ALLE ZEIT WACH 1842

Bioorganic Marine Chemistry Volume 6

Edited by Paul J. Scheuer

K.F. Albizati, V.A. Martin,
M.R. Agharahimi, D.A. Stolze

Synthesis of Marine Natural Products 2

Nonterpenoids

With 161 Structures and 263 Schemes

Springer-Verlag
Berlin Heidelberg New York London
Paris Tokyo Hong Kong Barcelona Budapest

Professor Paul J. Scheuer
University of Hawaii at Manoa, Department of Chemistry
2545 The Mall, Honolulu, Hawaii 96822, USA

ISBN-13:978-3-642-76840-8 e-ISBN-13:978-3-642-76838-5
DOI: 10.1007/978-3-642-76838-5

Library of Congress Catalog Card Number 89-649318

Softcover reprint of the hardcover 1st edition 1992

51/3020-54210–Printed on acid-free paper

Preface

Volumes five and six of *Bioorganic Marine Chemistry* differ from their predecessors in two respects – they deal exclusively with laboratory synthesis of marine natural products and they represent the effort of a single author and his associates.

The rationale for these departures is readily perceived. For several decades organic synthesis has without doubt been the most spectacular branch of organic chemistry. While the late R.B. Woodward's dictum – organic compounds can undergo only four basic reactions: they can gain electrons; they can lose electrons; they can be transformed with acid or with base – is still true, the wealth and variety of available reagents which will accomplish chemical transformations has reached staggering proportions. Little wonder then, that synthetic methodology has achieved a high degree of predictability and total synthesis of natural products has been successfully directed toward ever more challenging targets. As for the second point, that of single authorship, multiple authorship would invariably have led to gaps and overlaps, thus making it difficult to assemble and assess recent research in a systematic and comprehensive fashion.

These two volumes are significant not only as a testimonial to the productivity and versatility of marine biota and to the virtuosity of synthetic chemists. As the material is presented along biogenetic principles, it is ideally suited to support research into the biosynthesis of marine metabolites. The comprehensive nature of the work makes it an easy matter to compare and evaluate different synthetic approaches prior to any synthesis of labelled precursors.

The division into terpenoid (Vol. 5) and nonterpenoid (Vol. 6) compounds is a natural one not only because of bulk. Nonterpenoid, particularly amino acid-derived, metabolites have become the fastest growing group of marine natural products. As recently as a decade ago, this position was held by di-, and earlier by sesquiterpenoids. This change parallels the current trend in research emphasis. Much early work in marine natural products was the result of serendipitous collections and separations. By contrast, most of today's research is guided by biological activity, which in turn is skewed toward those activities – e.g. antitumor, antiviral, which receive funding in industrialized societies.

While reading and editing the manuscript I was struck by the large impact which marine natural product research has made on organic synthesis and indeed on contemporary chemistry. It occurred to me that these books could be valuable auxiliary texts for graduate courses in Organic Synthesis.

I am indebted to Dr. Albizati and his associates for the monumental task which this endeavor entailed. As before, I should like to express my appreciation to Springer Verlag for their prompt and expert cooperation. As always, I look forward to hearing from members of the scientific community how we can improve future volumes in the series.

August, 1991 Paul J. Scheuer

Table of Contents

Abstract

The growth and extent of chemical synthesis of marine natural products from the years 1960–1989 has been evaluated and reviewed in a near-comprehensive fashion for the first time. The rapid growth in the breadth and depth of this field in a comparatively short period of time mirrors the growth and interests of the synthesis community at large. Synthesis chemists are stimulated primarily by compounds which possess potential biomedical importance and/or provocative structures, of which there is an abundance among the metabolites from marine sources. Continued growth in this area is projected. The information in this review consists primarily of synthetic schemes and pathways which, after analysis, have been set to words. The metabolites synthesized have been organized according to broad biogenetic lines, including terpenes, alkaloids, fat-derived compounds, amino-acid-derived and miscellaneous.

3 Amino Acids and Peptides

A variety of unusual amino acids and small cyclic and acyclic oligopeptides have been isolated from marine organisms. These have been the subject of intense synthesis activity due to the potent and broad spectrum of activities exhibited by the various members of this class. Syntheses of a number of bromotyrosine-derived metabolites are also included in this section.

3.1 Various Amino Acids

3.1.1 Carnosadine

Carnosadine, a new cyclopropyl amino acid isolated [312] from the red alga *Grateloupia carnosa*, has been synthesized by Shiba [313] as shown in Scheme 192. Dipolar cycloaddition of diazomethane to acid **1212** gives pyrazoline **1213** which upon photolysis provides cyclopropane **1214**. Conversion of ester **1214** to the amide followed by Hofmann degradation gives amine **1215**. Resolution of the diastereomers and addition of guanidine to **1215** gives (−)-carnosadine in 13 steps and 6.1% overall yield.

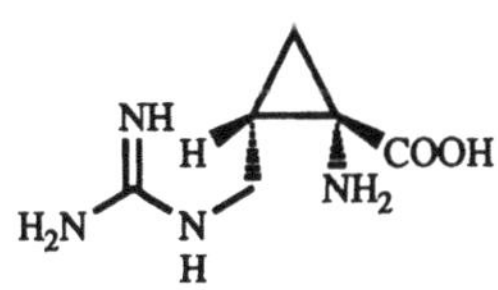

(-)-Carnosadine **1216**

3.1.2 Ovothiols A and C

Ovothiols A and C have been synthesized by Hopkins [314] from the parent heterocycle **1217** as shown in Scheme 193. Hydroxymethylation and chlorination of **1217** gives imidazole **1219** in two steps. Treatment of **1219** with α-glycinyl anion equivalent **1220** affords amino acid **1221**. Removal of the thiol

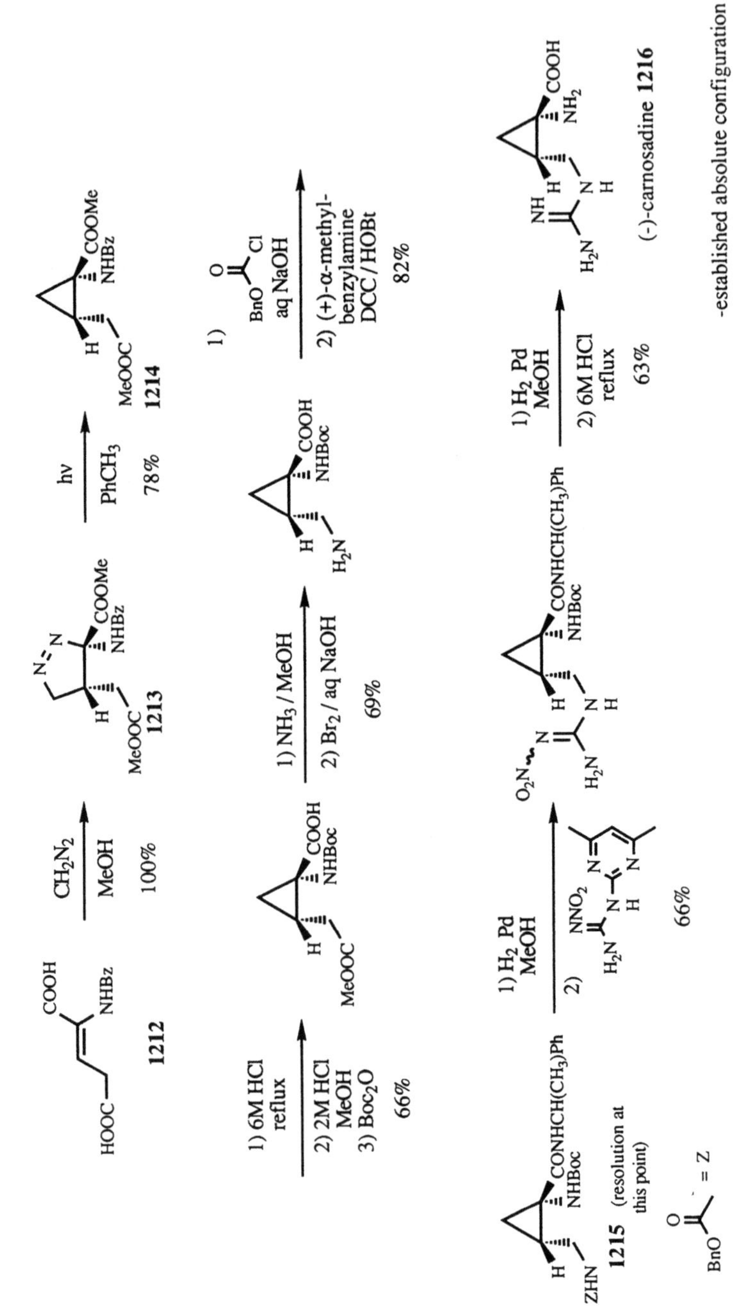

Scheme 192. Shiba Synthesis of (−)-Carnosadine

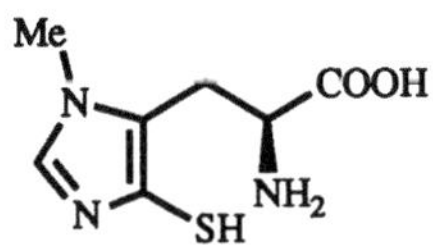

(+)-Ovothiol A **1222**

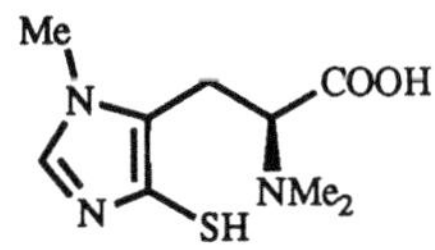

(+)-Ovothiol C **1223**

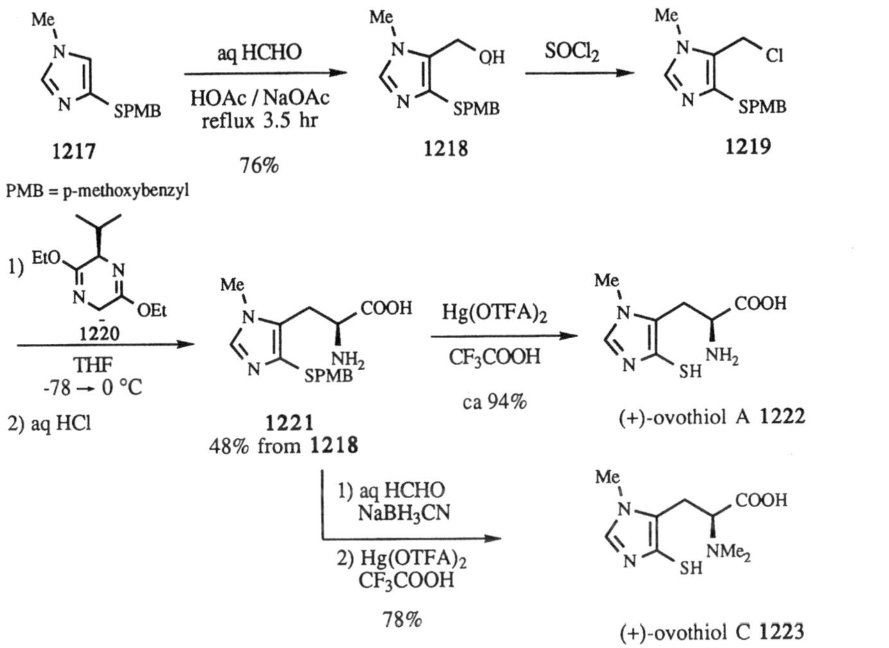

- established structure by localizing where the imidazole Me group was

Scheme 193. Hopkins Synthesis of (±)-Ovothiols A and C

protecting group yields (+)-ovothiol A in approximately 34% yield over 5 steps. Reductive methylation of **1221** prior to deprotection gives (+)-ovothiol C in 6 steps and 28% overall yield.

3.1.3 α-Kainic Acid and Allo-Kainic Acid

α-Kainic acid is the parent member of the kainoids, a group of structurally related pyrrolidine dicarboxylic acids. It has been isolated, along with its C4 epimer α-*allo-kainic* acid, from the alga *Digenea simplex* [315] and has also been

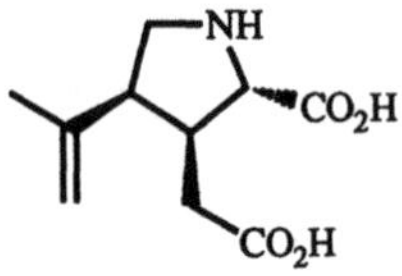

α-Kainic Acid **1234**

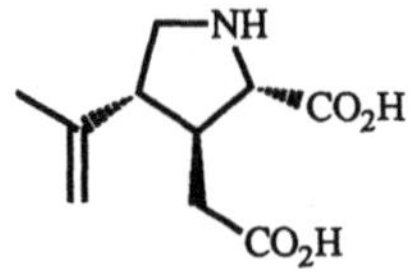

α-Allokainic Acid **1228**

found in the alga *Centrocerus clavulatum* [316]. Although α-kainic acid has been reported to possess anthelmintic and insecticidal attributes [317], it is of interest mainly due to its pronounced neuroexcitatory properties [318].

The diastereomeric α-*allo*-kainic acid was first prepared by Oppolzer [319] in 1978 as shown in Scheme 194. *N*-alkylation of the easily accessible *N*-trifluoroacetylaminomalonic ester **1225** affords compound **1226** which undergoes an ene reaction upon heating to give *trans*-substituted pyrrolidine **1227**. Hydrolysis and decarboxylation of **1227** yields (±)-α-*allo*-kainic acid in 6 steps and 53% overall yield. Crystallization of **1228** as its (−)-ephedrine salt provides enantiomerically pure (+)-α-*allo*-kainic acid.

Oppolzer's second synthesis [320] of (+)-α-*allo*-kainic acid (Scheme 195) utilizes (−)-8-phenylmenthol as a chiral auxiliary to promote a Lewis acid-catalyzed asymmetric intramolecular ene-type reaction that provides *trans*-pyrrolidine **1229** in 60% yield in 90% ee. Hydrolysis and decarboxylation gives (+)α-*allo*-kainic acid (**1228**).

Oppolzer's synthesis [321] of (±)-α-kainic acid appeared in 1979 (Scheme 196) and follows a similar strategy to that used in his syntheses of **1228**. Alkylation of ester **1231** via conjugate addition to a 2-methylthioacrylate ester

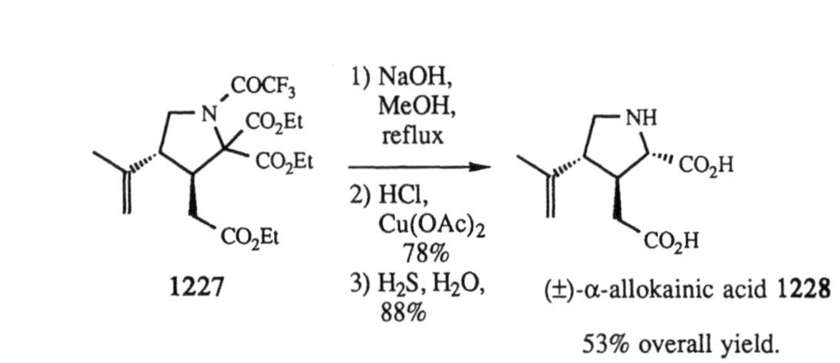

Scheme 194. Oppolzer Synthesis of (±)-α-Allokainic Acid

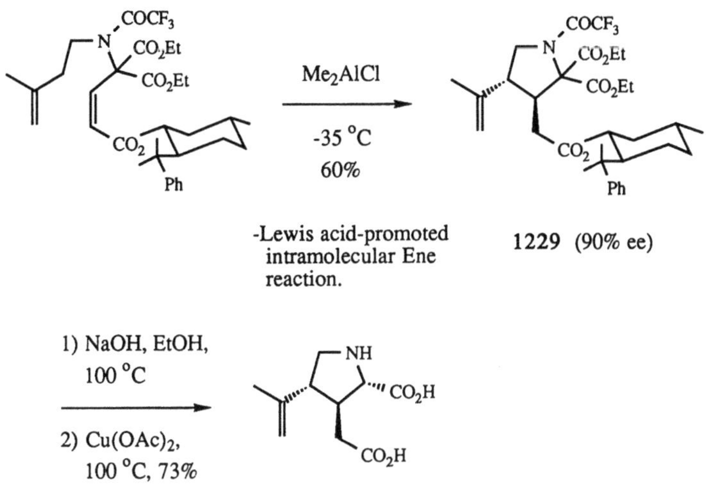

(+)-α-allokainic acid **1228**

Scheme 195. Oppolzer Synthesis of (±)-α-Allokainic Acid

followed by sulfoxide elimination gives diene **1232** as the $\Delta^{2,3}$ isomer. Heating this to 180 °C causes cyclization to *cis*-substituted pyrrolidine **1233**, presumably through an ene reaction of the thermally formed $\Delta^{3,4}$ isomer. Deprotection of the amine gives (±)-α-kainic acid **1234** in 7 steps and 41% overall yield.

An enantio- and diastereospecific synthesis of (−)-kainic acid by Oppolzer [322] is outlined in Scheme 197. Triester **1235** derived from (*S*)-glutamic acid is converted to 1,6-diene **1236** in six steps and 19% overall yield. Intramolecular ene cyclization of **1236** followed by desilylation, oxidation and hydrolysis provides (−)-α-kainic acid **1234** in 11 steps and 4.6% overall yield.

Kraus's entry into the arena of kainic acid synthesis [323] involved (±)-α-*allo*-kainic acid **1228** as outlined in Scheme 198. Stereospecific 1,3-cycloaddition of **1238** and **1239** provides functionalized pyrrolidine **1240** in 65% yield. Desulfurization of **1240** with one equivalent of Bu_3SnH leads to ketone **1241** which is converted to triester **1242** in four steps via standard transformations. Hydrolysis and deprotection occurs with epimerization, thus completing the synthesis to give (±)-α-*allo*-kainic acid **1228** in 11 steps and approximately 11% overall yield. In a full account of this work [324] it was disclosed that epimerization at C2 was performed after removal of the *t*-BOC protecting group.

The synthesis of (±)-α-*allo*-kainic acid **1228** by DeShong [325] involves the 1,3-dipolar cycloaddition of an azomethine ylid to an α,β-unsaturated ketone to give the functionalized pyrrolidine **1243** in one step along with < 1% of the C2 epimer (Scheme 199). Conversion of **1243** to diester **1244**, removal of the *N*-benzyl group with α-chloroethyl chloroformate, and treatment with sodium

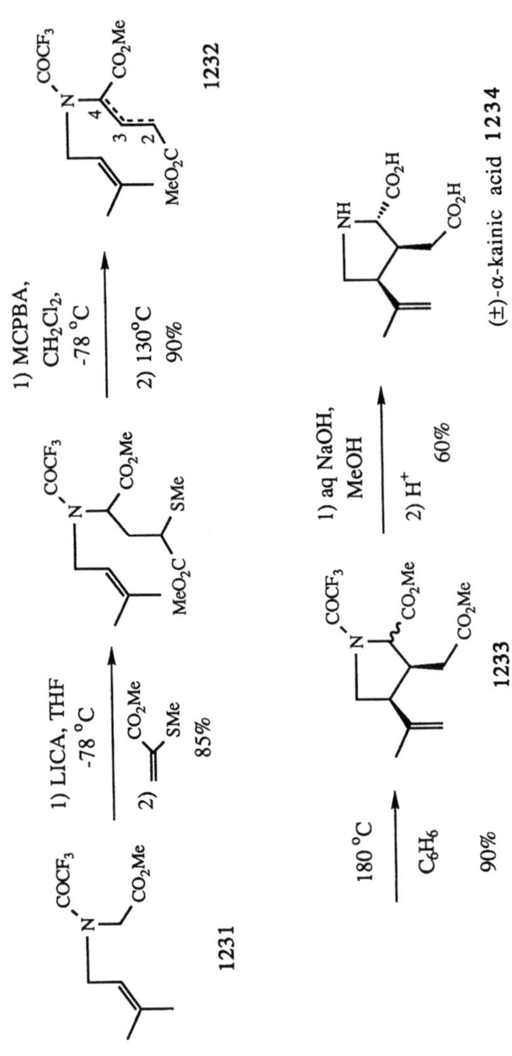

Scheme 196. Oppolzer Synthesis of (±)-α-Kainic Acid

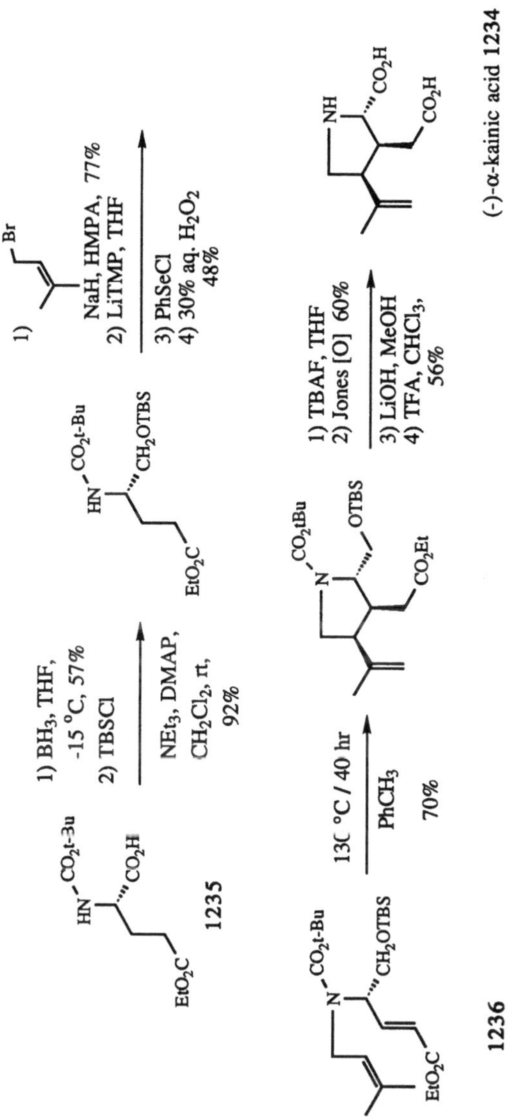

Scheme 197. Oppolzer Synthesis of (−)-α-Kainic Acid

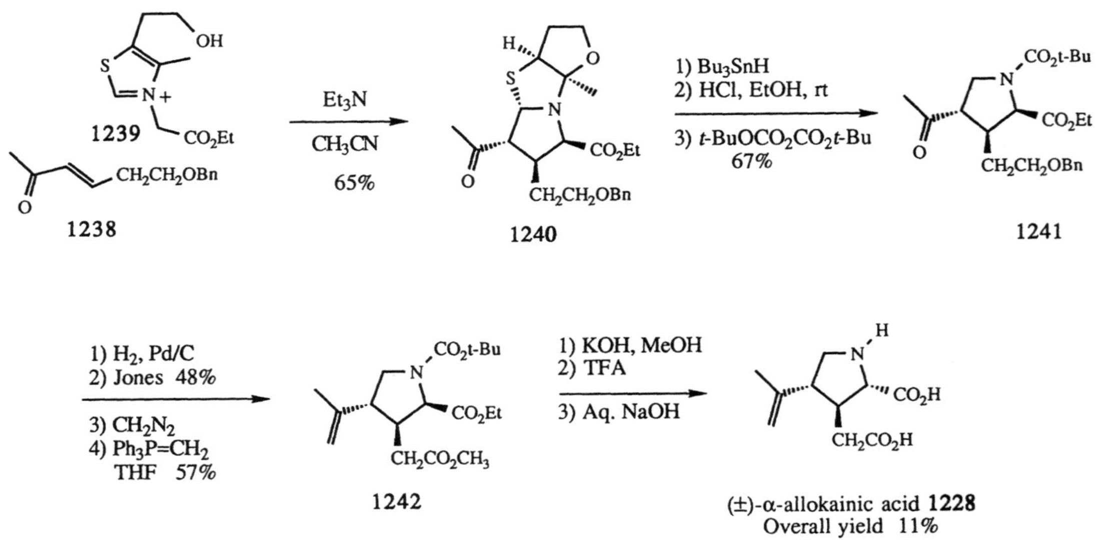

Scheme 198. Kraus Synthesis of (±)-α-Allokainic Acid

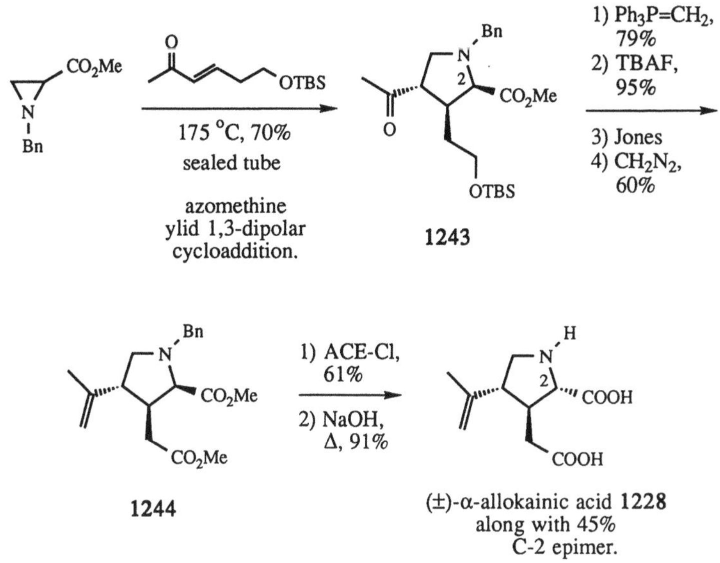

Scheme 199. DeShong Synthesis of (±)-α-Allokainic Acid

hydroxide gives a 1:1 mixture of (±)-α-*allo*-kainic acid (**1228**) and (±)-2-*epi*-α-*allo*-kainic acid. The natural product **1228** is obtained in 8.5% yield over 7 steps.

Knight's synthesis [326] of (−)-α-kainic acid (**1234**) is outlined in Scheme 200. The lithio anion of carbamate, derived from L-aspartic acid is *N*-alkylated with allylic chloride **1246** to give compound **1248**. Lactonization of **1248** gives nine-membered azalactone **1249**. The stereocontrolled enolate Claisen rearrangement of **1249** proceeds via a boat-like transition state to give substituted pyrrolidine **1250** as a single isomer. Homologation of the C3 carboxyl group, desilylation and oxidation gives diester **1251** which is converted to (−)-α-kainic acid **1234**. Fifteen steps were required with an overall yield of 3.3%.

The most recent offering in this area is a rather lengthy enantiospecific (but not diastereospecific) preparation of both (−)-α-kainic acid and (+)-α-*allo*-kainic acid by Baldwin [327] (Scheme 201). Reaction of optically pure epoxide **1252** with isonitrile **1253** provides an acyclic carbamate which upon treatment with base opens the epoxide in an intramolecular fashion to give amine **1254** after hydrolysis. Cobalt-mediated cyclization of the iodide derived from **1254** gives pyrrolidine **1255** as a 1.7:1 mixture of diastereomers. The mixture is separated, converted to acids **1256** and then to **1257** and **1258** in a total of 17 steps.

3.1.4 Domoic Acid

The closely related domoic acid (**1263**) isolated from the red alga *Chondria armata* (Okamura) [328] exhibits similar neurobiological activities and is in

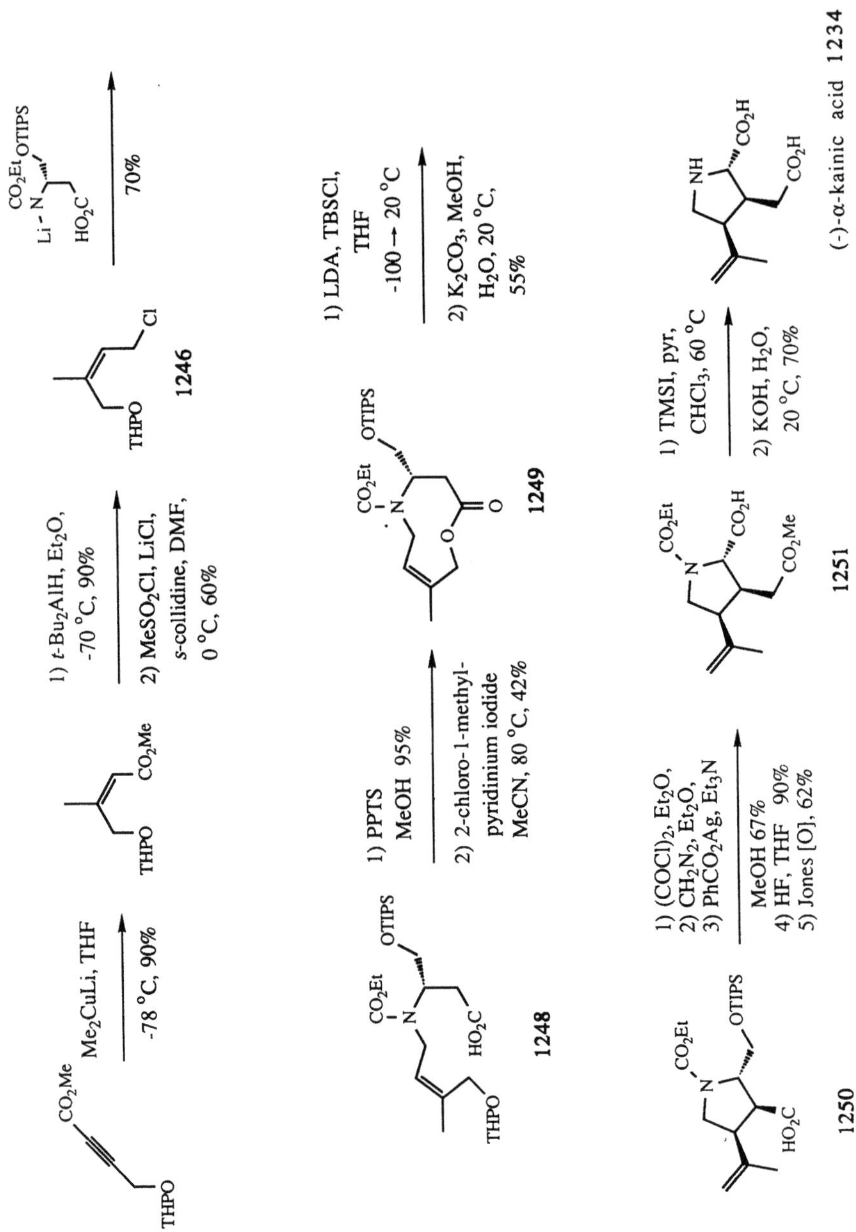

Scheme 200. Knight Synthesis of (−)-α-Kainic Acid

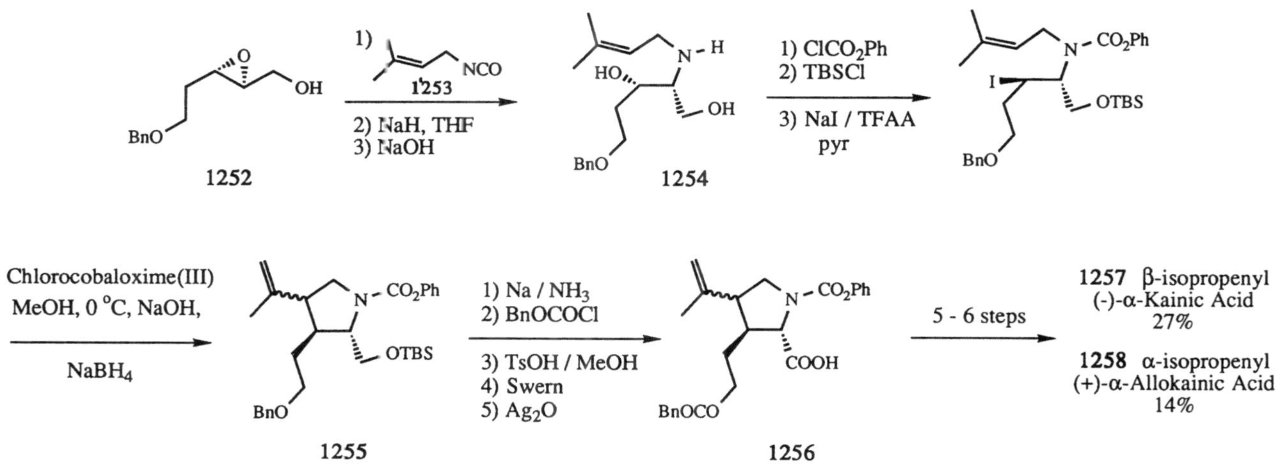

Scheme 201. Baldwin Syntheses of (−)-α-Kainic Acid and (±)-α-Allokainic Acid

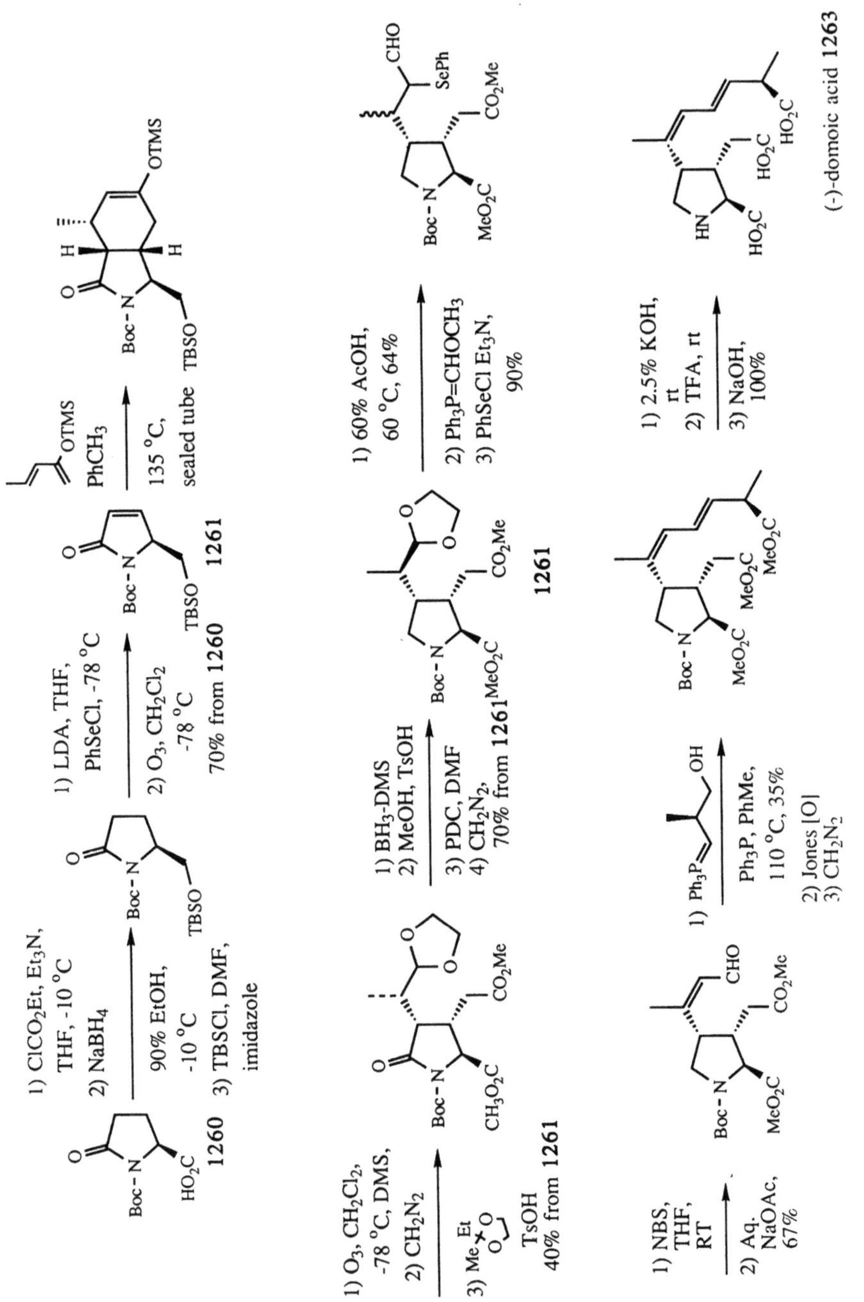

Scheme 202. Ohfune Synthesis of (−)-Domoic Acid

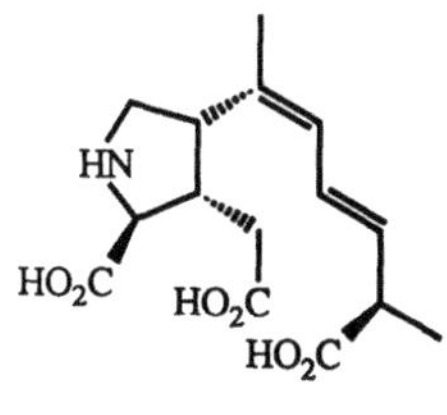

Domoic Acid **1263**

short supply due to the small amounts available from marine sources. Ohfune's synthesis of **1263** is presented in Scheme 202 [329]. *N-tert*-Butoxycarbonyl-L-pyrroglutamic acid (**1260**), available from L-glutamic acid, is converted to unsaturated lactam **1261**, thus setting the stage for the ensuing Diels-Alder reaction. Cycloaddition of **1261** with 2-trimethylsilyloxy-1,3-pentadiene sets the *cis*-stereochemistry of the nascent sidechains and provides the functionality needed for further elaboration. Ozonolysis and reductions complete the lower sidechain, which is protected as an ester, and removal of the amide carbonyl gives acetal **1262**. Elaboration of the upper sidechain via selenoxide elimination and Wittig olefination sets the proper geometry of the diene system and completes the synthesis to give (−)-domoic acid in 24 steps and 2.6% overall yield.

3.1.5 Tetraacetylclionamide

Ethanol extracts of the sponge *Cliona celata*, collected in the northwestern Pacific near British Columbia have been found to show in vitro antibiotic activity against *Staphylococcus aureus*. Unfortunately, all attempts to isolate and purify the active constituent of the ethanol extracts were unsuccessful. To circumvent this problem, the crude extracts were partially purified and acylated by treatment with acetic anhydride and sodium acetate. After isolation, the major component proved to be the 6-bromotryptophan enamide derivative tetraacetylclionamide (**1268**) [330]. Schmidt's synthesis [331] of **1268** is

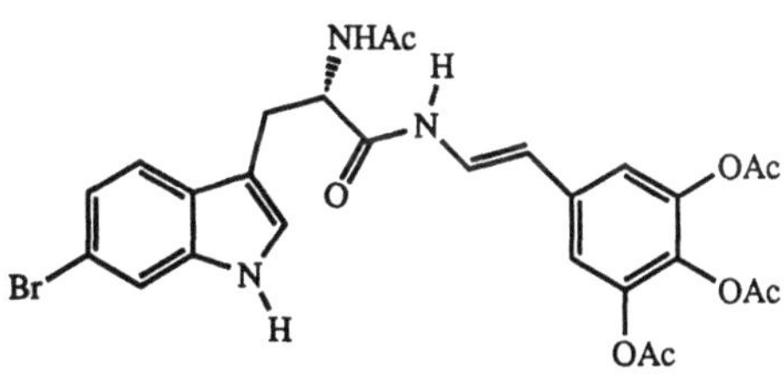

Tetraacetylclionamide **1268**

illustrated in Scheme 203. The preparation of **1268** is guided at all stages by the knowledge that care must be taken to avoid deacylation and transacylation reactions. The synthesis is initiated by the conversion of the acid chloride of triacetyl gallic acid (**1264**) to azido ketone **1265**. Azide **1265** is reduced to the amine and acylated with the pentafluorophenyl ester of *S*-BOC-6-bromotryptophan to give amide **1266**. Reduction of the ketone and elimination to the enamide **1267** followed by deprotection gives the acylated natural product **1268** in 9 steps and 25% overall yield.

3.1.6 Demethyldysidenin and Demethylisodysidenin

(+)-Demethyldysidenin (**1275**) and (−)-demethylisodysidenin (**1276**) (as well as dysidin *cf* 6.3.9) are among the very few naturally occurring compounds containing the trichloromethyl functionality. These three metabolites have all been isolated [332] from the Indo-Pacific sponge *Dysidea herbacea*. Demethylisodysidenin (**1276**) has been reported to exhibit antihypertensive activity when administered intravenously [333].

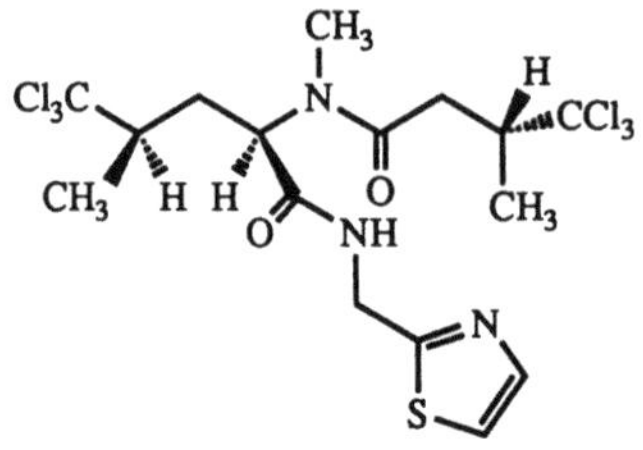

(+)-Demethyldysidenin **1275**

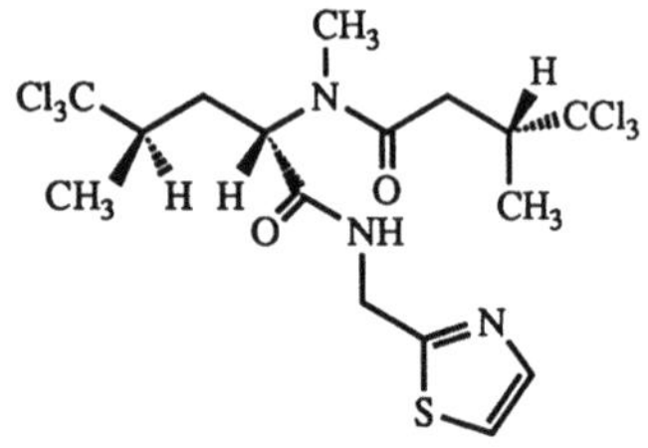

(-)-Demethylisodysidenin 1276

The chiral building block for Williard's syntheses [334] of (+)-demethyldysidenin and (−)-demethylisodysidenin (**1276**) (Scheme 204) is obtained by resolution of the half-acid ester of β-methylglutaric acid with either cinchonidine or quinine. Chiral acid **1270** is converted to the dichloroaldehyde **1271** and then to a key fragment, trichloromethyl aldehyde **1272**. Jones oxidation of **1272** gives a second key fragment, trichloromethyl acid **1273**. Utilizing the one-flask "four component peptide synthesis" described by Ugi [335], combination of **1272**, **1273**, isonitrile **1274** and methylamine gives optically pure **1275** and **1276** in 17% and 13% yields. Demethyldysidenin (**1275**) and demethylisodysidenin (**1276**) are produced in 9 steps and 2.1 and 1.6% overall yields respectively.

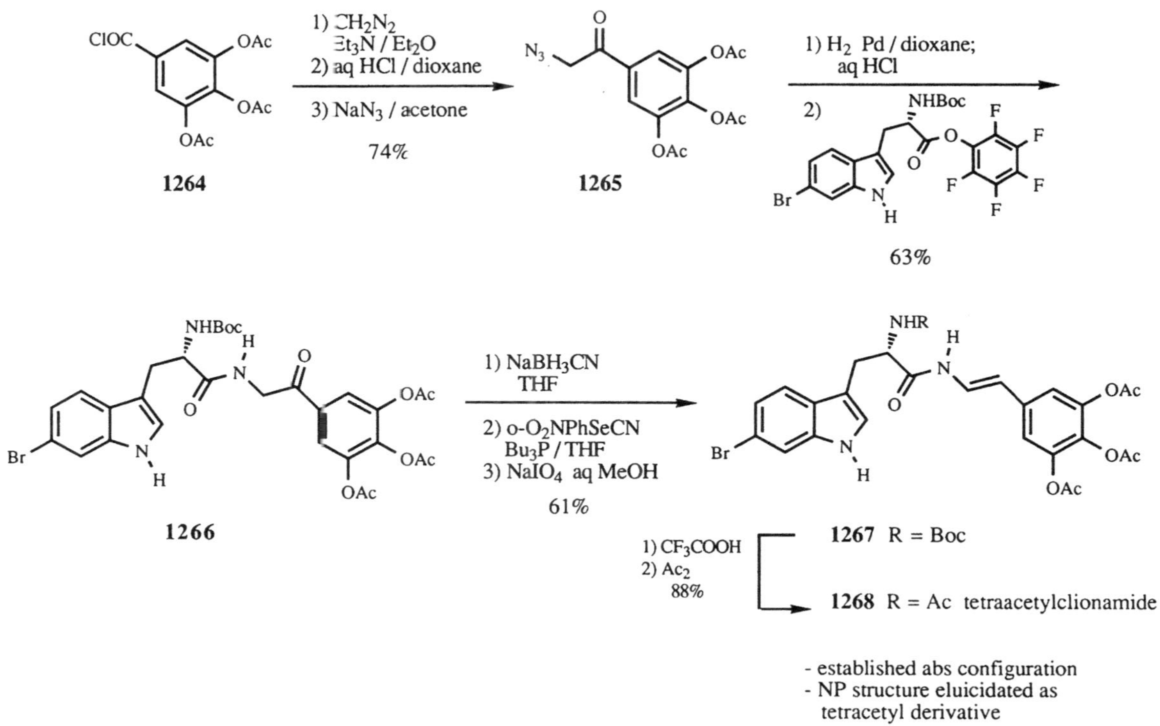

Scheme 203. Schmidt Synthesis of Tetraacetylclionamide

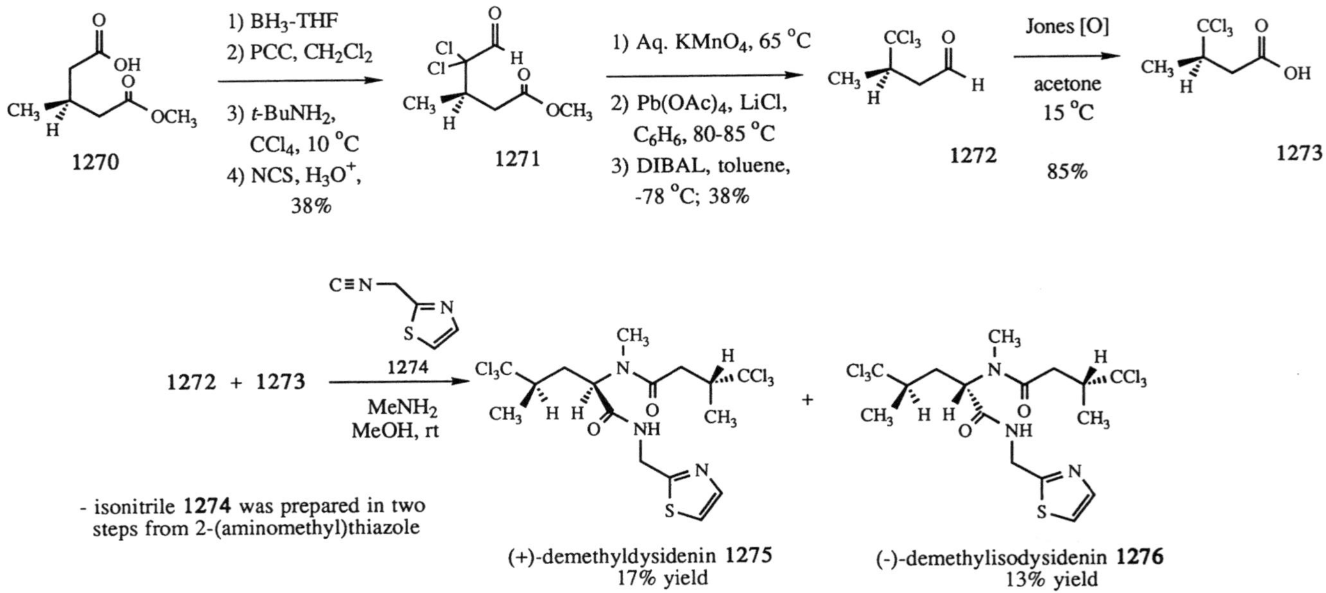

Scheme 204. Williard Syntheses of (±)-Demethyldysidenin and (−)-Demethylisodysidenin

3.2 Brominated Tyrosine Derivatives

3.2.1 Quinols

The quinol **1279** first isolated from the sponge *Verongia cauliformis* exhibits broad spectrum antibiotic properties [336]. A similar quinol **1283** has been isolated from the mollusc *Tylodina fungina* [337]. The first synthesis of **1279** is that of Sharma and Burkholder [338] in 1967 (Scheme 205). Bromination of acid **1277** and amide formation gives bromoamide **1278** (no description of these two steps was given). Oxidation of **1278** with nitric acid provides the natural product in three steps. An improvement on this methodology was developed by Yamada [339]. Formation of the amide by reaction with aqueous ammonia and diphenyl phosphite provides **1278** in 70% yield. The natural product is obtained by thallium(III) perchlorate oxidation of **1278** in 82% yield.

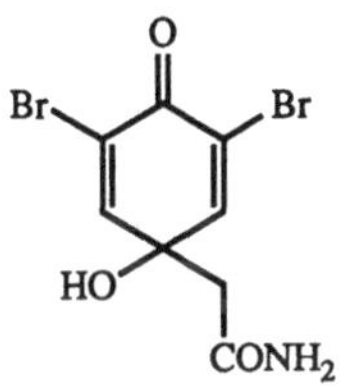

Verongia Metabolite **1279**

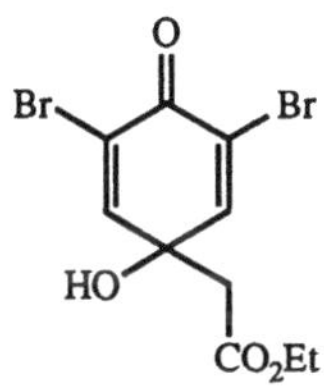

Tylodina Metabolite 1283

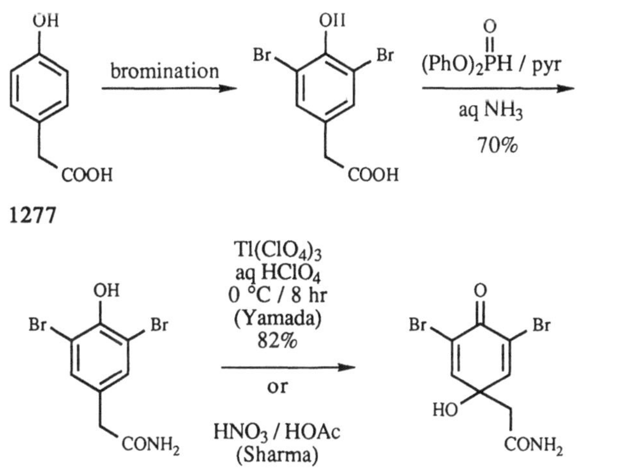

1278 *Verongia* metabolite 1279

Scheme 205. Syntheses of the *Verongia* Metabolite **1279**

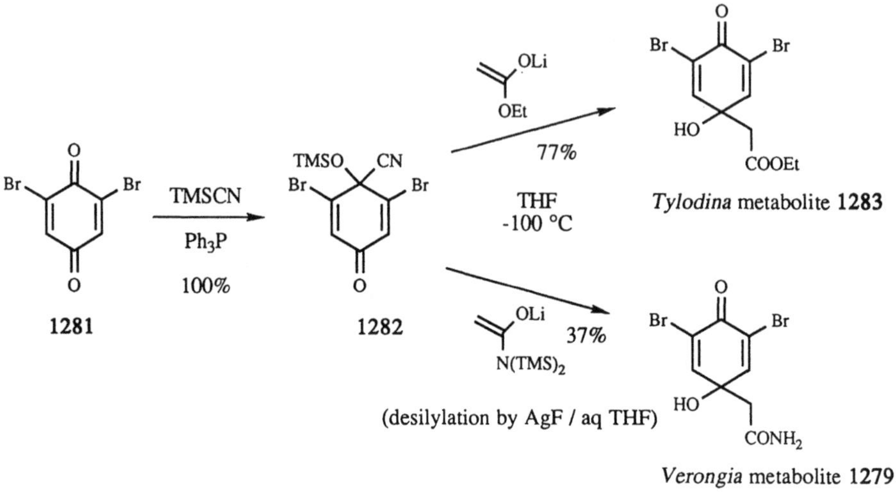

Scheme 206. Evans Synthesis of *Verongia* Metabolites **1279** and *Tylodina* Metabolite **1283**

Evans [340] has prepared both **1279** and **1283** from a common precursor (Scheme 206). Reaction of quinone **1281** with TMSCN affords dienone **1282** which upon reaction with the lithium enolate of ethyl acetate gives **1283** in 77% yield. Addition of the lithium enolate of *N*,*O*-bis(trimethylsilyl)acetamide to **1282** gives **1279** after desilylation with silver fluoride in 37% yield. Fischer [341], in conjunction with studies on functionalized organolithium reagents, has simplified the approach of Evans to do away with the need for monoprotection of the quinone **1281** (Scheme 207). At low temperatures, addition of ethyl

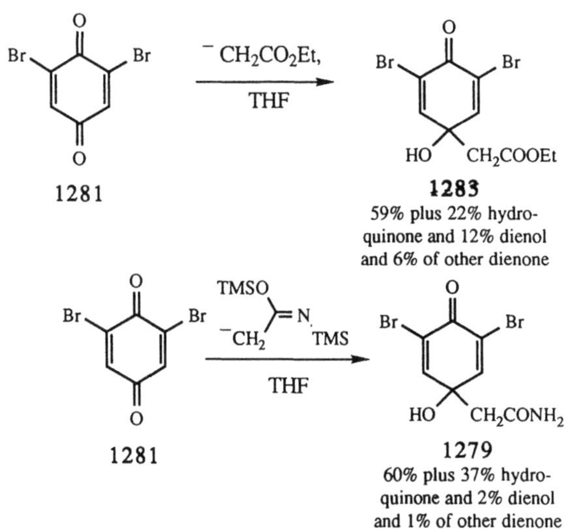

Scheme 207. Fischer Synthesis of *Verongia* Metabolite **1279** and *Tylodina* Metabolite **1283**

lithioacetate to brominated quinone **1281** provides *Verongia* metabolite **1283** in 59% yield. In a similar manner, addition of the lithium anion of *N*,*O*-bis(trimethylsilyl)acetamide to **1281** gives **1279** in 60% yield.

3.2.2 Aplysinadiene

The rearranged dibromotyrosine derivative aplysinadiene (**1287**) was first isolated from the sponge *Aplysina aerophoba* collected near Graciosa Island, Canary Islands. The isolation and synthesis (Scheme 208) of (**1287**) is described by Norte [342]. Benzylic oxidation of **1284** followed by Baeyer-Villiger oxidation and methylation provides **1285** in 43% yield. Chloromethylation of **1285** sets up the transformation to lactone **1286** which is converted to aplysinadiene via aldol condensation and elimination. Aplysinadiene **1287** is obtained in 9 steps and 15% overall yield. Other isomers of **1287** were prepared in a similar manner to confirm the identity of the natural product.

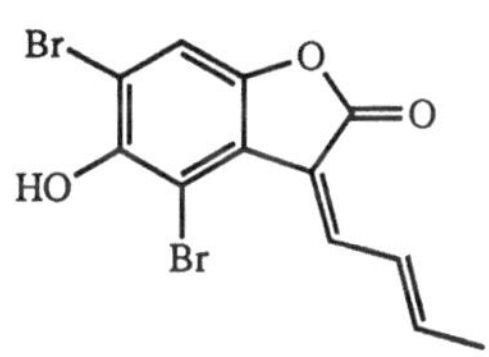

Aplysinadiene 1287

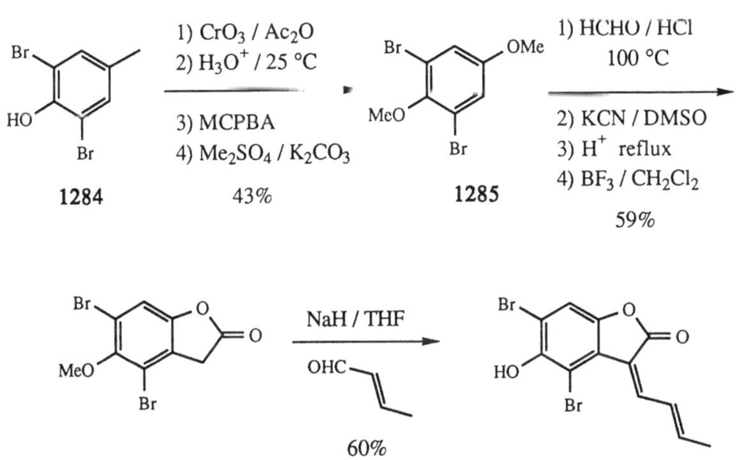

1286

aplysinadiene 1287

Scheme 208. Norte Synthesis of Aplysinadiene **1287**

3.2.3 Aerothionin, Homoaerothionin and Aerophobin-1

Aerothionin (**1295**) and homoaerothionin (**1296**) are novel tyrosine-derived metabolites containing spiroisoxazoline moieties. These substances have been isolated from the sponges *Aplysina aerophoba*, *A. fistularis* and *Verongia thiona*. Their relative and absolute configurations have been determined by X-ray crystallographic analysis and circular dichroism studies [343]. Also isolated [344] from these sponges are the related spiro compounds aerophobin-1 (**1294**) and -2. Yamamura [345] has prepared aerothionin (**1295**), homoaerothionin (**1296**) and aerophobin-1 (**1294**) as shown in Scheme 209. Azalactone **1290** is prepared from aldehyde **1289** via known chemistry. Conversion of **1290** to **1291** occurs in five steps and sets the stage for formation of the spiroisoxazoline substructure. Oxidation of oxime **1291** with $Tl(OTF)_3$ affords the spiro compound **1292** in 27% yield. Zinc borohydride reduction of **1292** affords the key intermediate **1293** in 29% yield along with 40% of the *cis* isomer. Reduction with sodium borohydride gives only the *cis* isomer. Condensation of **1293** with histamine affords aerophobin-1 in 8 steps and 1.7% overall yield. Condensation of **1293** with 1,4-butanediamine or 1,5-pentanediamine yields **1295** and **1296** in 0.37 and 0.089% yields respectively, over 8 steps. Although it is not known with certainty, it is possible that diastereomers of **1295** and **1296** may have been formed due to the coupling of racemic fragments.

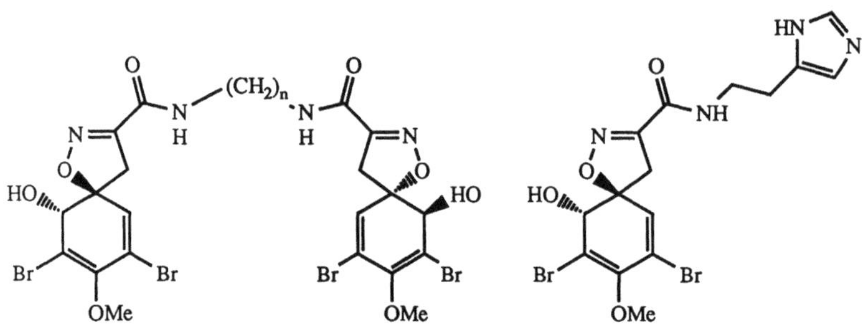

Aerothionin (n = 4) **1295**
Homoaerothionin (n = 5) **1296**

Aerophobin-1 **1294**

3.2.4 Bastadins-1, -2, -3 and -6

Bastadins-1, -2 and -3 are dimeric tyrosine derivatives composed of four Tyr units. Bastadin-6 is a 28-membered ring lactam ether that is structurally related to bastadin-2. These metabolites were isolated from the Verongid sponge *Ianthella basta* by Wells [346] in 1980. They possess potent in vitro and some in vivo activity against Gram-positive bacteria. Yamamura [347] has completed syntheses of all four of these compounds in a highly efficient manner. 3-Bromotyramine (**1297**) and the oximes **1299** and **1300** were required for bastadin

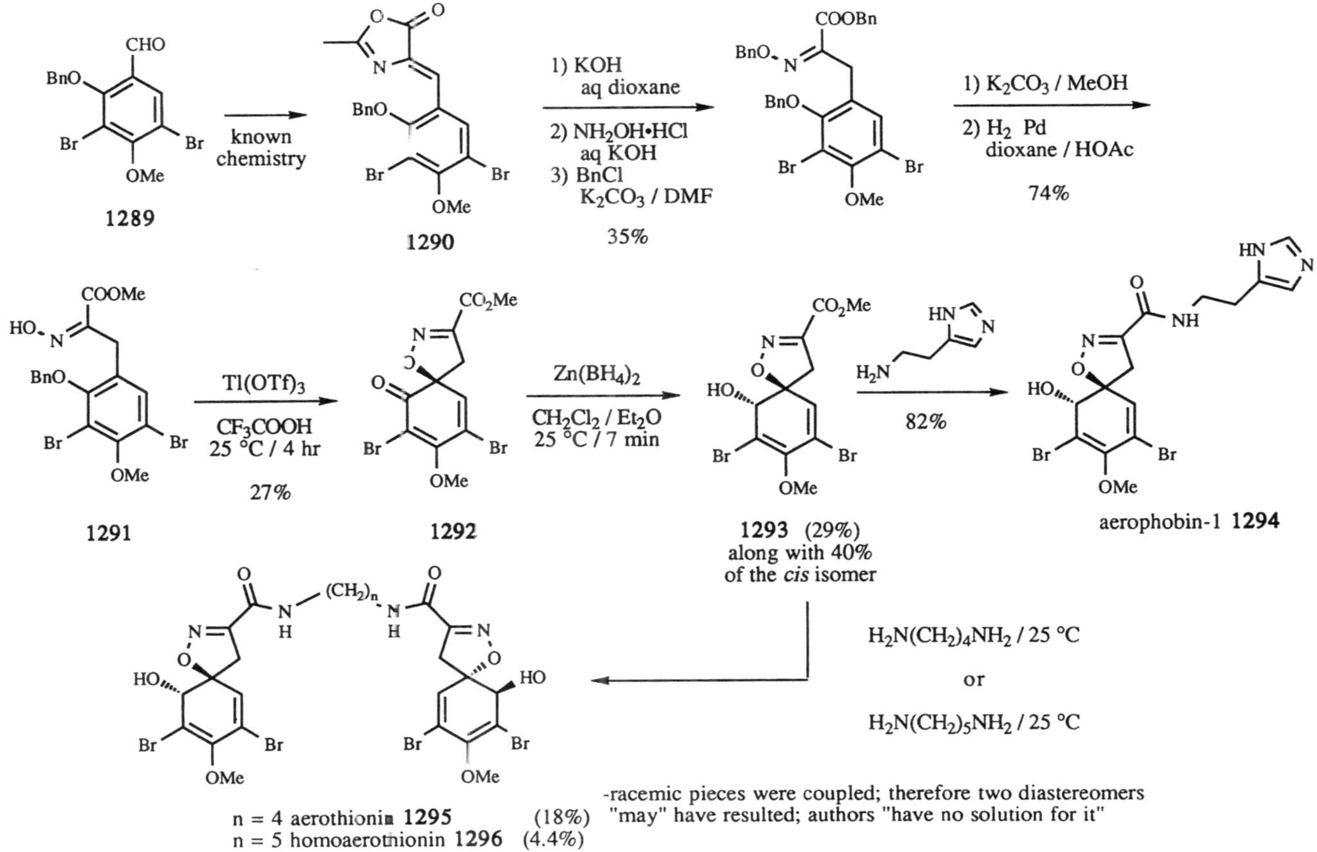

Scheme 209. Yamamura Syntheses of (±)-Aerothionin, (±)-Homoaerothionin and (±)-Aerophobin-1

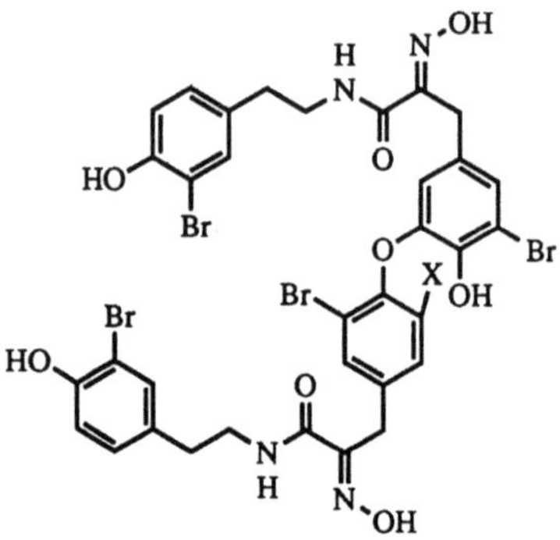

Bastadin-2 (X = Br) **1304**
Bastadin-1 (X = H) **1309**

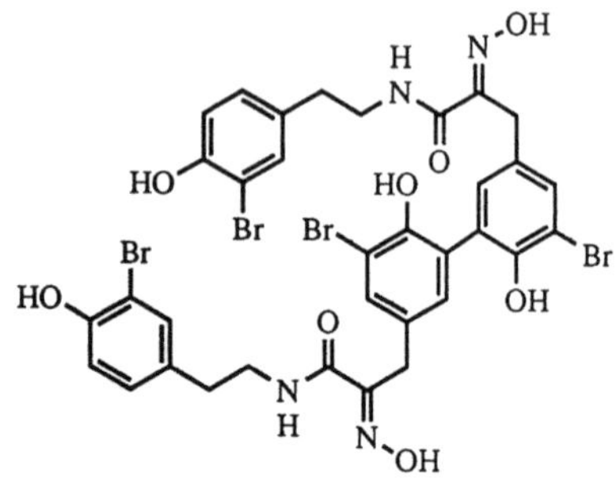

Bastadin-3 1311

synthesis. The oximes were produced in a manner similar to analogous compounds used in Yamamura's aerothionin synthesis. 3-Bromotyramine (Scheme 210) was prepared from 3-bromo-4-hydroxybenzaldehyde. Phenolic dimerization of oxime **1299** using $Tl(NO_3)_3$ (Scheme 211) produced isoxazoline **1301** in 44% yield which was quantitatively reduced to the phenolic dimer **1310**. Reaction with **1302** followed by removal of the PMB group gave bastadin-2 (**1304**). In a similar fashion, dimerization of **1300** with $Tl(OTFA)_3$ gave rise to a mixture of products in low yield, including **1301**, **1305** and **1306**. Compound **1301** was reduced to the analog **1302** to provide bastadin-1 (**1309**) after deprotection. To produce bastadin-3, the bis-spirooxazoline **1307** was reduced to **1310** in 8% overall yield from **1300**. Reaction of this substance as before with protected 3-bromotyramine followed by phenolic deprotection led to bastadin-3 (**1311**). Assuming that no $E \rightarrow Z$ oxime equilibration took place during the procedures, these syntheses establish the *E*-oxime configurations in these four metabolites and presumably others in the series as well.

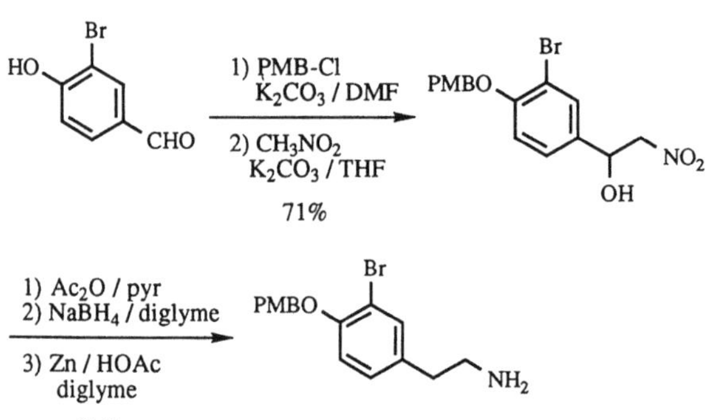

3-bromotyramine **1297**
(PMB ether)

Scheme 210. 3-Bromotyramine Synthesis (Yamamura)

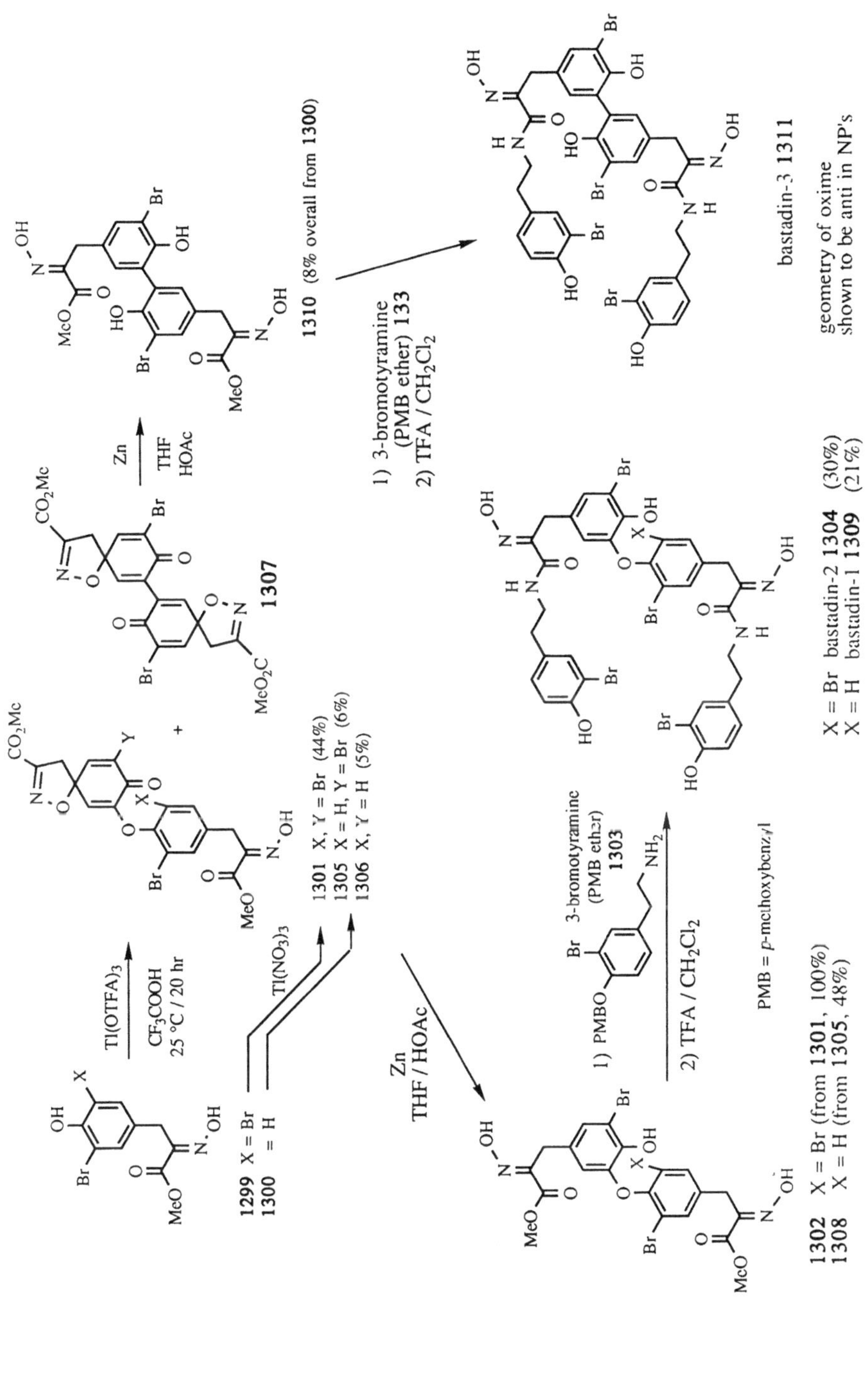

Scheme 211. Yamamura Synthesis of Bastadins-1, -2 and -3

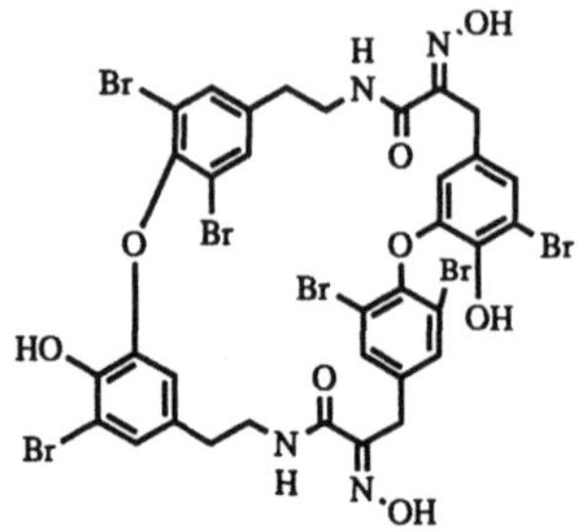

Bastadin-6 **1316**

Bastadin-6 was synthesized [348] from the penultimate intermediate in the bastadin-2 synthesis and requires an intramolecular, and perhaps biomimetic, phenol oxidation (Scheme 212). Benzylation of the oxime hydroxyls of **1312** and removal of the PMB groups gave **1313**. Addition of two bromines was accomplished in high yield providing **1314**. Cyclization via $Tl(NO_3)_3$ oxidation gave macrocycle **1315** in 13% yield along with an undesired isomeric substance. Zn reduction of the quinol ether to the phenol and debenzylation yielded bastadin-6. Previously, this group had prepared bastadin-6 trimethyl ether via a similar route [349]. Although several of the yields in these processes were low, the directness of this biomimetic approach is laudable.

3.3 Peptides

A number of cyclic and acyclic peptides [350] isolated from marine organisms exhibit a variety of biological activities. In particular, the anticancer properties of dolastatin-3 and -10 and various didemnins have attracted many groups to participate in this area. Synthesis has been particularly important in this area since many of the compounds are available in only very small amounts in amorphous form, making them untenable substrates for X-ray crystallography. Indeed, several of the originally proposed structures have been reassigned by synthesis. Although much of the work is comprised of well-known peptide synthesis techniques, the presence of several unusual amino acid units and non-amino acid sequences in many metabolites required more than classical amino acid coupling chemistry. We have used generally accepted acronyms in referring to the various reagents commonly used in peptide synthesis. Syntheses of marine polypeptides containing only common amino acids or >15 residues have been omitted [351].

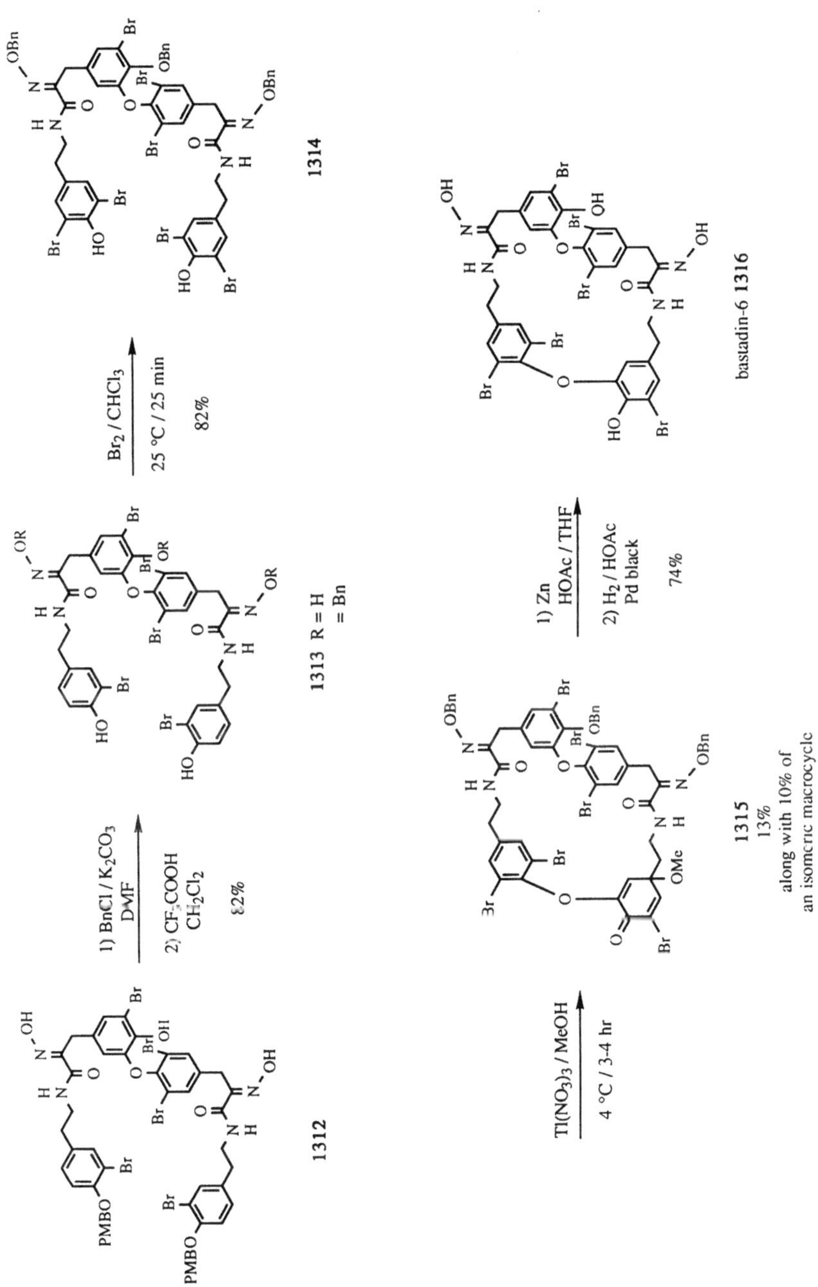

Scheme 212. Yamamura Synthesis of Bastadin-6

3.3.1 Didemnins

The cyclic depsipeptides didemnins A, B and C were originally isolated by Rinehart from the tunicate *Trididemnum solidum* and have subsequently been found in other *Trididemnum* sp. Didemnin B was the first marine natural product to enter clinical trials as an anticancer agent, hence its synthesis has engendered much interest. The original structural assignment has been corrected by synthesis. Rinehart [352] has described a route to didemnins A, B and C using classical peptide synthesis techniques to couple common amino acid and uncommon moieties which make up this metabolite. Dipeptides **1318** and **1319** were constructed (Scheme 213) via standard chemistry and then connected to give the larger fragment **1320**. The mixed fragment **1323** was prepared from optically pure acid chloride **1321** which was prepared from L-Val (Scheme 214). Reaction with the half-acid **1322** gave **1323** as an inevitable mixture of isomers. It is reasonable to assume that such a stereo-labile center would exist in the thermodynamically preferred orientation in the natural product and this was borne out in the Rinehart and the later Shioiri and Schmidt work as well. The mixture of diastereomers was carried on through the synthesis. Coupling with leucine provided the larger unit **1324** which was further coupled to **1320** providing the seco-peptide **1325**. Deprotection and cyclization via macrolactamization led directly to didemnin A (**1326**) which served as the precursor to both didemnins B and C. Modification of the leucyl side chain gave rise to didemnins B and C.

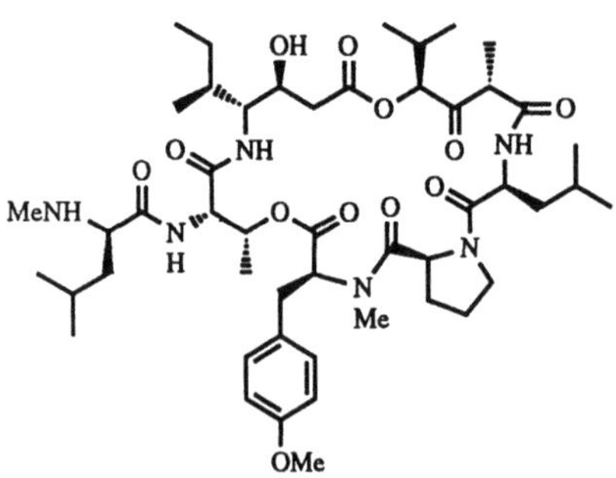

Didemnin A

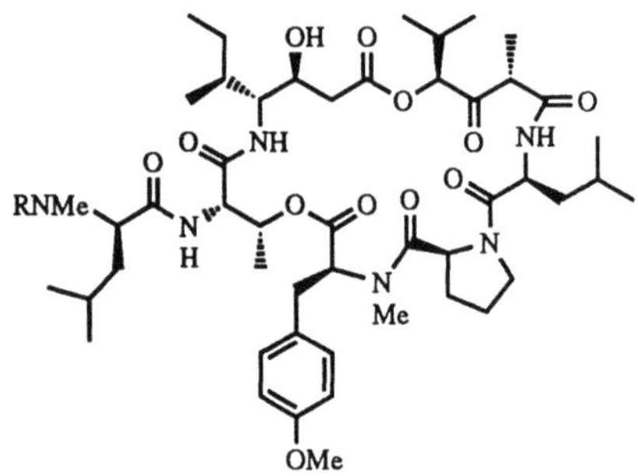

Didemnin B R =

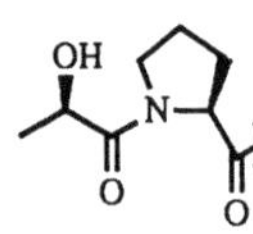

Didemnin C R =

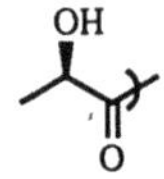

Schmidt has also synthesized didemnins A, B and C using a different strategy [353]. The major fragment **1329** is equivalent to fragment **1324** in the Rinehart synthesis and was also built up from (*S*)-α-hydroxyisovaleric acid (Schemes 215 and 216). However, the similarity ends here, as the remainder of the synthesis

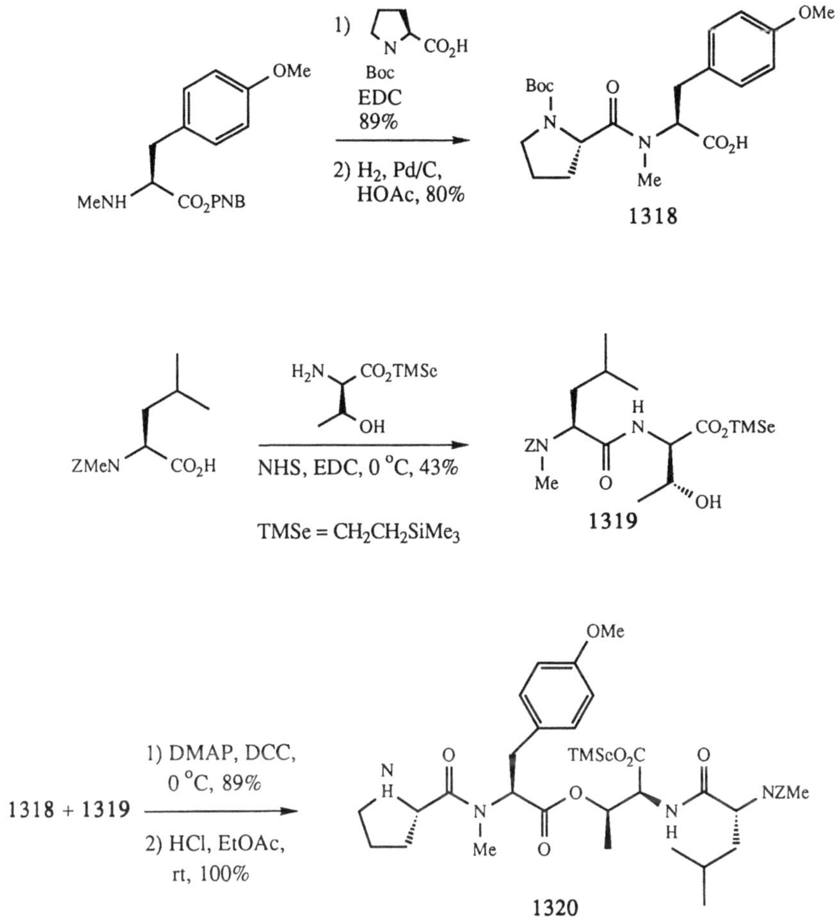

Scheme 213. Syntheses of Didemnin Fragments (Rinehart)

takes a different strategic, if not operational approach. Macrolactamization was carried out between the Pro and Leu residues on a substrate which carried an unadulterated Thr amino group to provide **1331** (Scheme 217). As before, ring closure was accompanied by slow epimerization of the methyl group at C2 of the 1st unit to provide a single peptolide isomer. Simple modification of the Thr amino group led to didemnin A and from there to didemnins B and C.

More recently, Shioiri has described a highly efficient convergent approach to didemnins A and B [354]. Two large eastern and western fragments (Schemes 218 and 219) were produced by standard chemistry. As in the Rinehart syntheses, a mixture of diastereomers at the C2 position of the Hip residue was generated (**1334**). Thus, methylation of the acylated Meldrum's acid **1333** (Scheme 218) proceeded without selectivity to produce **1334** as a mixture which

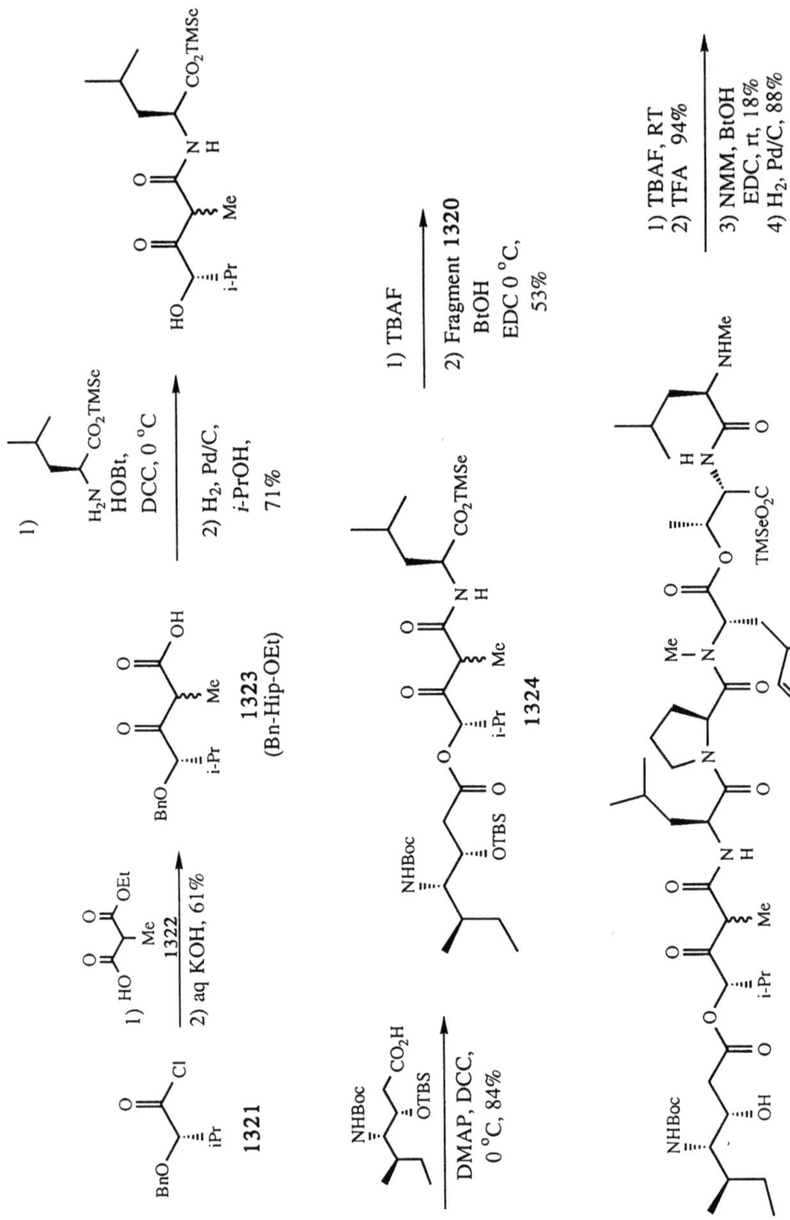

1321
1) 1322
2) aq KOH, 61%
1323
(Bn-Hip-OEt)
1) H_2N CO_2TMSe HOBt, DCC, 0 °C
2) H_2, Pd/C, i-PrOH, 71%
DMAP, DCC, 0 °C, 84%
1324
1) TBAF
2) Fragment 1320 BtOH EDC 0 °C, 53%
1325
1) TBAF, RT
2) TFA 94%
3) NMM, BtOH EDC, rt, 18%
4) H_2, Pd/C, 88%

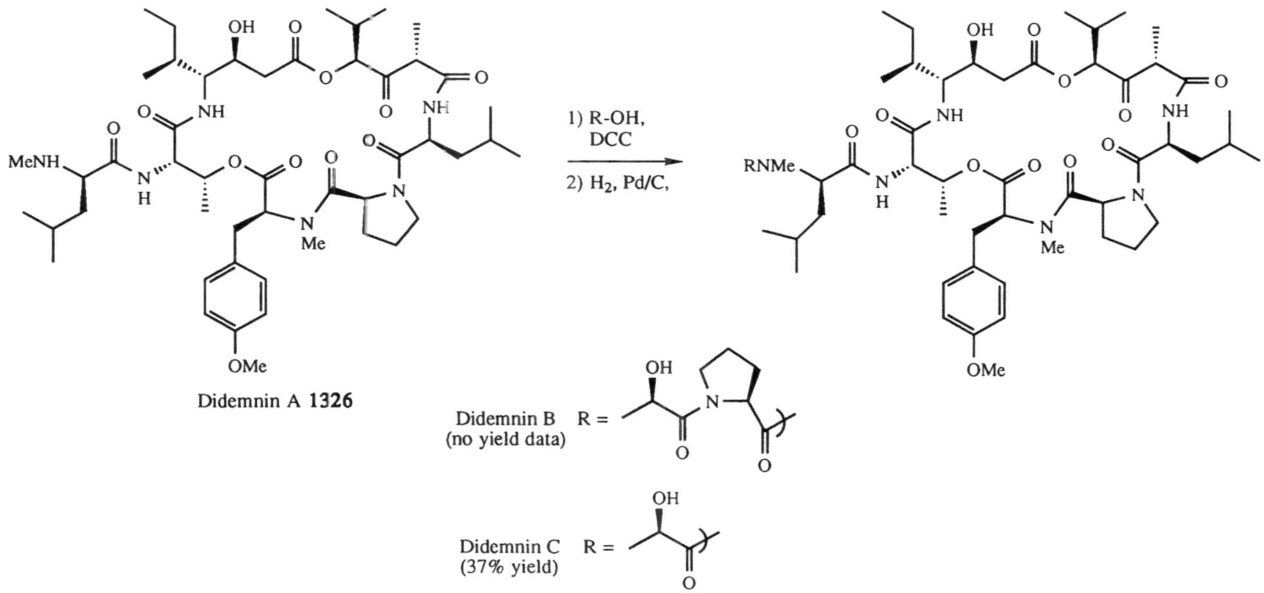

Scheme 214. Rinehart Syntheses of Didemnins A, B and C

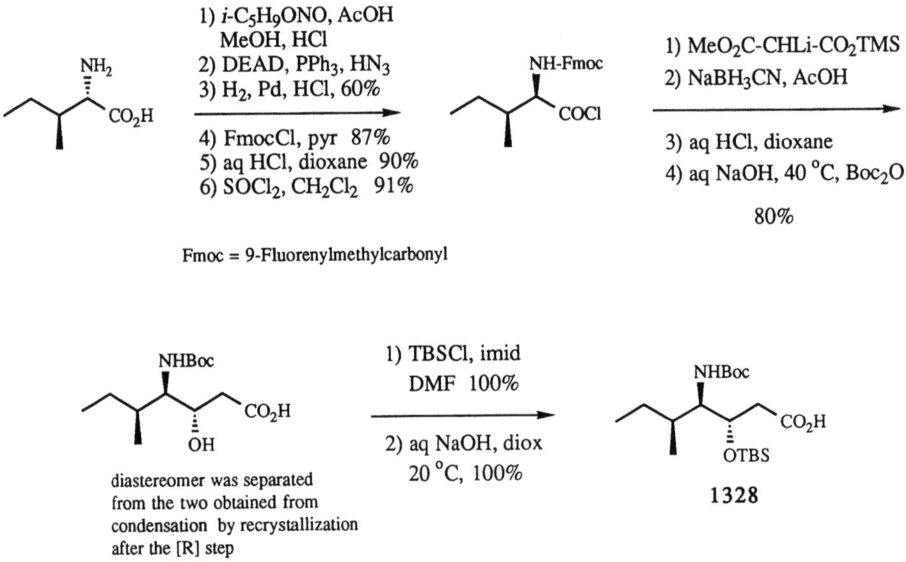

1328

Scheme 215. Synthesis of Didemnin Fragments (Schmidt)

was carried through the synthesis. Coupling of the large fragments via ester formation followed by macrolactamization gave didemnin A directly (Scheme 220). As in the Rinehart synthesis, didemnin B was prepared by coupling with a Lac-Pro unit followed by deblocking.

Nordidemnin B is a minor component of *Trididemnum cyanophorum*, differing from didemnin B by lacking a methyl group on the isostatine (1st) residue. Jouin [355] has synthesized this minor metabolite by assembling three large fragments as shown in Schemes 221 and 222. Combination of **1340** and **1341** provided the seco peptide **1342** (Scheme 223). Palladium-promoted deallylation was followed by removal of the acetonyl group and BOP-assisted macrocyclization and deblocking gave the peptolide ring **1343**. Coupling with fragment **1338** then provided a synthetic product which was identical using several criteria to naturally-derived nordidemnin B.

3.3.2 Jaspamide

The cyclodepsipeptide jaspamide (jasplakinolide) was isolated by Faulkner [356] and by Crews [357] from *Jaspis* sp. (sponge). Two unusual amino acid units, β-tyrosine and 2-bromoabrine, are incorporated into this metabolite as well as a polypropionate stretch containing methyl branching at alternating carbons. This metabolite exhibits a broad spectrum of activities including insecticidal, antiparasitic and antifungal properties. An enantiospecific synthesis of **1354** has been reported by Grieco [358]. A large fragment containing the two

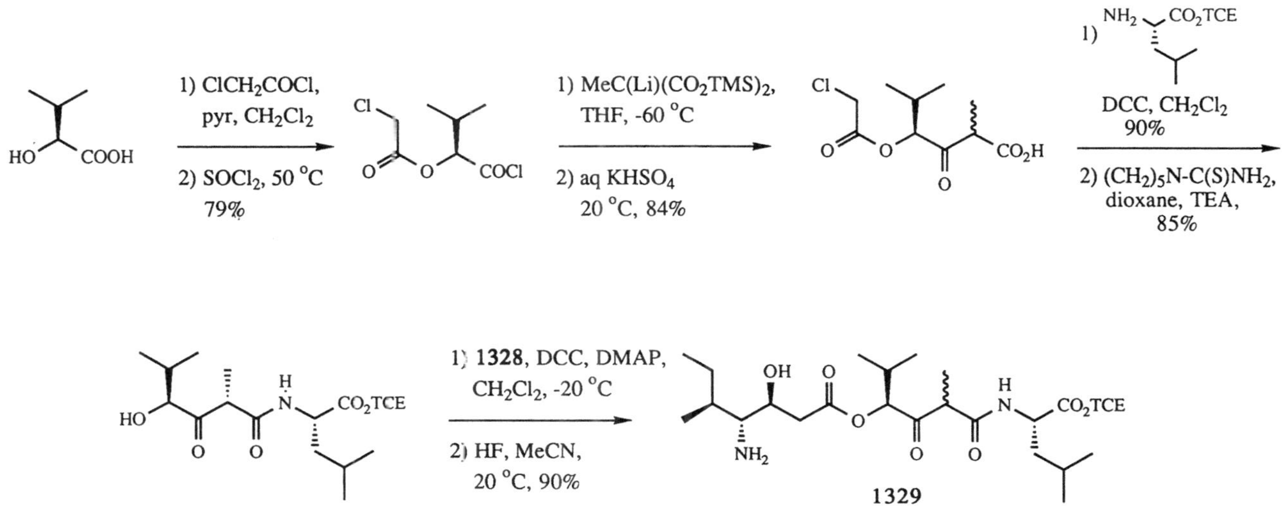

1329

Scheme 216. Synthesis of Didemnin Fragment **1329** (Schmidt)

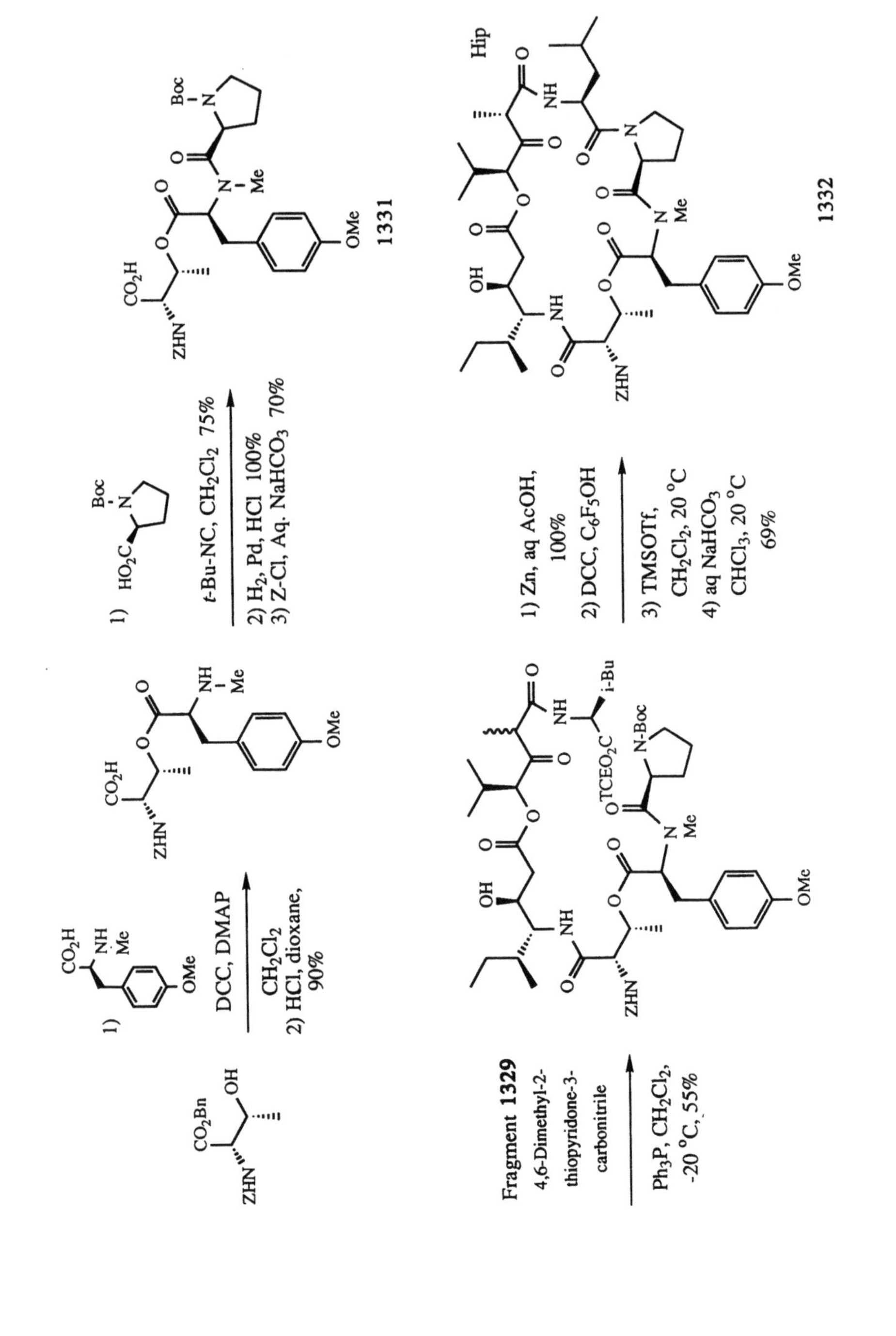
ZHN
CO2Bn
OH
1)
CO2H
NH
Me
OMe
DCC, DMAP
CH2Cl2
2) HCl, dioxane,
90%
1) HO2C
Boc
t-Bu-NC, CH2Cl2 75%
2) H2, Pd, HCl 100%
3) Z-Cl, Aq. NaHCO3 70%
1331
Fragment 1329
4,6-Dimethyl-2-thiopyridone-3-carbonitrile
Ph3P, CH2Cl2,
-20 °C, 55%
TCEO2C
N-Boc
i-Bu
1) Zn, aq AcOH,
100%
2) DCC, C6F5OH
3) TMSOTf,
CH2Cl2, 20 °C
4) aq NaHCO3
CHCl3, 20 °C
69%
Hip
1332

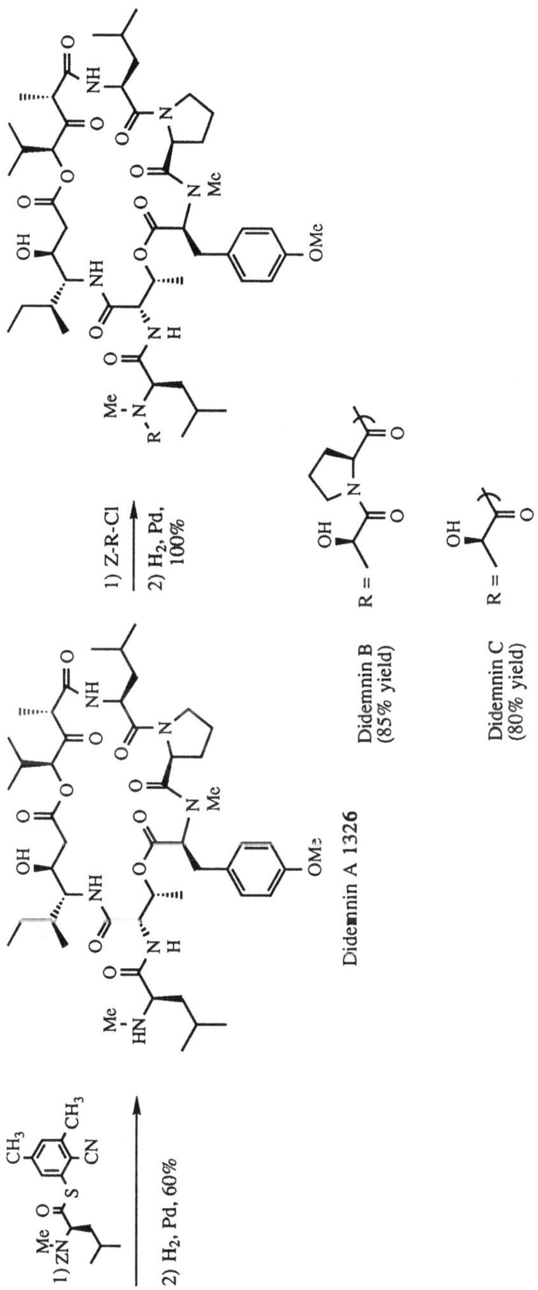

Scheme 217. Combination of Didemnin Fragments (Schmidt)

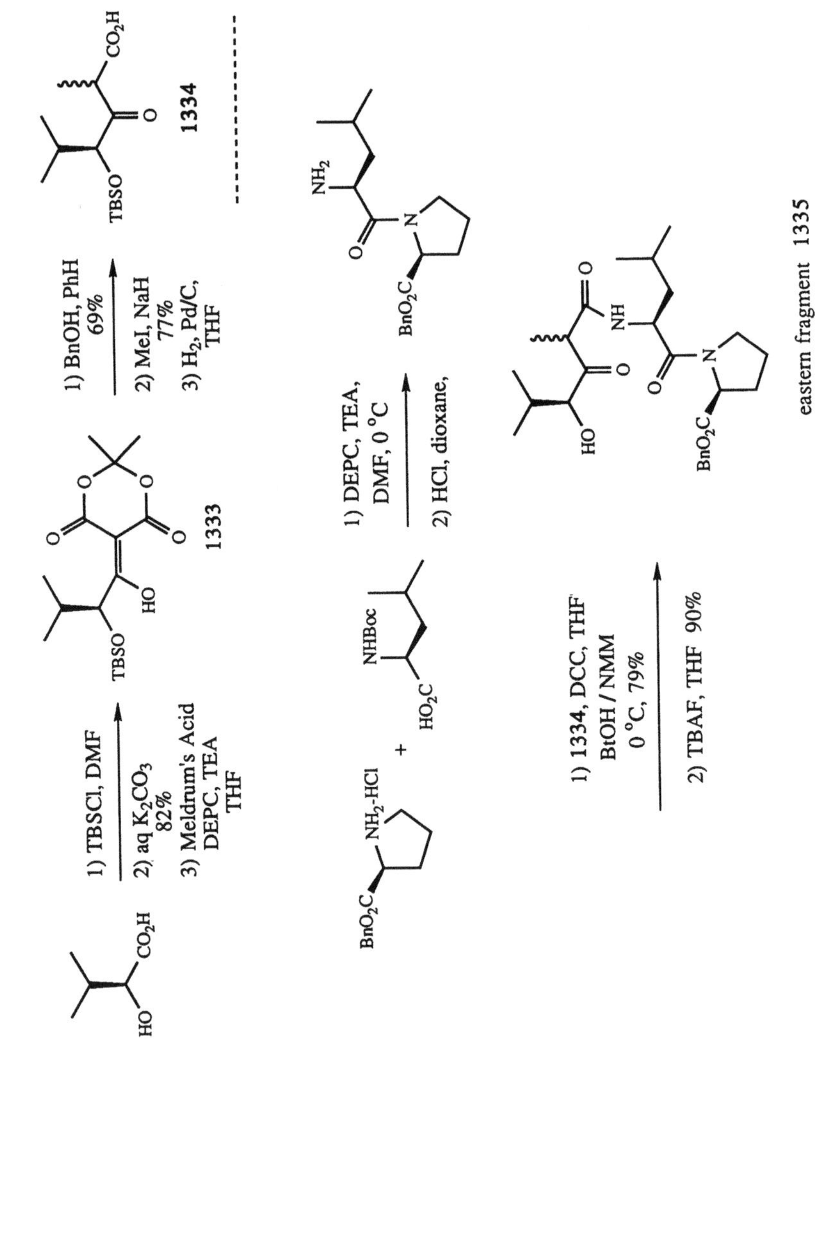

Scheme 218. Synthesis of Didemnin "Eastern" Fragment (Shiori)

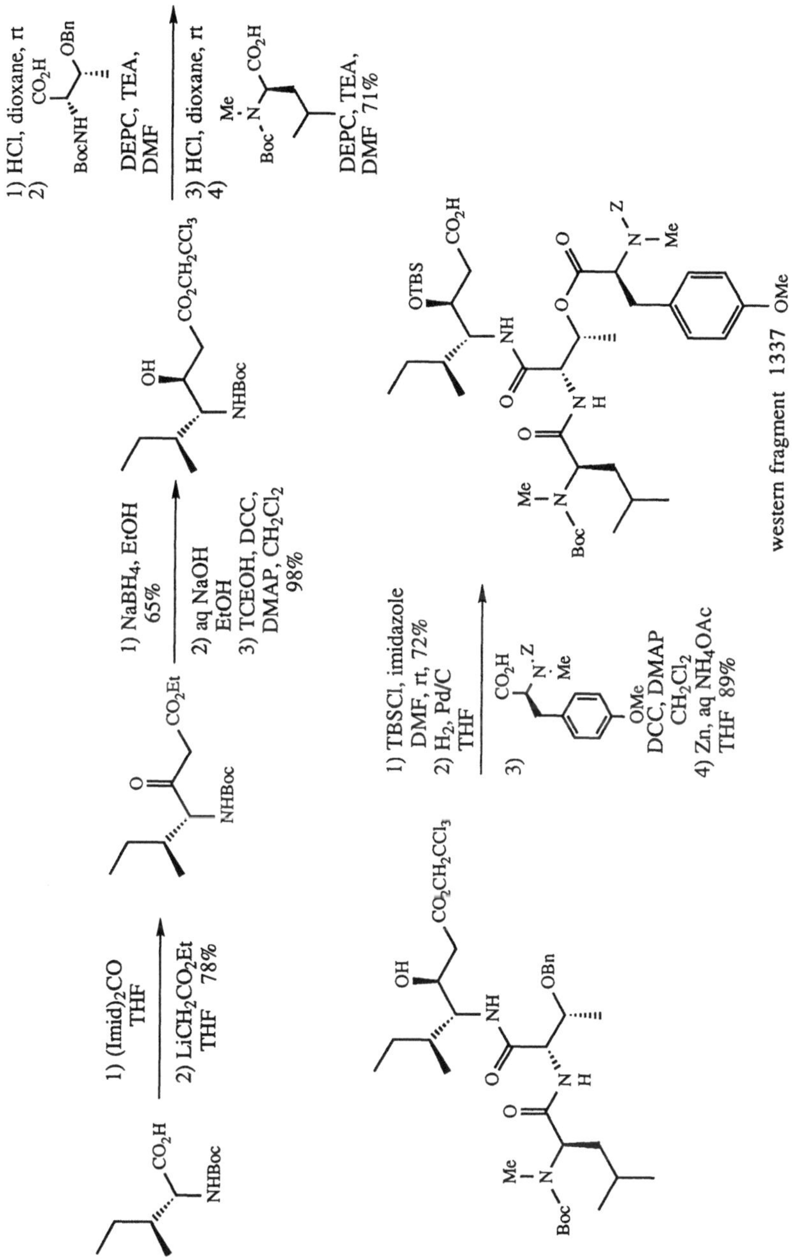

western fragment 1337

Scheme 219. Synthesis of Didemnin "Western Fragment" (Shioiri)

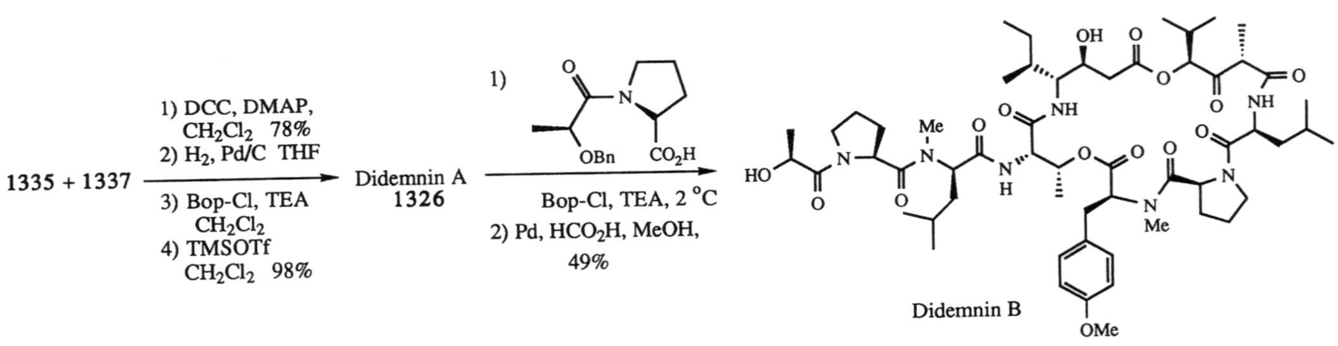

Scheme 220. Combination of Eastern and Western Fragments: Synthesis of didemnins A and B (Shioiri)

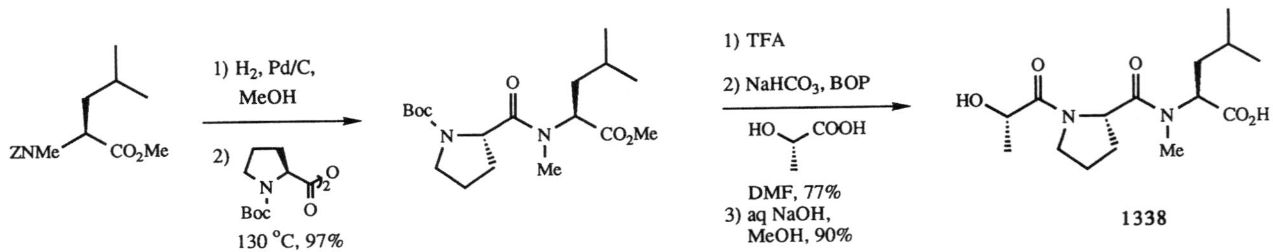

Scheme 221. Jouin Synthesis of Nordidemnin B Fragment **1338**

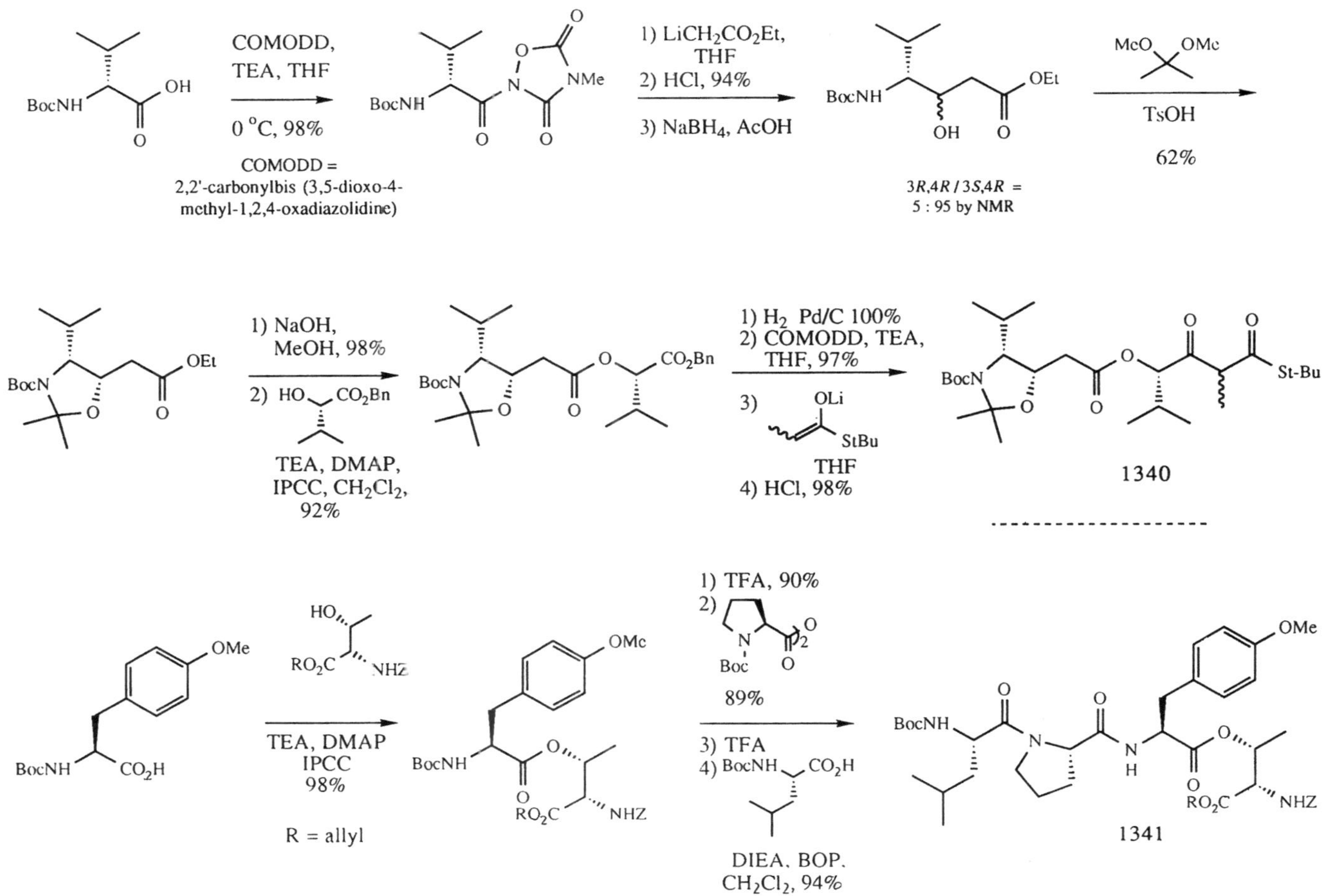

Scheme 222. Jouin Synthesis of Nordidemnin B Fragments

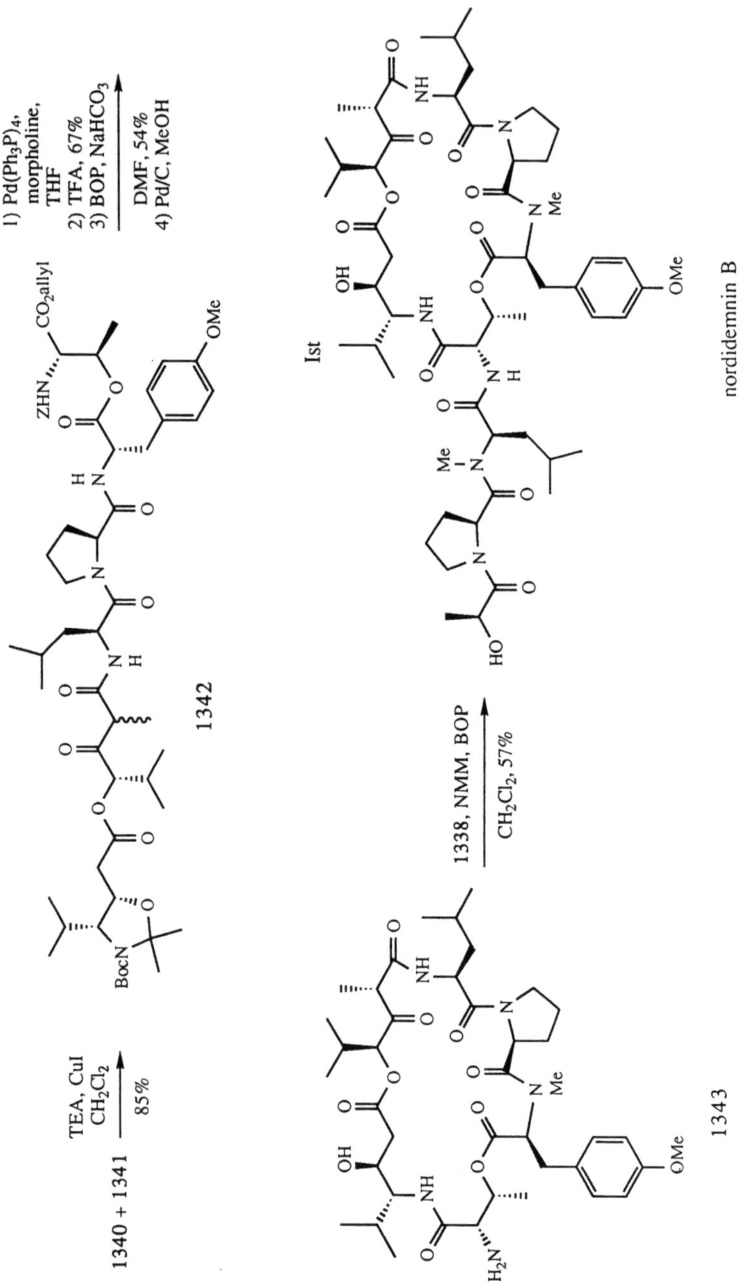

Scheme 223. Union of Nordidemnin B Fragments (Jouin)

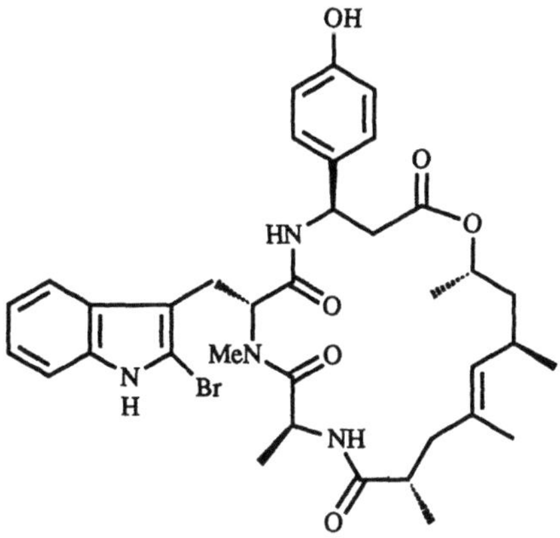

Jaspamide (Jasplakinolide) **1354**

unusual amino acids was prepared (Scheme 224). Homologation of tyrosine **(1345)** produced the blocked β-tyrosine **1346** while blocking and bromination of tryptophan gave bromoindole **1347**. Coupling of these pieces and modification led to the primary amino acid fragment **1348**. Preparation of the remaining fragment and coupling to **1348** are shown in Scheme 225. The optically active acid **1349** was routinely converted to the allylic alcohol **1350**. Orthoester Claisen rearrangement was used to establish the correct *E* olefin geometry which was followed by connection to the Evans chiral auxiliary producing **1351**. Methylation of **1351** via the sodium enolate provided the desired stereochemistry at the α carbon in 71% yield. Conversion to the pyridinethiol ester and coupling with alanine led to **1352**. Connection of the large fragment **1348** at this point produced the seco compound **1353**, which was deprotected and cyclized with DCC/DMAP to yield (+)-jaspamide identical to the natural product.

3.3.3 Geodiamolides A and B

Geodiamolides A and B [359], isolated from a sponge of the genus *Geodia* contain the same nonenoic acid fragment as jaspamide and differ only in the identity of the halogen on the tyrosine unit. Grieco has also synthesized (+)-geodiamolide B [360] (Scheme 226). An appropriately substituted tyrosine unit **(1356)** was prepared and brominated to give **1357**. Coupling to alanine at both the carboxyl and amino groups gave **1358**, ready for connection of the nonenoic acid piece. This was accomplished via formation of the amide bond which was followed by deblocking the hydroxyl and carboxylic acids prior to lactonization, which was carried out using the DCC/DMAP method as in their jaspamide synthesis. Desilylation of the phenolic ether gave (+)-geodiamolide B.

Geodiamolide A has been produced by two groups. White [361] followed an approach similar to Grieco in the assembly of fragments (Scheme 228).

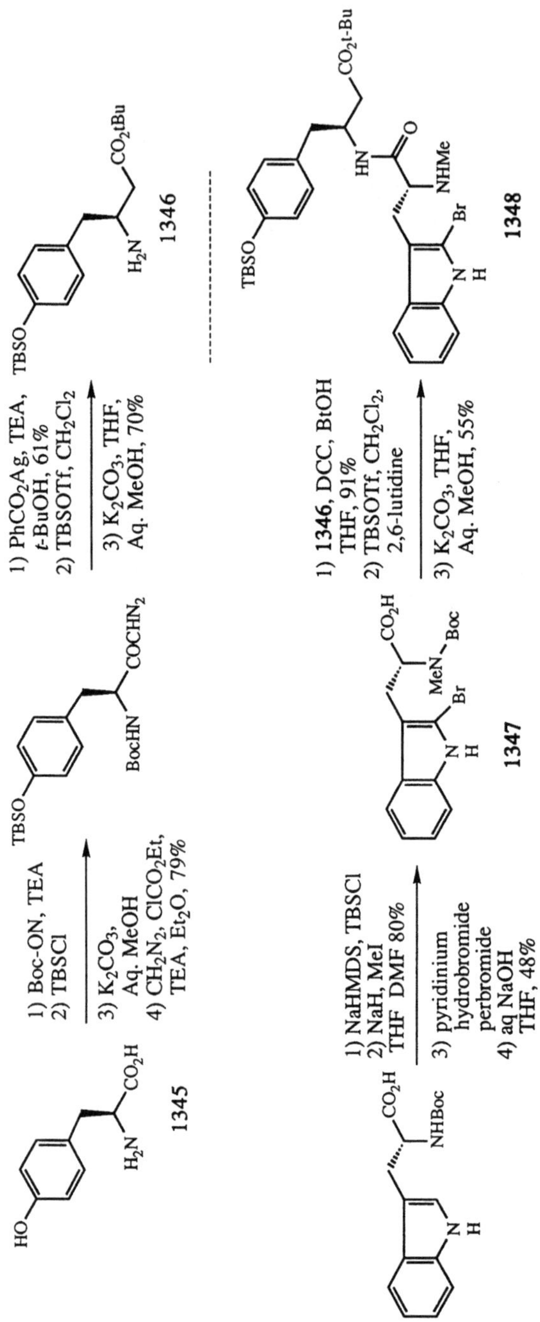

Scheme 224. Synthesis of Jaspamide (Jasplakinolide) Fragment **1348** (Grieco)

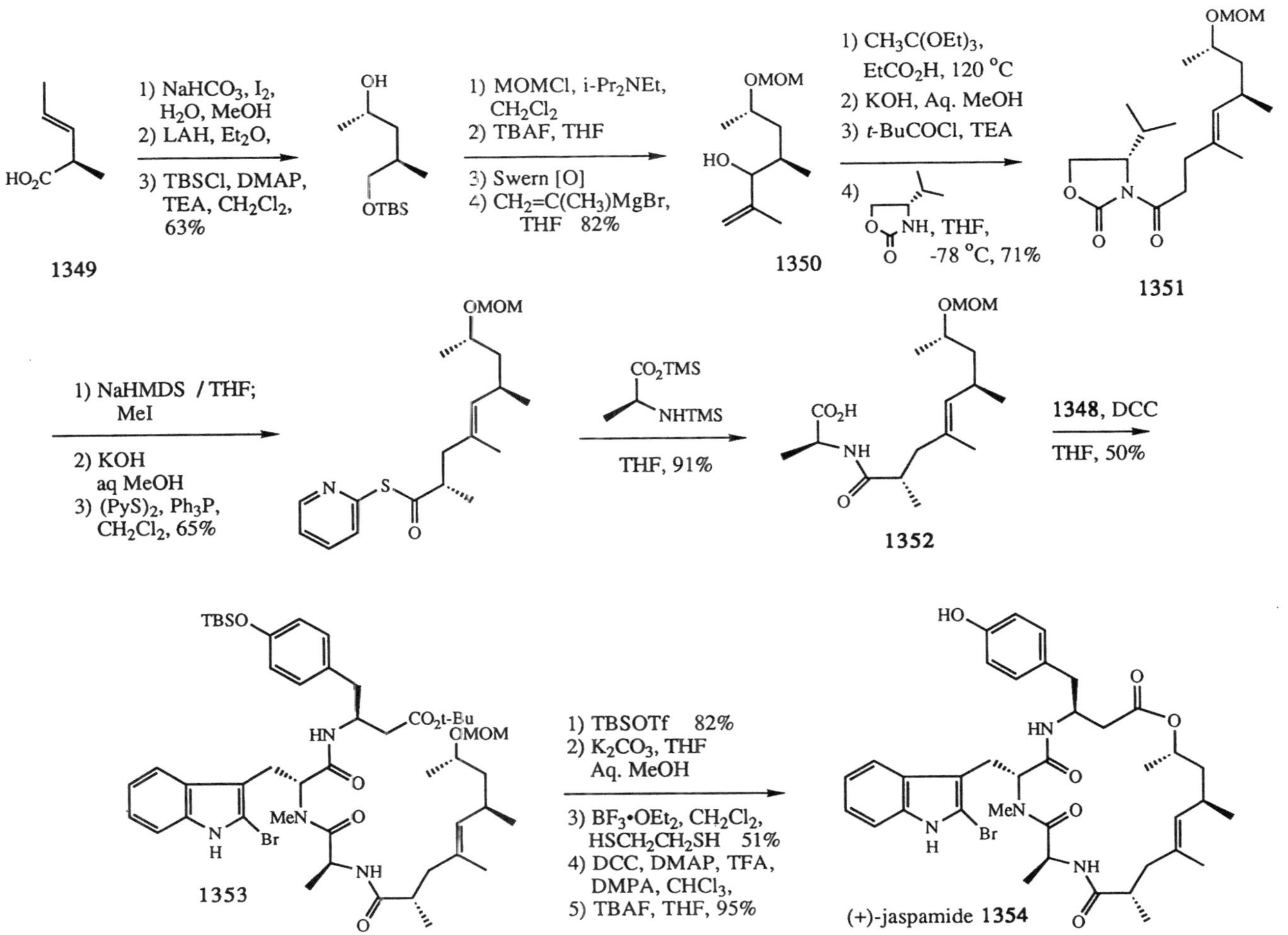

Scheme 225. Grieco Synthesis of (±)-Jaspamide (Jasplakinolide)

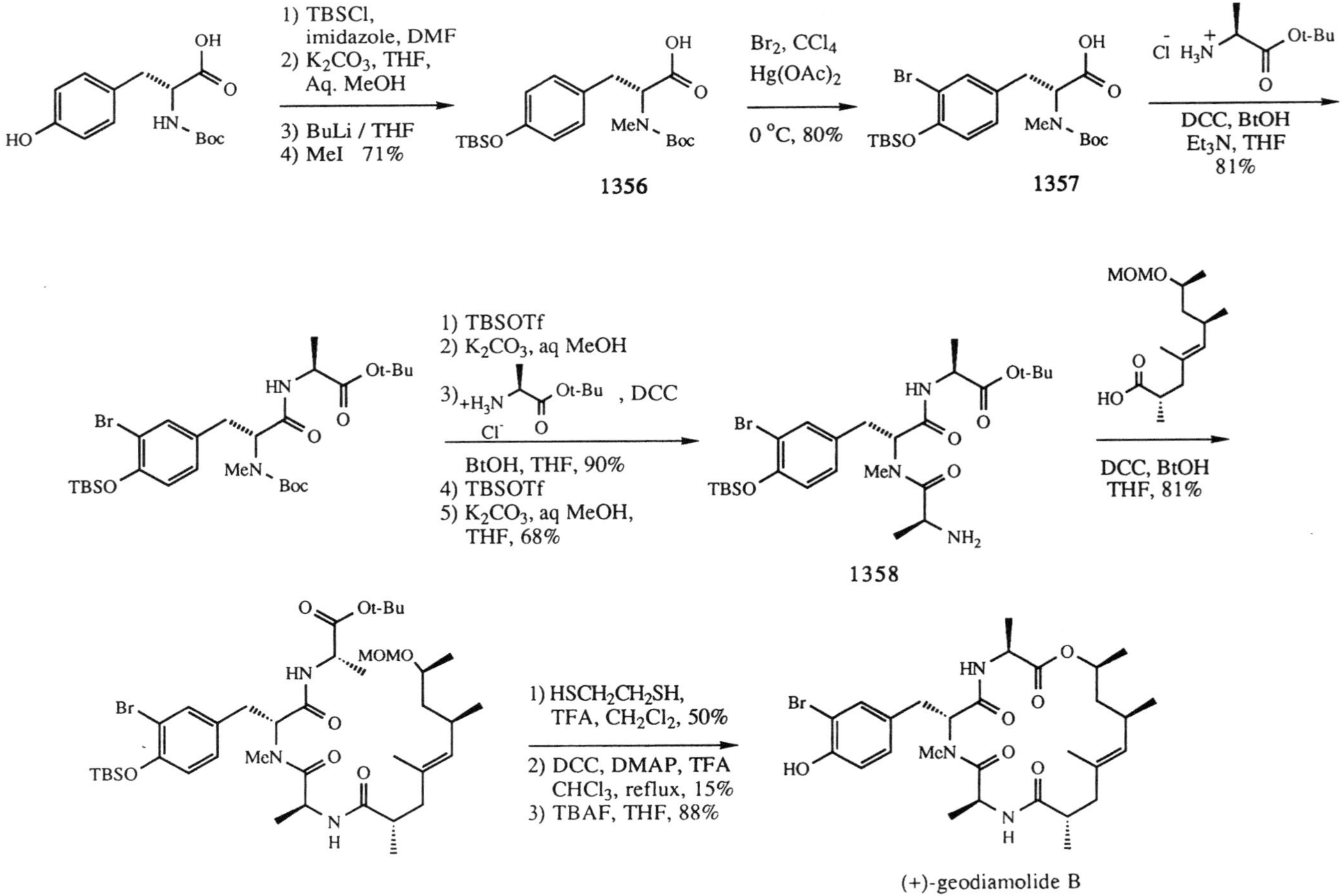

Scheme 226. Grieco Synthesis of (±)-Geodiamolide B

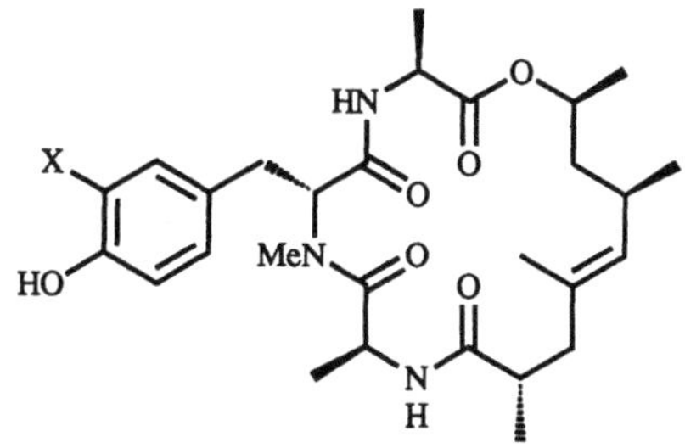

X = I Geodiamolide A
= Br Geodiamolide B

Tyrosine was iodinated and converted to the appropriately blocked substance **1363**. *N*-methylation and coupling to two alanine units using standard chemistry produced tripeptide **1364**. Connection of the nonenoic acid fragment via amide formation and macrolactonization via the DCC/DMAP method led to the natural product. The construction of the nonenoic acid fragments shown in Scheme 227 also features a [3,3] sigmatropic rearrangement strategy for setting the geometry of the trisubstituted olefin. The allylic alcohol **1360** was generated from the known optically active lactone **1359** via standard transformations. Orthoester Claisen rearrangement in this case was effected with triethyl orthopropionate, such that the methyl group at C2 did not have to be added subsequently. However, the Claisen rearrangement only proceeded with 1.5:1 selectivity and required an HPLC separation of C2 epimers.

The Momose synthesis [362] of (+)-geodiamolide A followed only a slightly different pathway from that of White and Grieco (Scheme 230). An analogous ala-tyr-ala tripeptide fragment was constructed, but iodination was deferred until the last stages, eventually producing **1368**. Connection of the nonenoic acid fragment (Scheme 229) was analogous to the other syntheses, proceeding once again with amide formation first. Macrolactonization was accomplished using the Yamaguchi procedure [363] involving a DMAP mediated cyclization. The nonenoic acid fragment was again assembled using [3,3] sigmatropic rearrangement, as the Ireland variant and was reported to give a 77% yield of the desired isomer **1366** and only 6% of the C2 epimer. The synthesis was initiated with the known optically active (*S*)-propylene oxide **1365**.

3.3.4 Thiazole-Containing Peptides

3.3.4.1 Patellamides A, B and C

A number of cancer-active cyclic peptides containing thiazole and oxazoline rings have been described from sponges, as exemplified by patellamides A, B and C [364], ulicyclamide [365] and ulithiacyclamide [366]. Shioiri and Schmidt have been most active in this area, with several syntheses being accomplished.

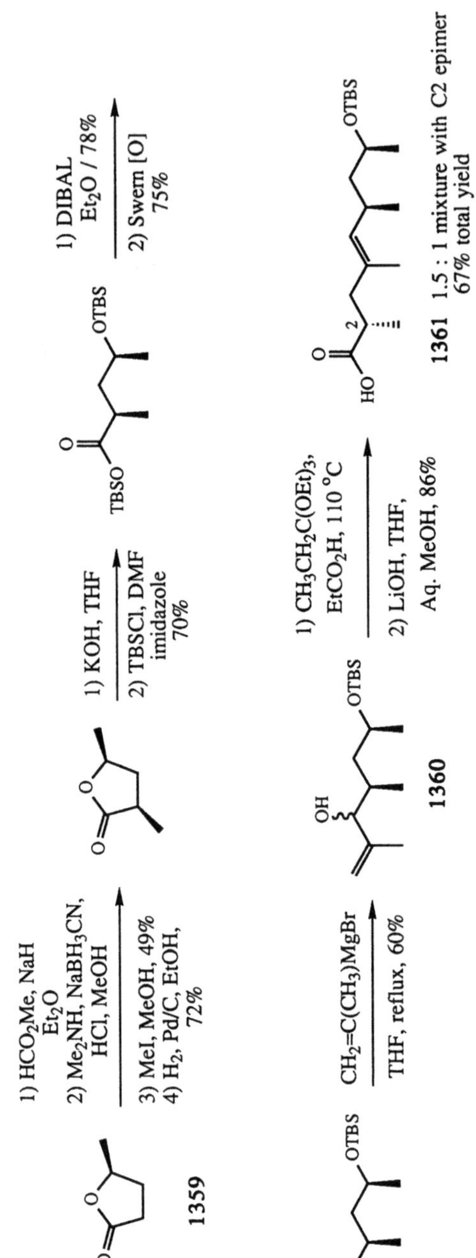

Scheme 227. White Synthesis of Geodiamolide A Nonenoic Acid Fragment

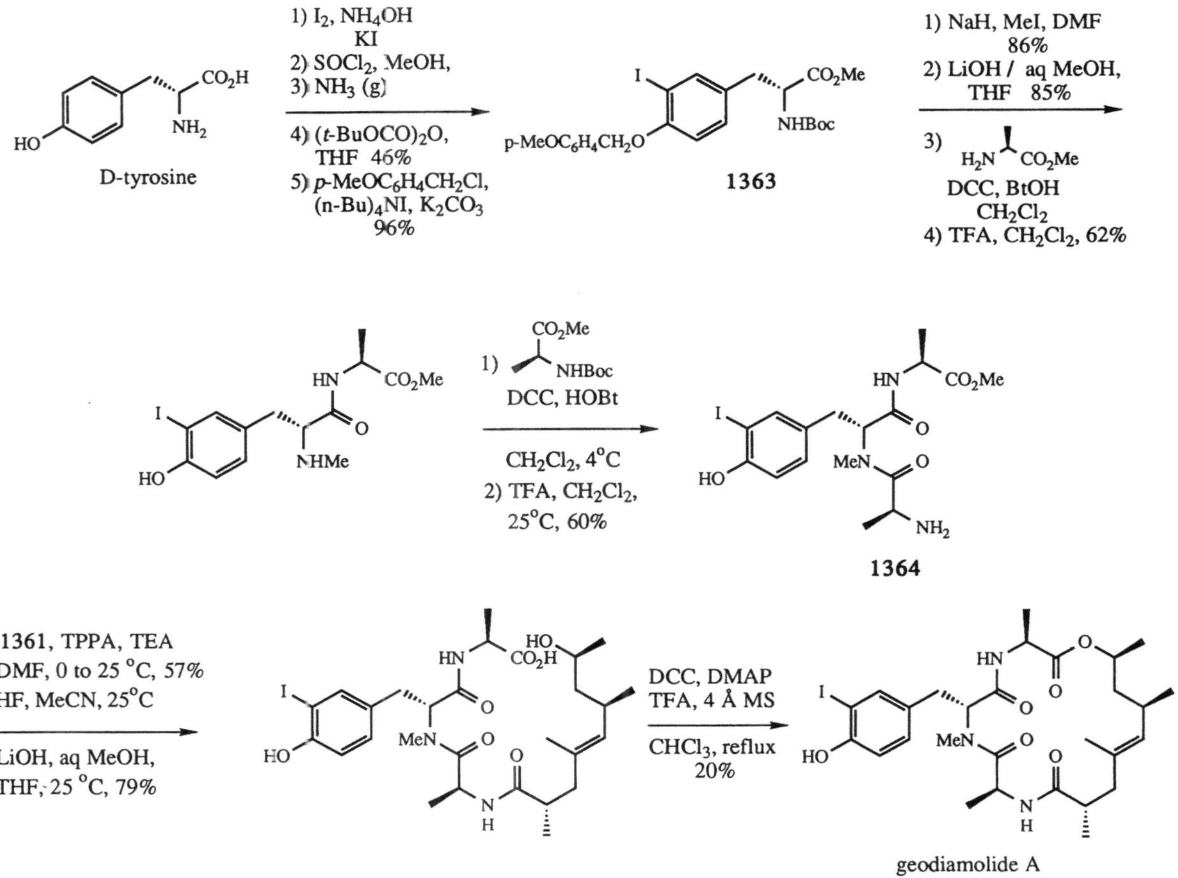

Scheme 228. Assembly of Geodiamolide A Fragments (White)

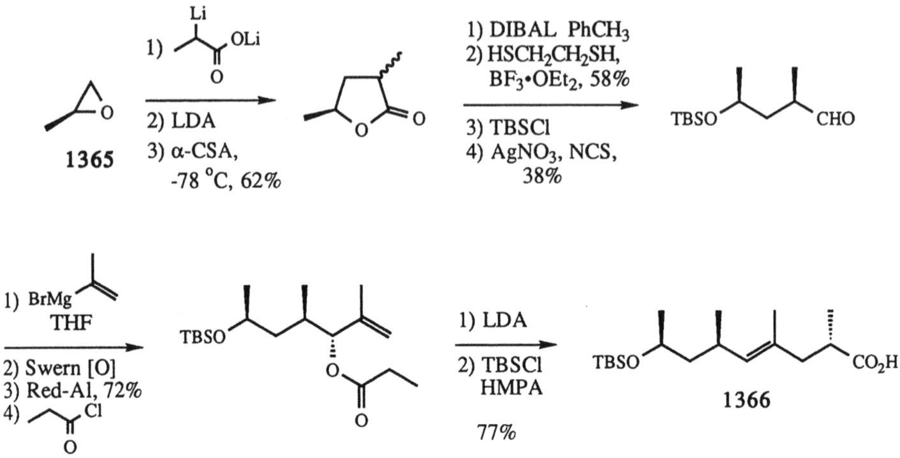

Scheme 229. Synthesis of Geodiamolide A Nonenoic Acid Fragment (Momose)

Shioiri has constructed patellamides A [364], B, and C [368] using common fragments and a common strategy. Scheme 231 shows a patellamide A synthesis using the strategy of constructing two tripeptides **1370** and **1371** from a common thiazole **1369**. Condensation of **1370** and **1371** provides the seco compound **1372** which is closed just prior to bis-thiazole formation by formal cyclization of hydroxyls onto amide carbonyls. Patellamides B and C, differing only in the non-thiazole units were assembled using essentially the same approach (Schemes 232 and 233). These three syntheses, along with syntheses of several incorrect isomers, finally led to the correct formulation of these metabolites.

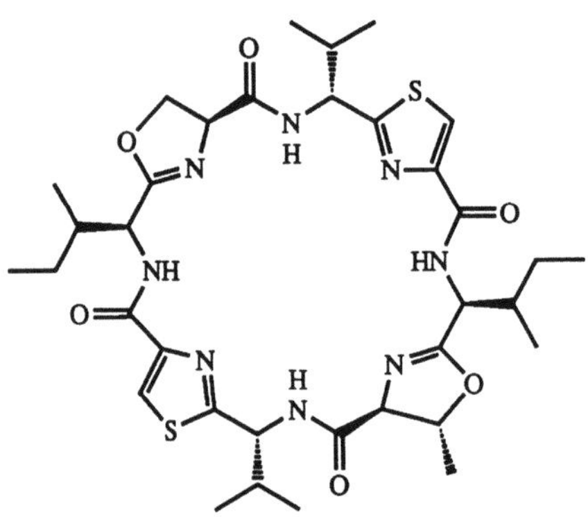

patellamide A

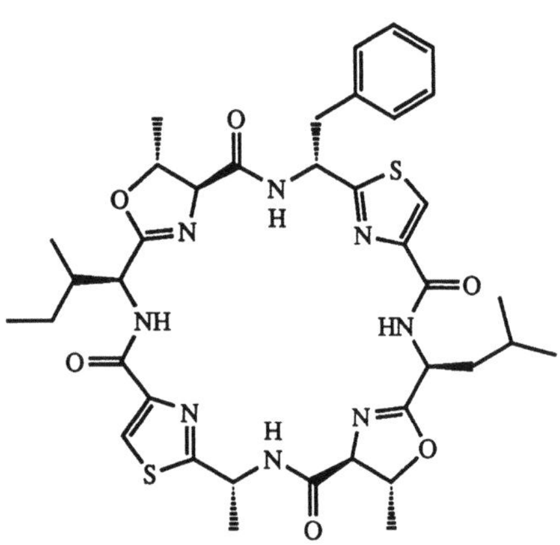

patellamide B R = isobutyl
patellamide C R = isopropyl

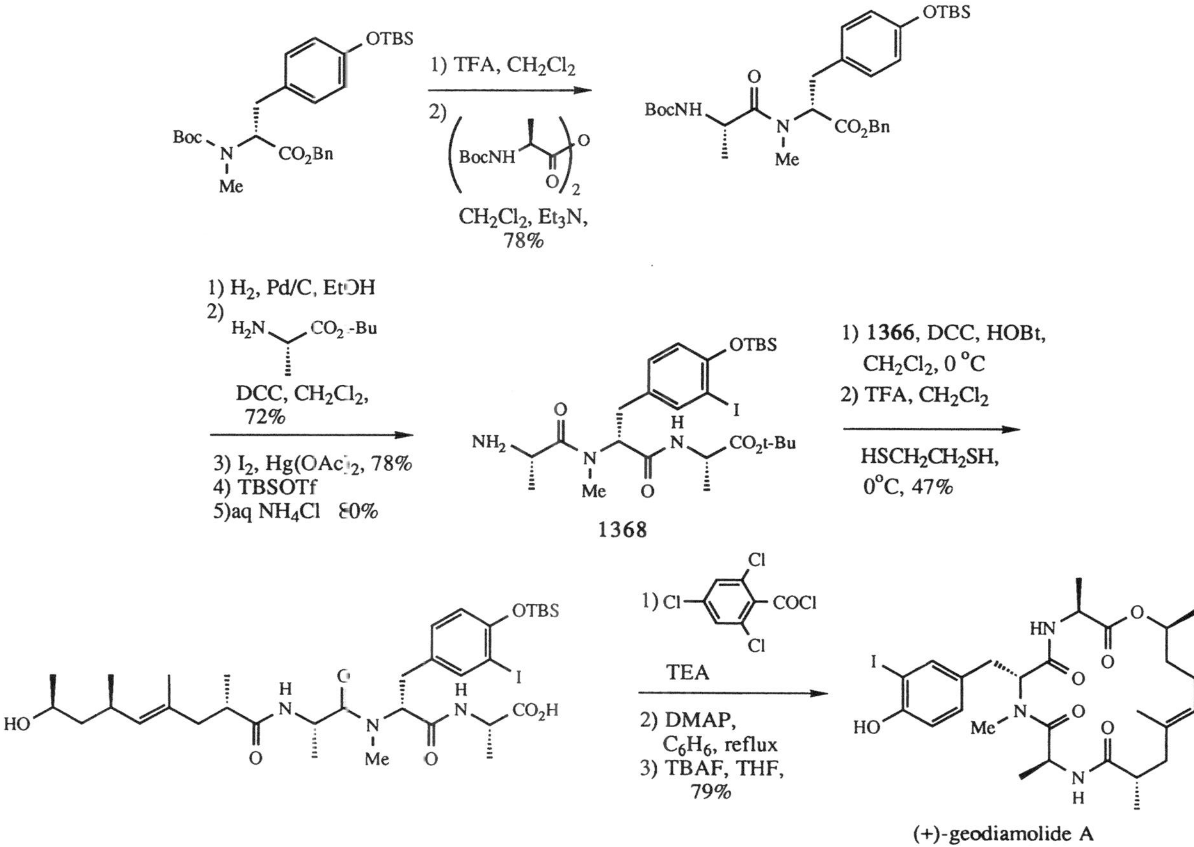

Scheme 230. Momose Synthesis of (±)-Geodiamolide A

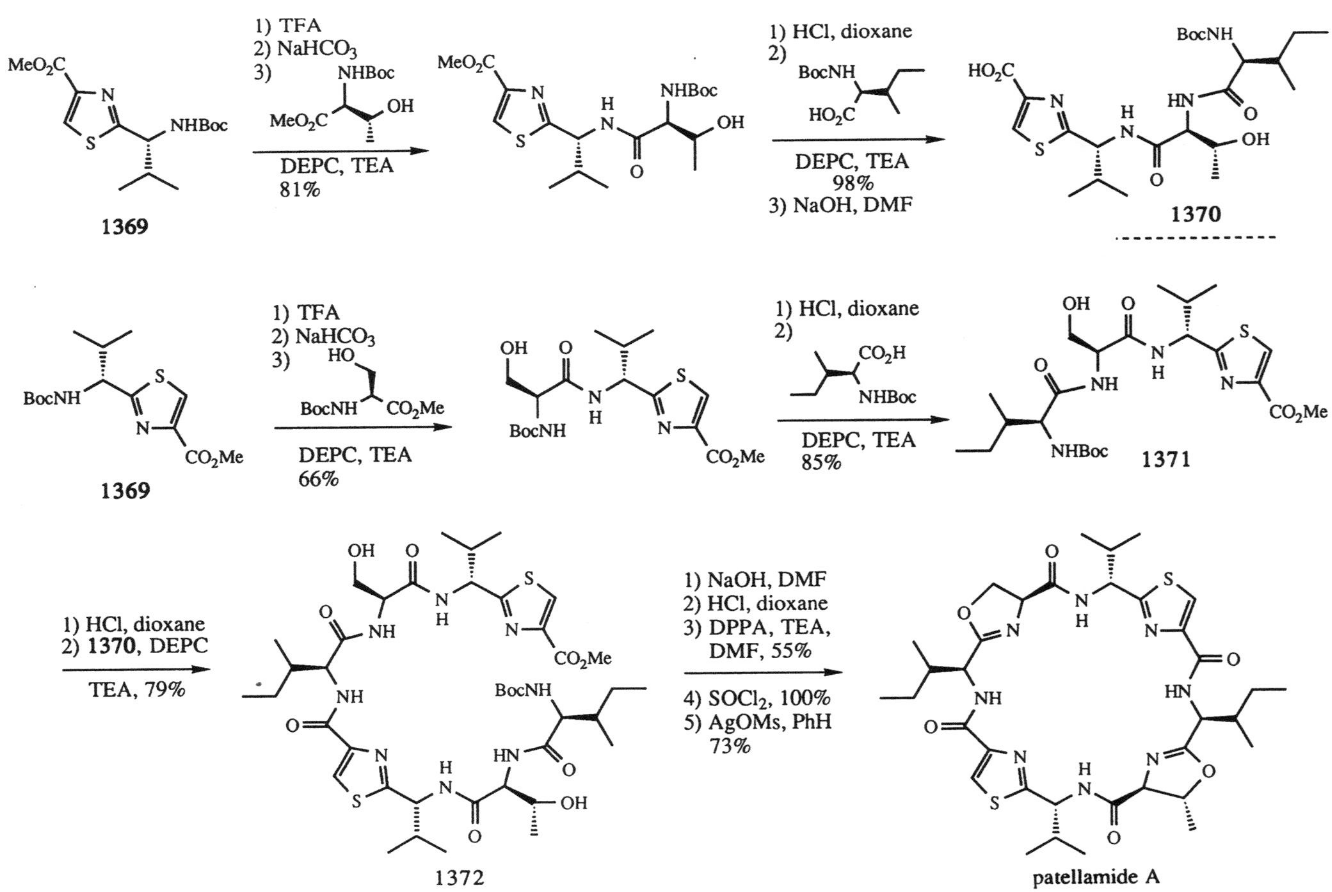

Scheme 231. Shioiri Synthesis of Patellamide A

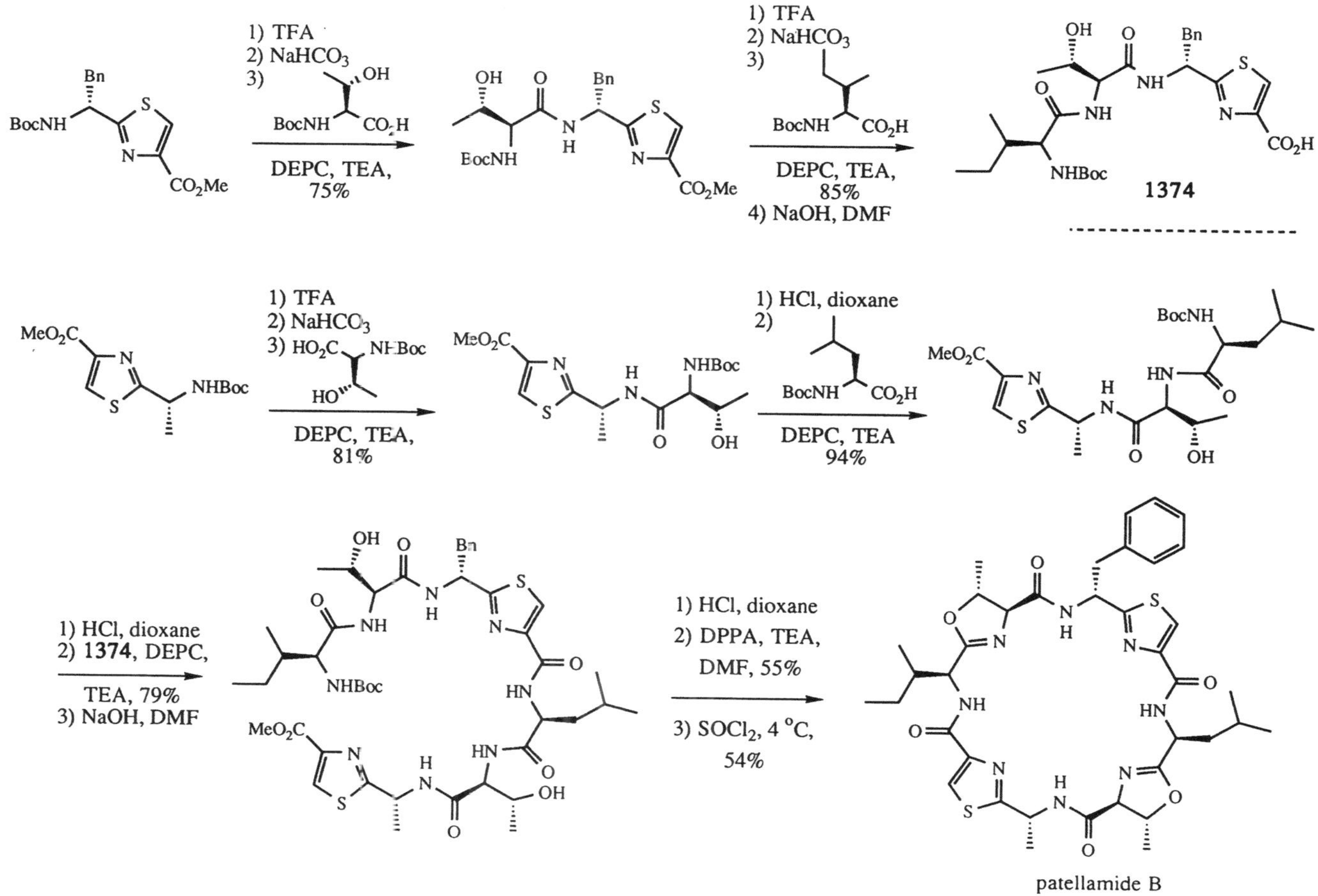

Scheme 232. Shioiri Synthesis of Patellamide B

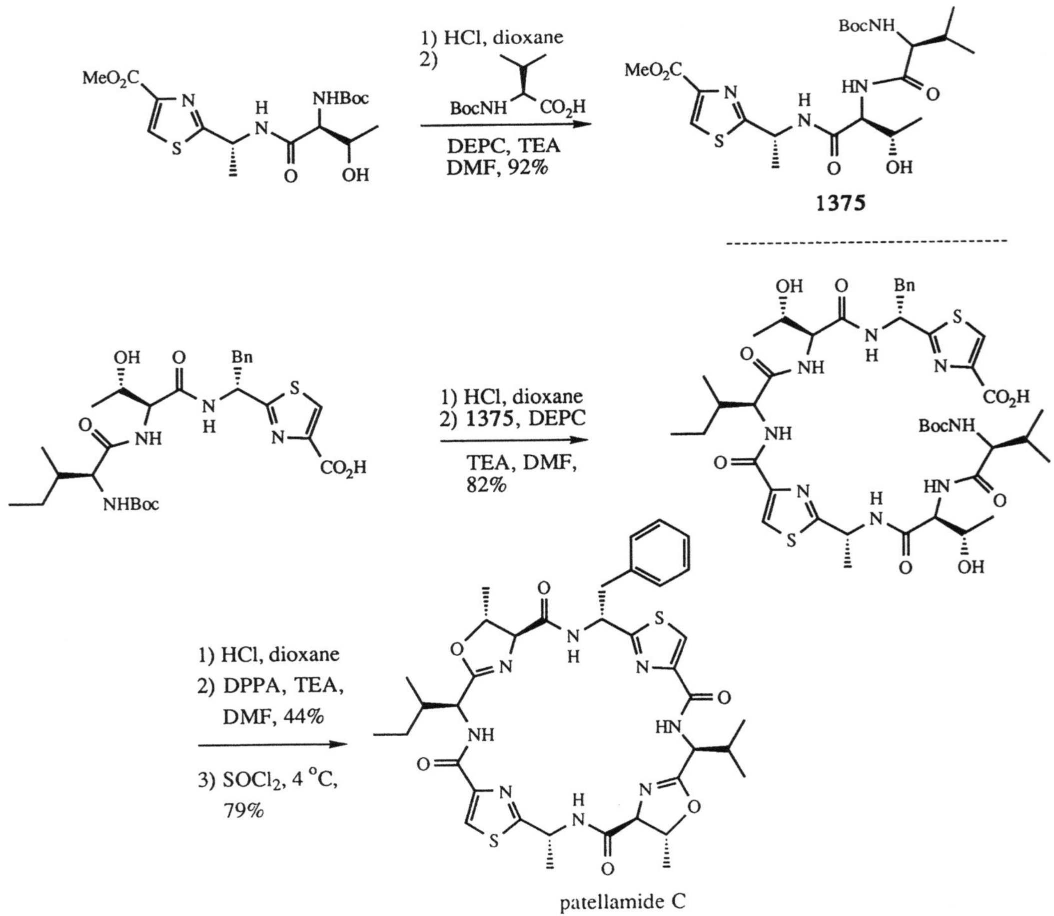

Scheme 233. Shioiri Synthesis of Patellamide C

Schmidt also recorded a synthesis of patellamide B [369] using the thiazole **1376** as the primary building block (Scheme 234). Mitsunobu inversion of the alcohol to the azide and reduction led to an amine which was DCC-condensed with a blocked threonine to give **1377**. Deblocking of the threonine was followed by reaction of the amino-alcohol with the imidate **1380** generating the oxazoline ring **(1381)** *prior* to final cyclization. A similar set of reactions was used to convert **1379** to a second tripeptide fragment **1381**. Combination of these two led to the seco compound **1382** which was deblocked, activated and closed to patellamide B.

3.3.4.2 Ulicyclamide and Ulithiacyclamide

In their syntheses of ulicyclamide and ulithiacyclamide Shioiri and Schmidt used chemistry analogous to their respective patellamide syntheses. The Shioiri ulicyclamide synthesis [370] (Scheme 235) used a proline anchored to a polystyrene base and sequentially added two thiazole-containing units and a threonine to give **1384**. At this point the resin was removed and the synthesis completed by macrolactamization and formation of the oxazoline ring. The Schmidt synthesis of ulicyclamide [371] proceeded as shown in Scheme 237 and features the coupling of two thiazole units to give the intermediate bis-thiazole **1387**. Mitsunobu inversion of the alcohol to the amine and coupling with a blocked threonine unit followed by deblocking gave **1388**. Coupling of **1388** with an Ala-Pro unit via oxazoline formation led eventually to the penultimate intermediate **1389** which was macrolactamized to ulicyclamide. An example of the synthesis of a thiazole-containing unit is shown in Scheme 236.

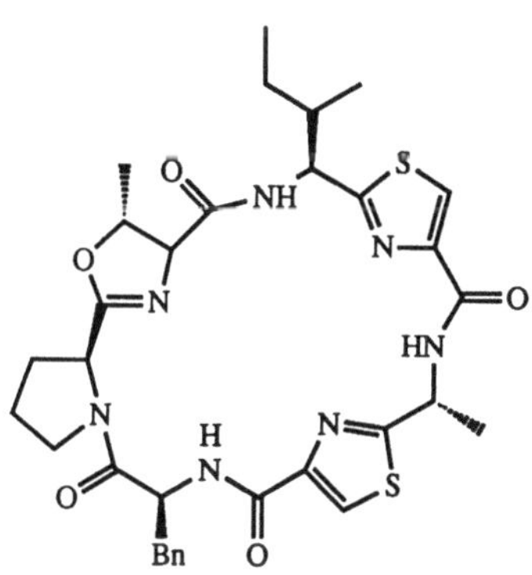

Ulicyclamide

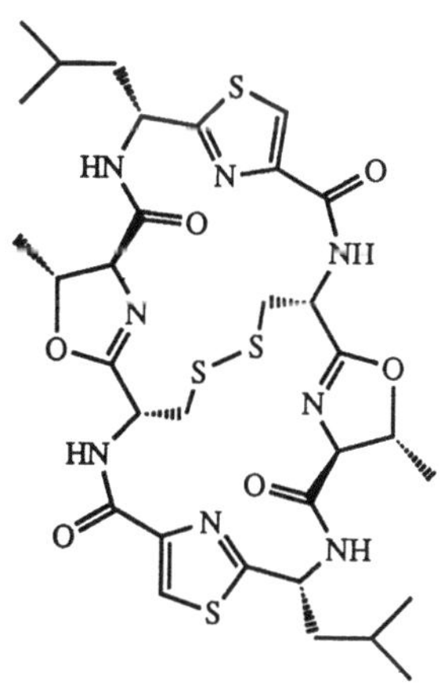

Ulithiacyclamide

Ulithiacyclamide contains a disulfide bridge of Cys units and, once again, has been synthesized by Schmidt [372] and Shioiri [373]. The Schmidt approach (Scheme 238) to this symmetric substance involved the coupling of two

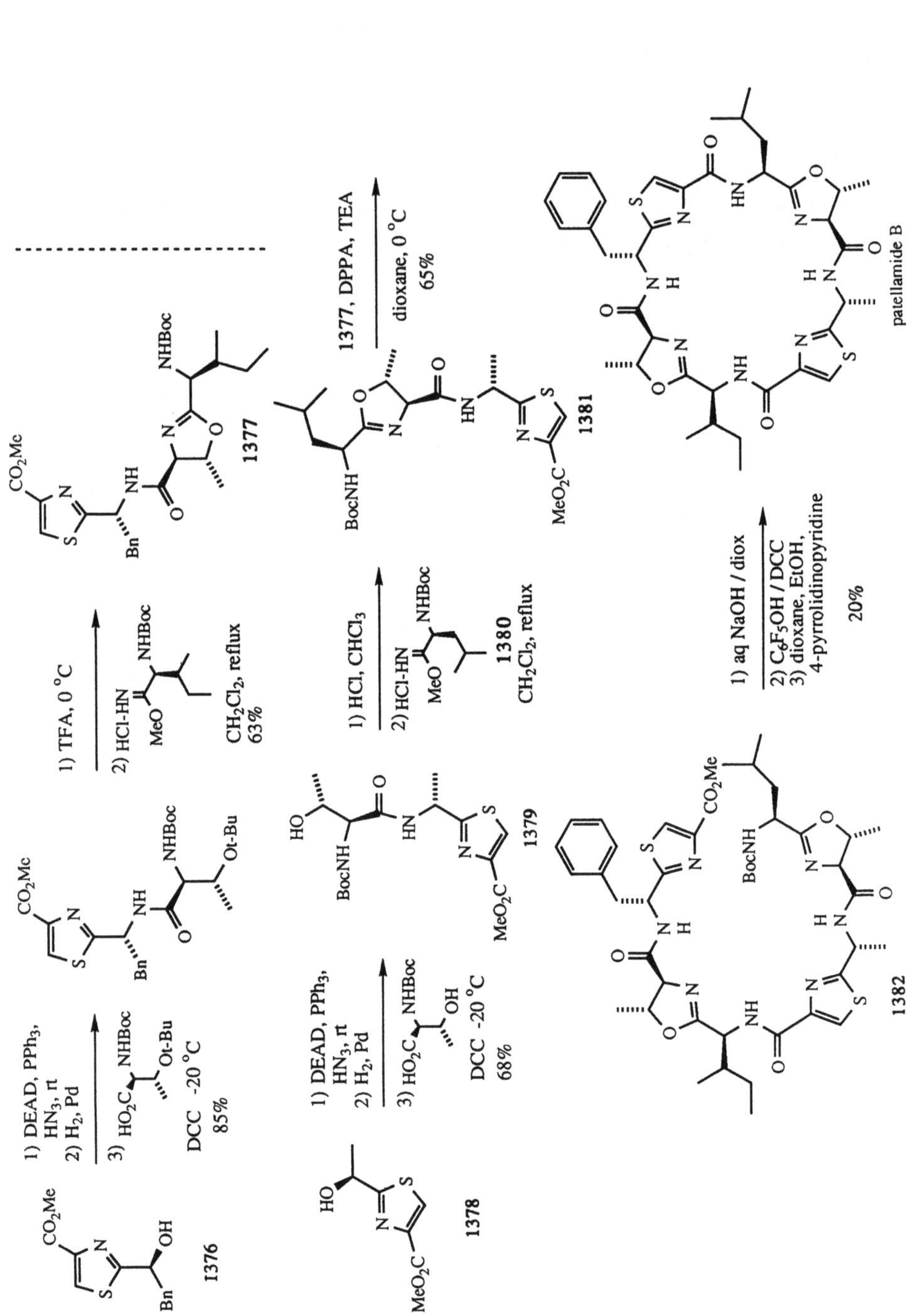

Scheme 234. Schmidt Synthesis of Patellamide B

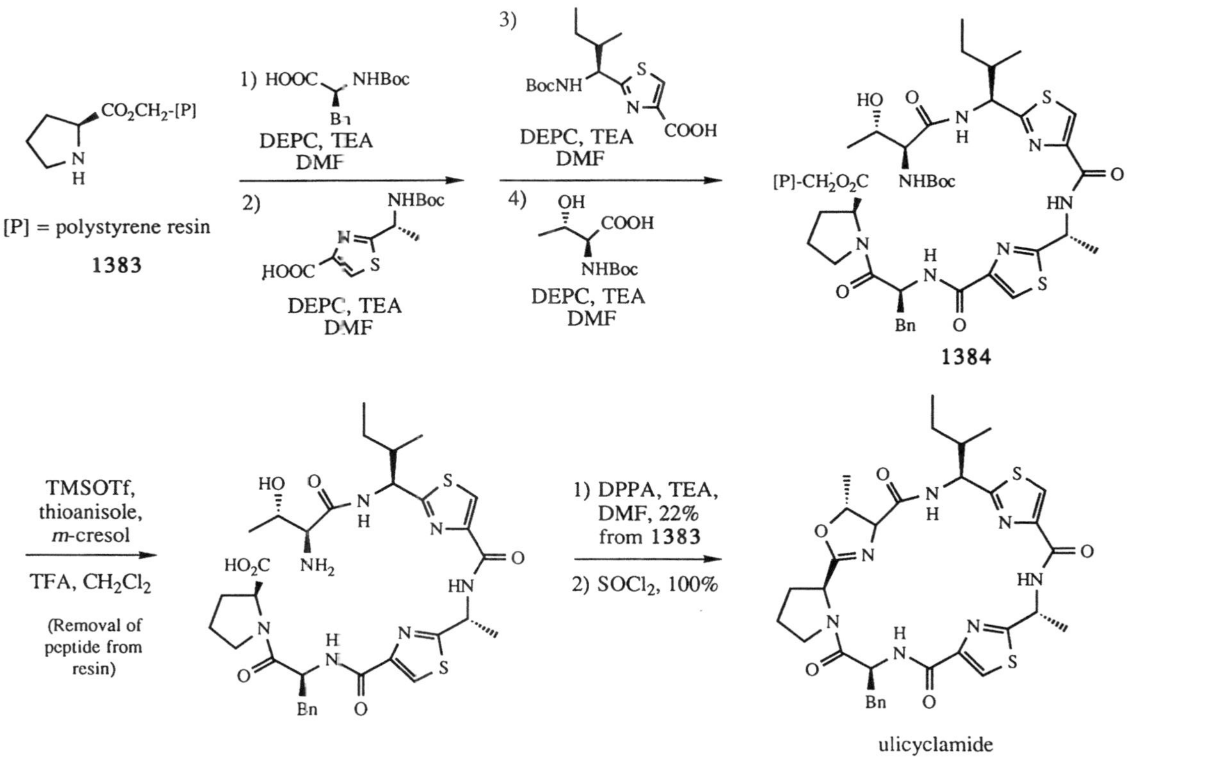

- an extensive deblocking/washing procedure was used between each coupling reaction

Scheme 235. Shioiri Synthesis of Ulicyclamide

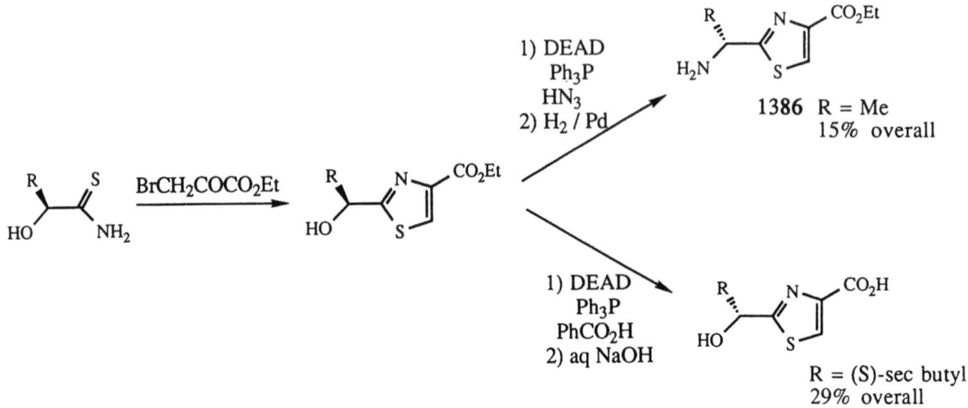

Scheme 236. Synthesis of Thiazole Units (Schmidt)

identical thiazoles to a sulfur-linked Cys dimer **1390**. After functional group manipulation, the bis-seco compound **1391** was obtained. Sequential cyclization of **1391** gave rise to **1392** which was submitted to $SOCl_2$ to complete an efficient pathway. The Shioiri approach is equally efficient (Scheme 239), involving dimerization of the tripeptide fragment **1394**. This was followed by intramolecular disulfide formation and finished off with closure of the oxazoline rings.

3.3.4.3 Ascidiacyclamide and Dolastatin-3

Ascidiacyclamide (**1395**) [374] is a simpler member of this class, being essentially a cyclic dimer. Shioiri [375] assembled **1395** using previously outlined technology as shown in Scheme 240, involving as the key step the cyclodimerization. The synthesis required only 9 steps and established the absolute configuration of this metabolite.

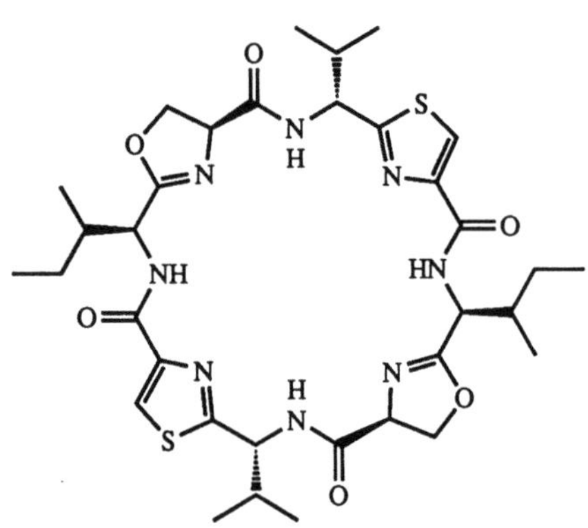

Ascidiacyclamide **1395**

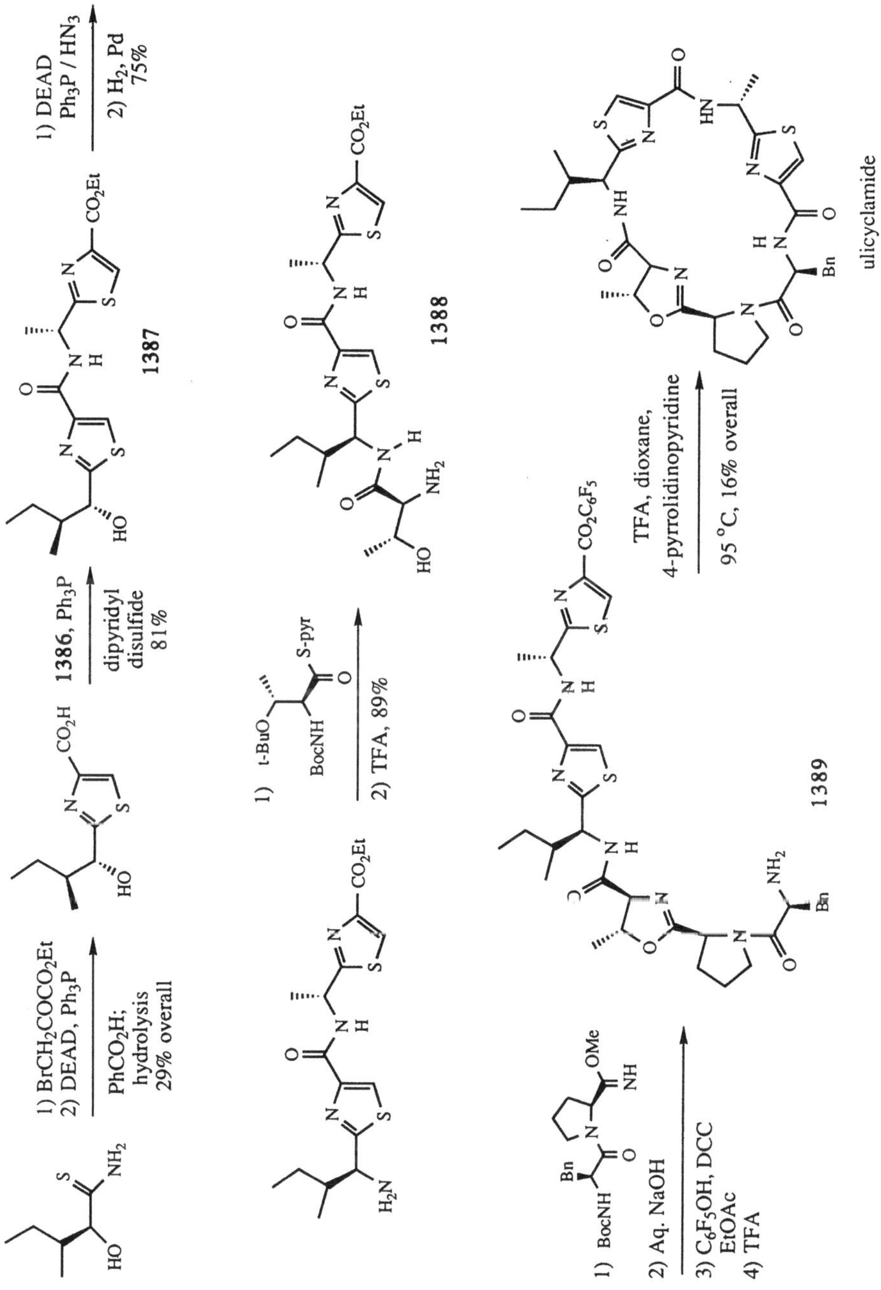

Scheme 237. Schmidt Synthesis of Ulicyclamide

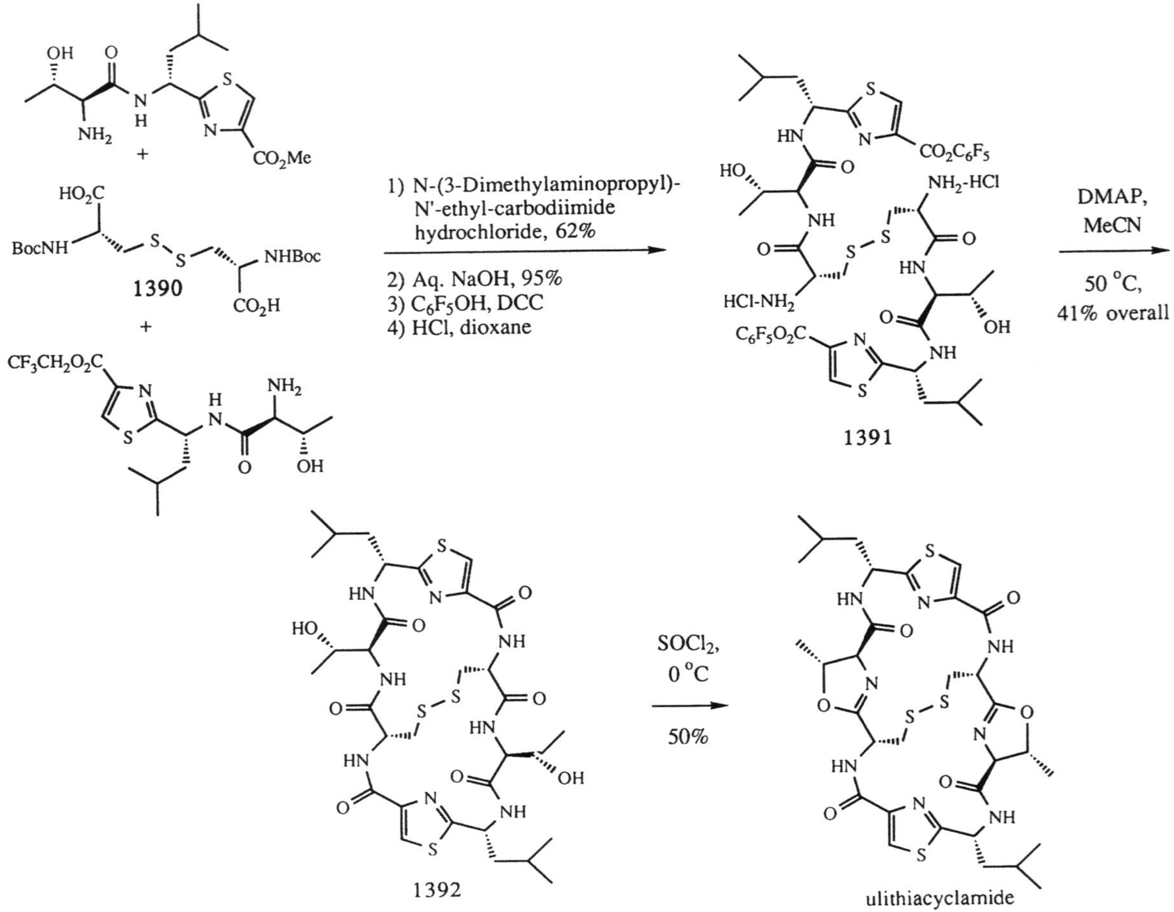

Scheme 238. Schmidt Synthesis of Ulithiacyclamide

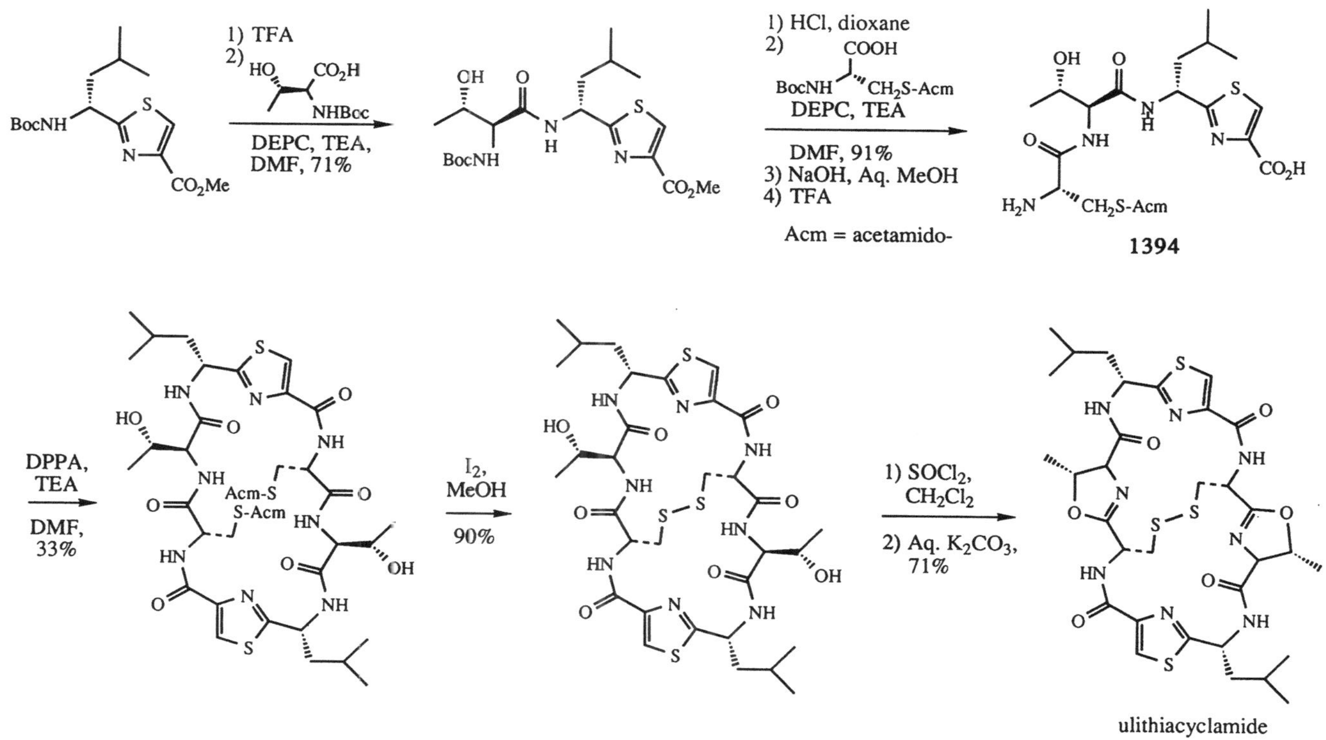

Scheme 239. Shioiri Synthesis of Ulithiacyclamide

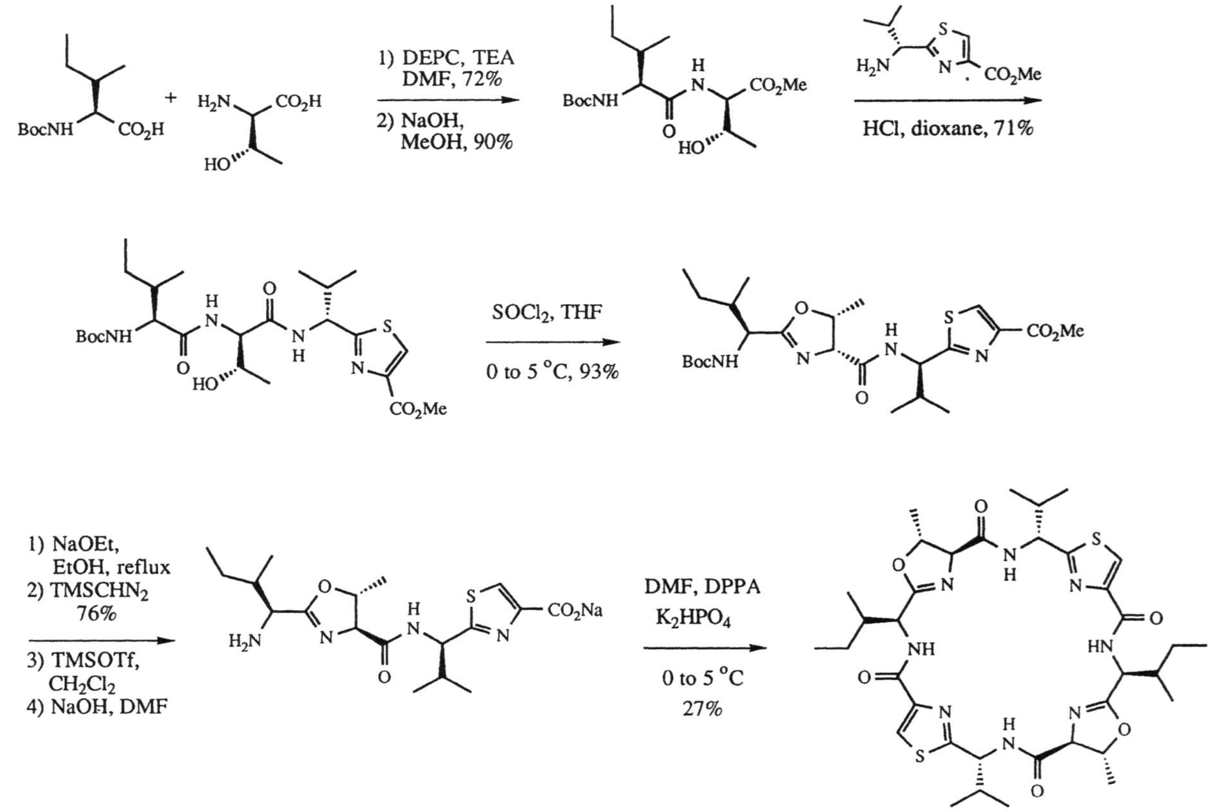

ascidiacyclamide 1395

Scheme 240. Shioiri Synthesis of Ascidiacyclamide

Elucidating the structure of dolastatin-3 has required extraordinary effort due to the minute amounts obtainable from the natural source (*Dolabella auricularia*) and the chemical instability of this substance. Several synthetic ventures failed to produce the natural peptide. Pettit finally succeeded in pinning down the structure of this cyclic pentapeptide as **1398**. This was accomplished by a standard linear sequence of condensations as shown in Scheme 241. Beginning with proline, a valine unit was followed by two thiazole containing amino acids and a leucine producing the seco peptide **1397**. Closure was effected in good yield to produce a substance identical to the natural (−)-enantiomer.

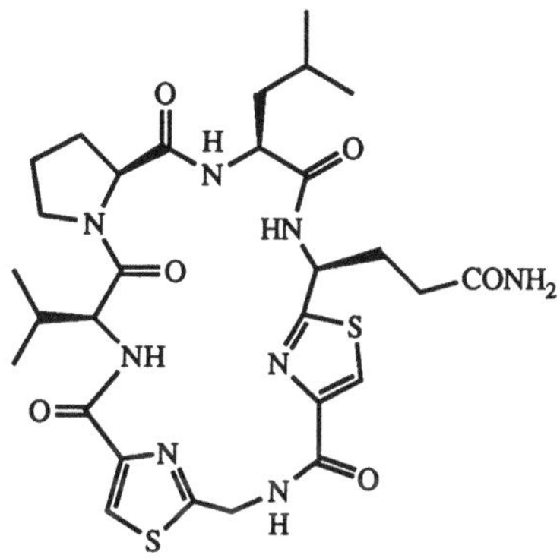

(-)-Dolastatin 3 **1398**

3.3.5 Dolastatin-10 and Hexaacetylcelenamide

Several acyclic marine peptides have attained importance due to their anticancer properties. The dolastatin family, in particular, seems to show much promise in the development of new antineoplastic agents. Isolated by Pettit from the sea hare *Dolabella auricularia*, dolastatin-10 [376] **(1408)** is a very promising cancer-active substance because of its *in vivo* assay performance. Pettit has also synthesized **1408** confirming the absolute stereochemistry of this important substance (Scheme 243) [377]. The two uncommon amino acids and the aminothiazole units were prepared as shown in Scheme 242. Dolaisoleuine was

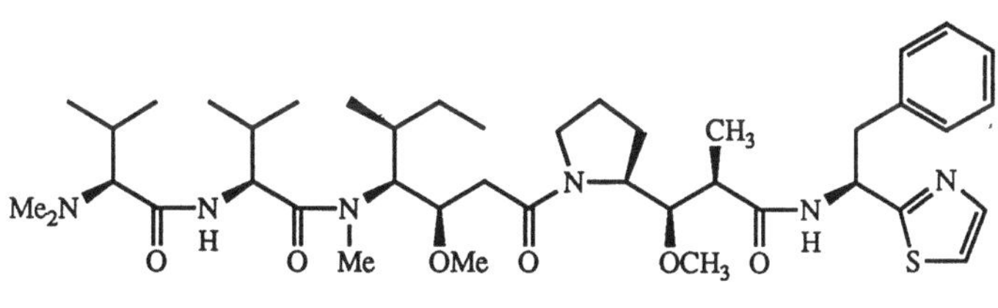

(-)-Dolastatin 10 **1408**

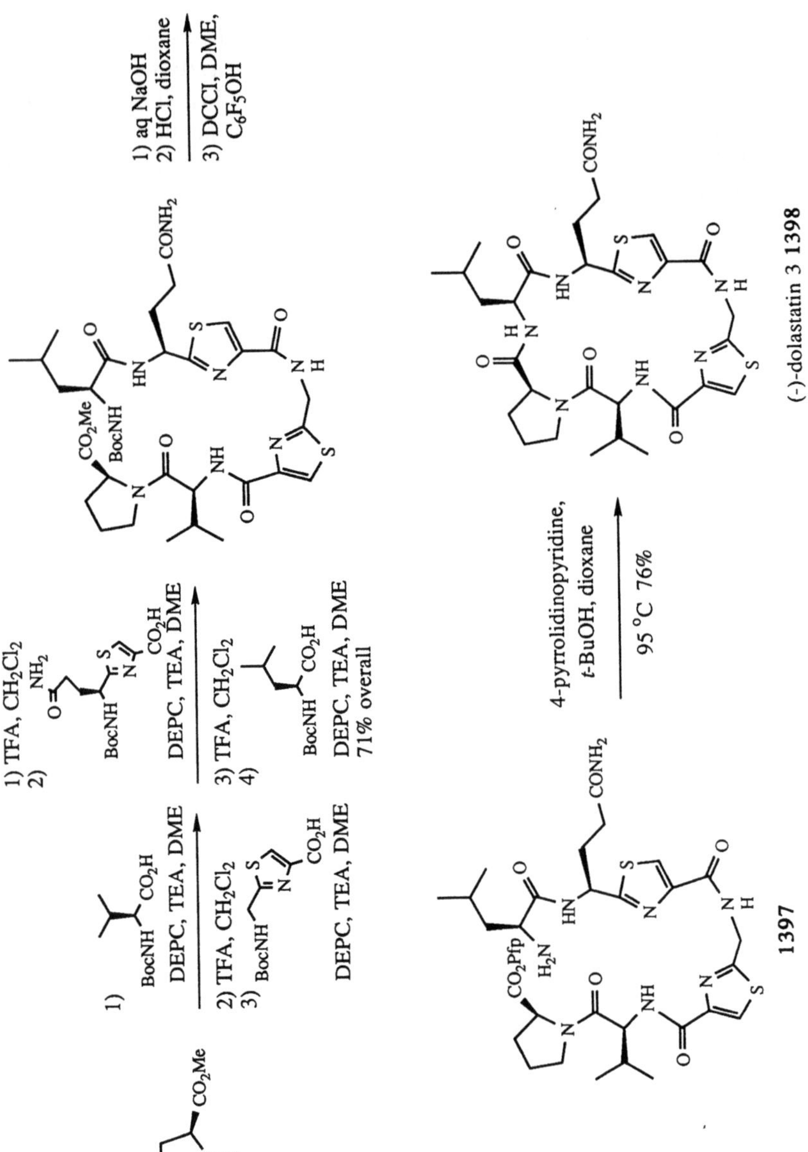

Scheme 241. Pettit Synthesis of (−)-Dolastatin 3

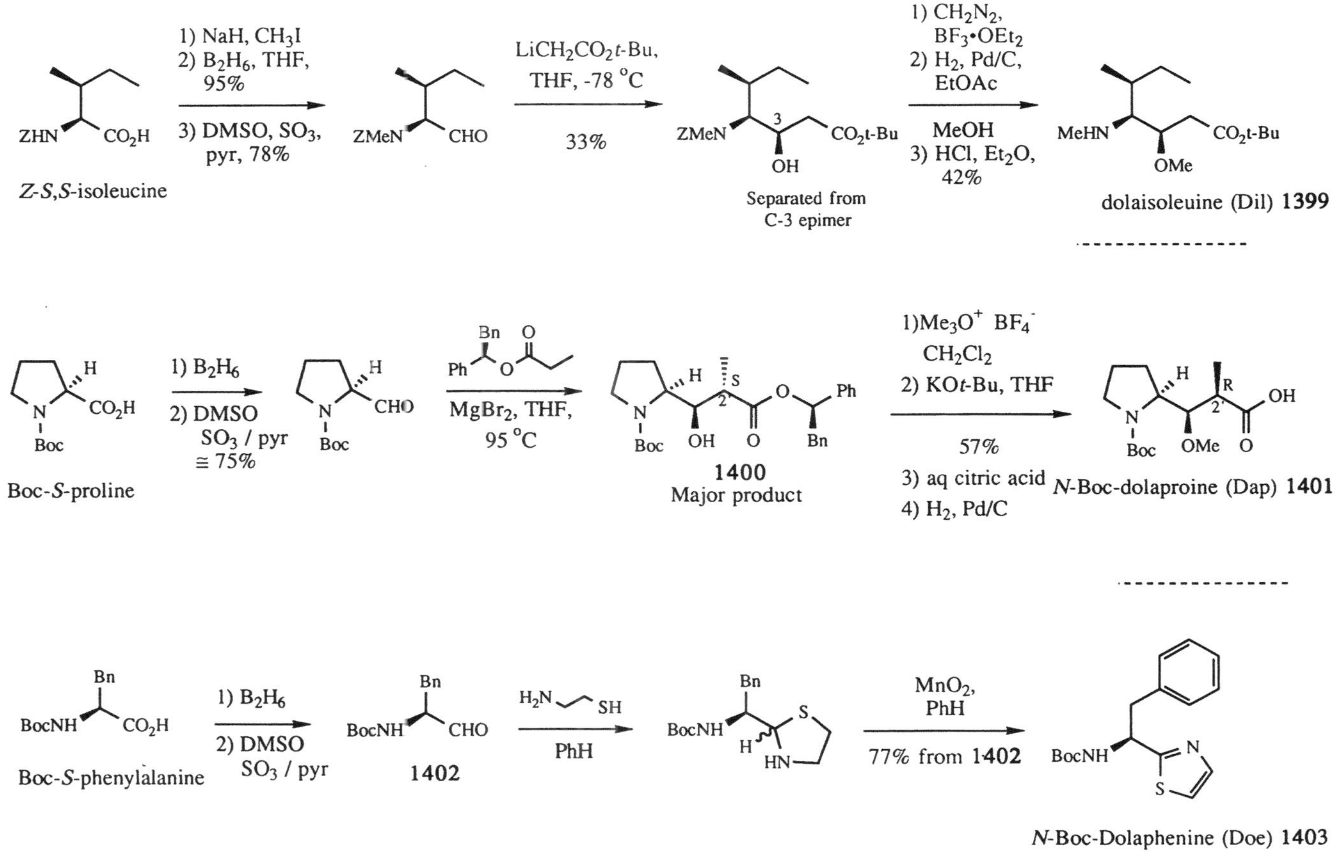

Scheme 242. Pettit Synthesis of (−)-Dolastatin 10 Fragments

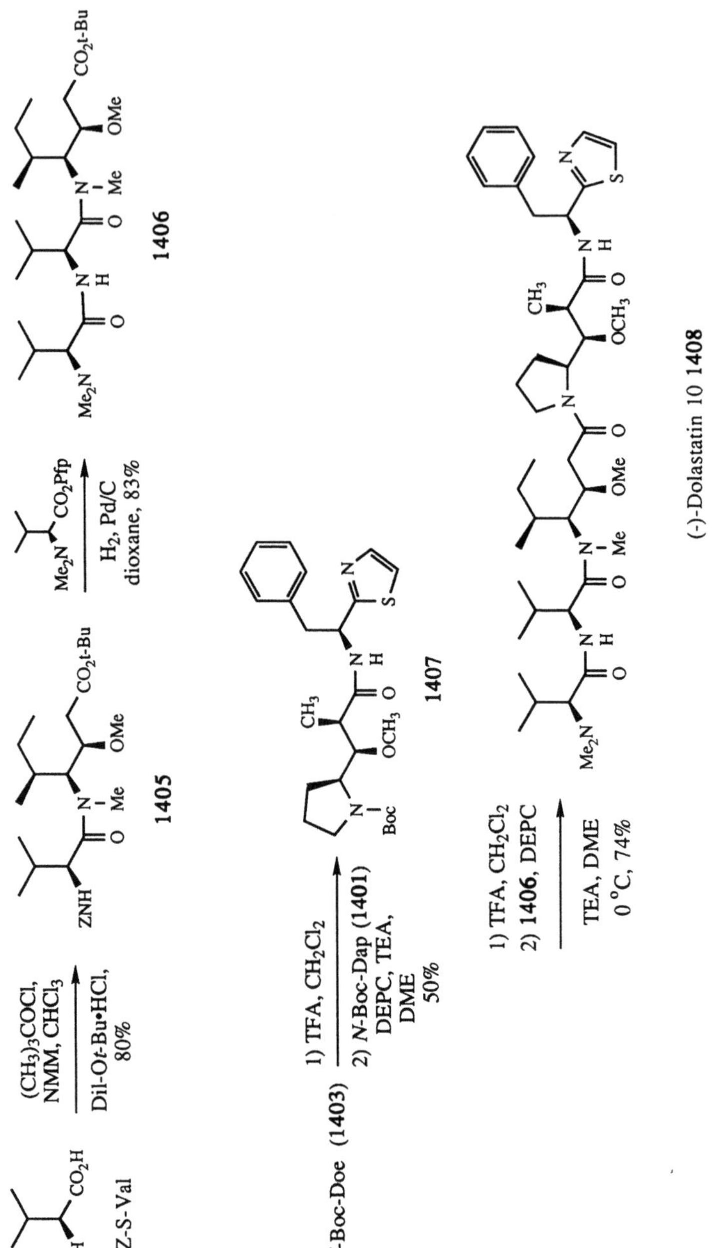

Scheme 243. Pettit Assembly of (−)-dolastatin 10 Fragments

made by 2-carbon aldol chain extension of a leucine derivative from which the C3 epimer **1399** was separated. Blocked dolaproine **1401** was similarly produced by chain extension of *N*-Boc-(*S*)-prolinal. An epimerization of the C2′ center in **1400** was required in this process. The thiazole-containing dolaphenine **1403** was prepared from (*S*)-Phe. The remaining amino acids were common. Assembly of the fragments took a standard course. (*S*)-Val was connected to the extended leucyl fragment producing **1405** which was extended at the *N*-terminus with a second Val unit. The units **1403** and **1406** were connected by condensation producing **1407** which was then hooked on to fragment **1406** producing (−)-dolastatin-10 **(1408)**.

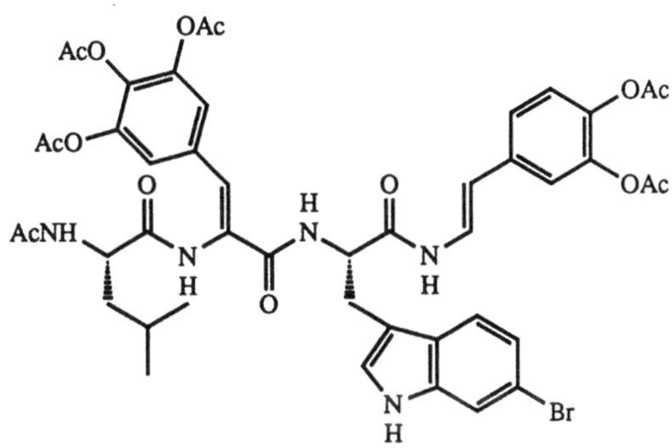

Hexaacetylcelenamide A **1417**

The pseudo-tetrapeptide hexaacetylcelenamide A was isolated from the Pacific sponge *Cliona celata* by Andersen [378] as its hexaacetyl derivative and possesses a relatively rare and fairly reactive enamide functionality. A synthesis has been reported by Schmidt [379] in which this functional group was generated in the final step of a synthesis of hexaacetylcelenamide A **1417** (Schemes 244 and 245). The (*S*)-6-bromotryptophan fragment **1410** was prepared by Wadsworth-Emmons reaction of the aldehyde **1409** and subsequent hydrogenation with an asymmetric catalyst producing **1411** in 98% ee. The propenamide containing group **1412** was constructed using standard chemistry, with the olefin eventually being generated by selenide oxidation and elimination. The fragments were assembled as shown in Scheme 245. The interesting phosphonate **1413** was condensed with the bromotryptophan fragment to give **1414**. Wadsworth-Emmons reaction with 2,3,4-triacetoxybenzaldehyde gave the *E* isomer **1415** which was connected to fragment **1412** by DCC condensation to provide **1416**. Final oxidative elimination generated hexaacetylcelenamide A **(1417)**.

3.3.6 Teleocidin A-1 (Lyngbyatoxin A) and Teleocidin A-2

The teleocidins are metabolites of *Streptomyces mediocidicus* [380]. One of these substances, teleocidin A-1, was shown to be identical with lyngbyatoxin A,

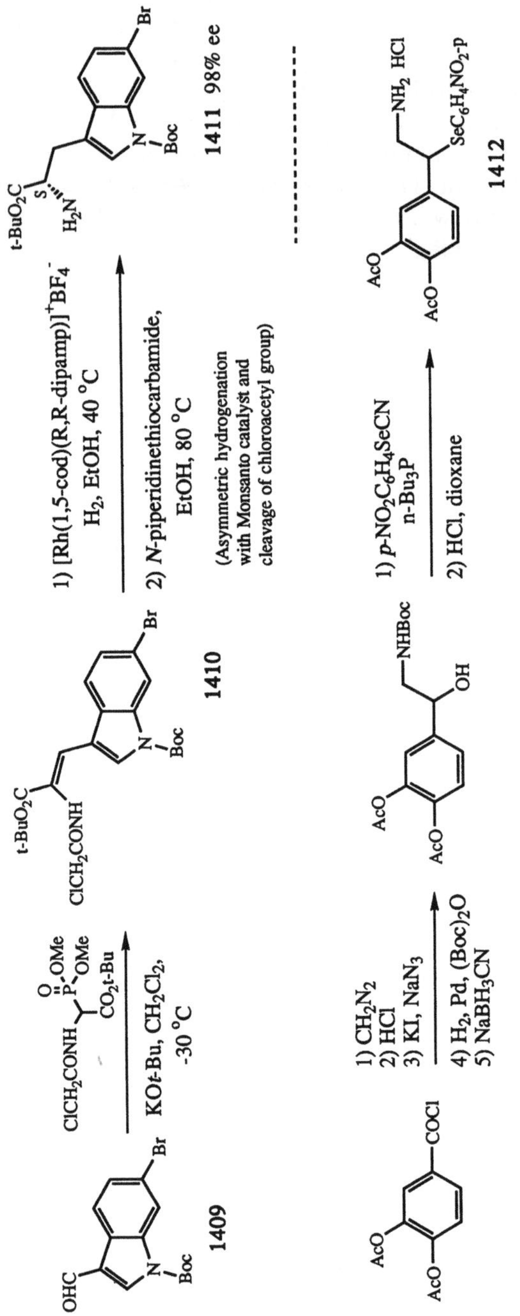

Scheme 244. Schmidt Synthesis of Hexaacetylcelenamide A Fragments

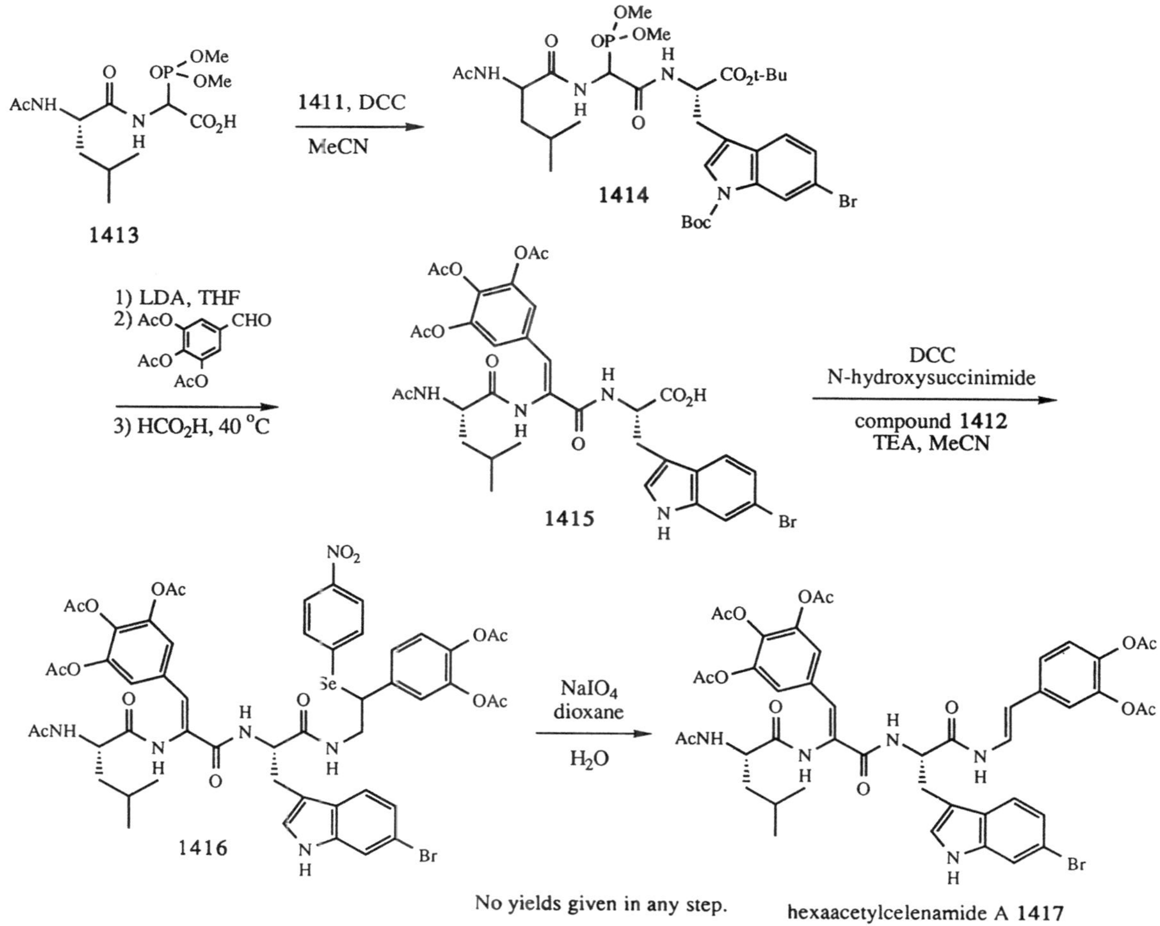

Scheme 245. Assembly of Hexaacetylcelenamide A Fragments (Schmidt)

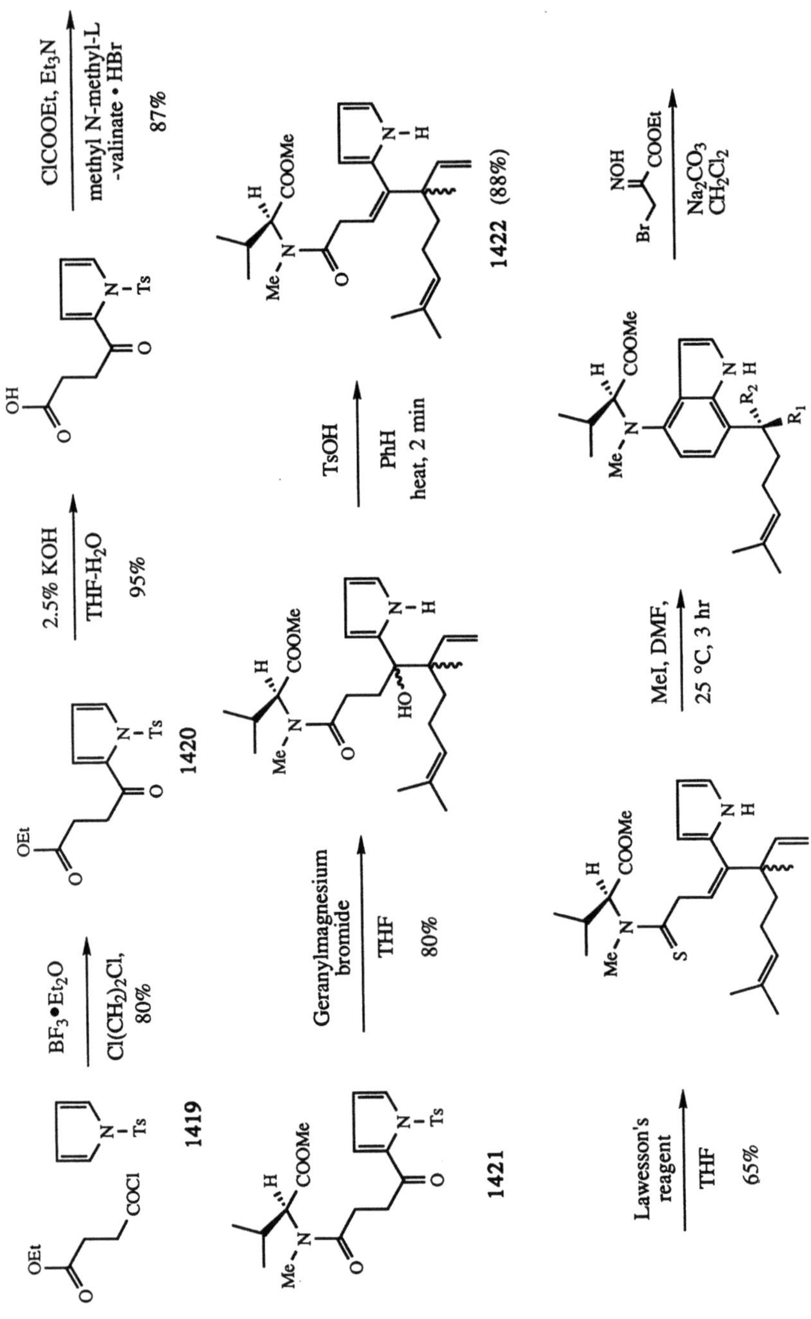

1423

1424 series **a** R_1= vinyl, R_2= Me (37%) plus 3 side products
1424 series **b** R_1= Me, R_2= vinyl (24%) plus 3 side products

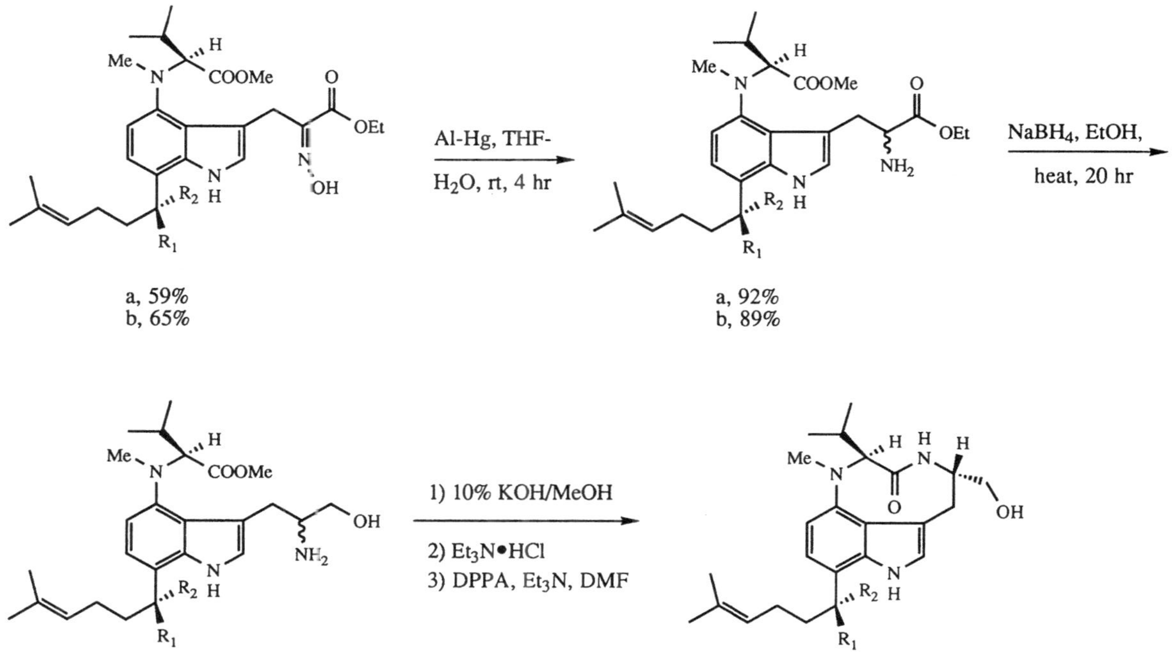

a, 59%
b, 65%

a, 92%
b, 89%

1425a, 51%
1425b, 53%

teleocidin A-1 (Lyngbyatoxin A) **1426**
23% + 31% of two other diastereomers

teleocidin A-2 **1427**
21% + 29% of two other diastereomers

Scheme 246. Natsume Synthesis of Teleocidins A-2 and A-1 (Lyngbyatoxin A)

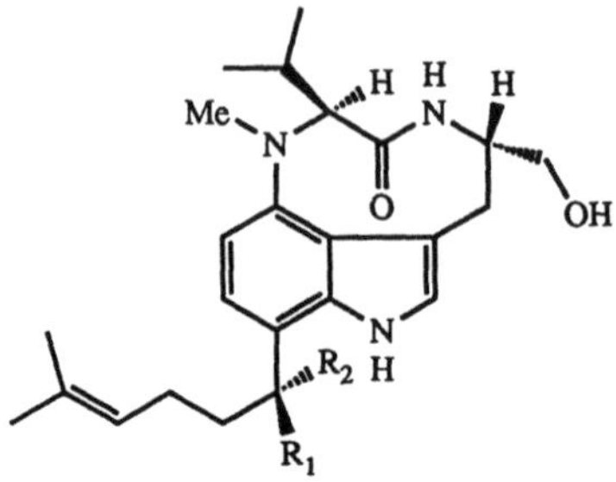

Teleocidin A-1 (Lyngbyatoxin A) **1426**
Teleocidin A-2 **1427**

isolated from the blue-green alga *Lyngbya majuscula* in Hawaii [381]. Teleocidin A-1 is a highly inflammatory and vesicatory compound that exhibits very potent tumor promoting properties. Teleocidins A-1 and A-2 have been prepared enantiospecifically by Natsume [382] in a synthesis that was very much lacking in diastereoselectivity (Scheme 246). The synthesis is initiated by the Friedel-Crafts acylation of *N*-tosylated pyrrole **1419** to afford ketoester-**1420**. Saponification of the ester, activation of the carboxyl and condensation with methyl *N*-methyl-L-valinate provides optically active ketoamide **1421**. Addition of geranylmagnesium bromide to **1421** and elimination with TsOH affords **1422** as an inseparable mixture of diastereomers. Direct conversion of **1422** to indole **1424** is difficult, therefore **1422** is transformed into thioamide **1423** and treated with methyl iodide to give **1424a** and **1424b** in 37 and 42% yields respectively. Addition of the C3 substituent is accomplished by alkylation with ethyl 3-bromo-2-hydroxyiminopropanoate, reduction of the oxime, and selective reduction of the less hindered ester to give indoles **1425a** and **1425b** as mixtures of diastereomers. Formation of the 9-membered lactam occurs via saponification of the ester, neutralization of the excess alkali, and treatment with diphenylphosphoryl azide to give **1426** in 23% yield along with 31% of two other diastereomers and **1427** in 21% yield along with 29% of two other diastereomers. Teleocidins A-1 (**1426**) and A-2 (**1427**) are obtained in 0.71 and 0.47% yields respectively over 12 steps.

4 Fatty Acid Derived Metabolites

A large number of metabolites appear to have been derived at least partially from fatty acid metabolism. Included in this section are metabolites which appear to be related to this metabolic pathway, or in any case, which contain unbranched or nearly unbranched carbon skeleta.

4.1 Haloethers of *Laurencia* sp.

There have been a number of halogenated cyclic ethers based on a linear pentadec-3-en-1-yne skeleton isolated from red algae of the genus *Laurencia.* A number of related compounds that are presumed to be biogenetic precursors of these haloethers have also been described. Total syntheses of several of these have been performed. Of the biogenetic precursors, laurencenyne (**1433**) and neolaurencenyne (**1434**) were isolated [383] from *Laurencia okamurai* and their structures confirmed by total syntheses. The laurediols have been isolated [384] from *L.nipponica* (Yamada) by Irie, along with several related compounds. Another related compound, although possibly of different biogenetic origin, is the trienyne acetate **1457** isolated from *L. pinnatifida* (Gmal) Lamour by Gonzalez [385]. Of the halogenated cyclic ethers, laurencin was first isolated [386] by Irie from *L. glandulifera* (Kutzing) in 1968 and the similar oxocin laurenyne **(1478)** was obtained from *L. obtusa.* The tricyclic maneonenes (e.g. **1489**) are chloroethers [387] isolated from *L. pinnatifida.* Panacene **(1497)**, a brominated allene, has been isolated [388] from the sea hare *Aplysia brasiliana,* a large sluglike gastropod mollusc indigenous to the Gulf coast of Florida. Although not yet found in *Laurencia sp.*, it has been suggested that panacene is derived from dietary sources. The *cis-* and *trans*-maneonenes are bicyclic halogenated diethers isolated [389] from the bright green variety of the Hawaiian alga *L. nidifica.*

4.1.1 Laurencenyne

Yamada's synthesis [390] of laurencenyne is shown in Scheme 247. Copper-catalyzed Grignard coupling of diyne **1428** with magnesium acetylide **1429**

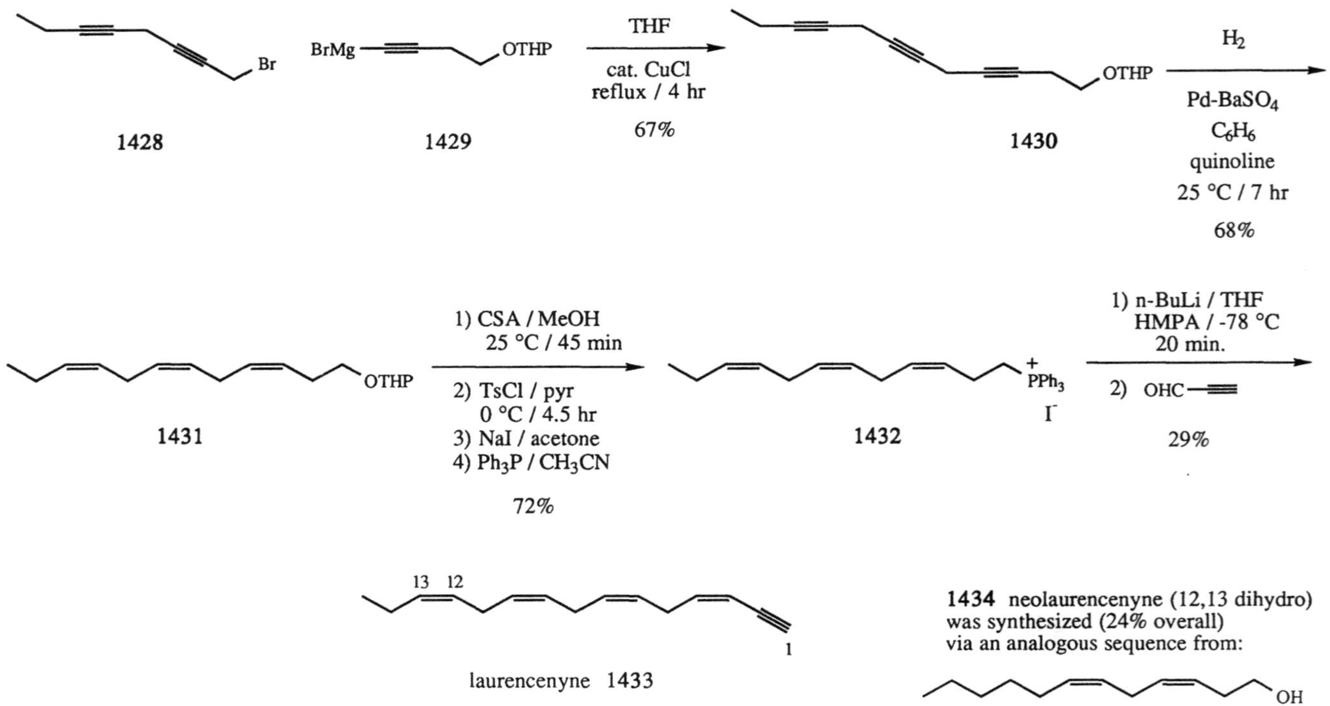

Scheme 247. Yamada Synthesis of Laurencenyne

Laurencenyne 1433

affords triyne **1430**. Lindlar reduction of **1430** yields the all *cis*-triene **1431**. Triene **1431** is then converted in four steps to phosphonium salt **1432** which is coupled in a Wittig reaction with propynal to afford laurencenyne **(1433)** in 7 steps and 9.5% overall yield. In a similar manner, neolaurencenyne (12,13-dihydrolaurencenyne) **(1434)** was synthesized in 24% overall yield.

4.1.2 Laurediols

A variety of laurediols have been prepared by Martin [391]. The preparation of *cis*-(6*S*,7*S*)-laurediol is shown in Scheme 248. Lindlar reduction of alkyne **1436** followed by Sharpless asymmetric epoxidation enantiospecifically provides epoxide **1437**. Rearrangement of epoxide **1437** to the terminal epoxide and Mitsunobu inversion of the secondary alcohol gives epoxide **1438**. In a series of standard transformations, the terminal epoxide is converted to ketal **1439**. Allylic oxidation of alcohol **1439** and transformation to the methylidene dibromide **1440** allows formation of the terminal acetylene by treatment with *n*-butyllithium. Methanolysis under acidic conditions produces *cis*-(6*S*,7*S*)-laurediol in 16 steps and 16% overall yield. Using similar technology, the Δ3,4-*cis*-isomer and the (6*R*,7*R*)-Δ3,4 *cis* and *trans* isomers were also prepared.

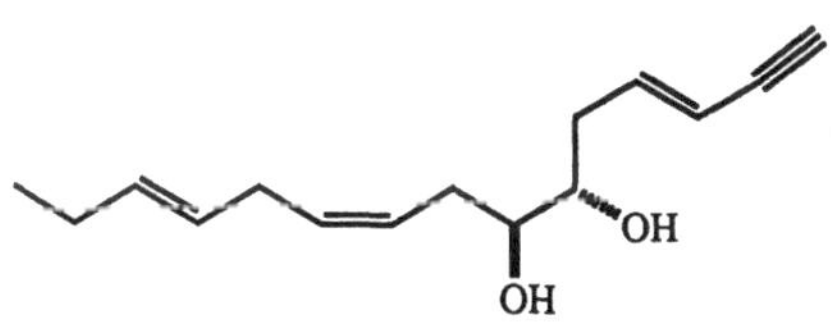

6*S*, 7*S*-*trans*-Laurediol **1441**

Masamune's synthesis of 6*S*,7*S*-*trans*-laurediol is outlined in Scheme 249 [392]. The key starting material is optically pure epoxide **1442** obtained from (+)-tartaric acid. Acetylide opening of the epoxide provides the required *trans* diol geometry required in the natural product. From this point all that is needed to obtain **1441** is elaboration of the sidechains. The diene sidechain is obtained by addition of *cis*-allylic bromide **1443** and Lindlar reduction to afford alcohol **1444**. The *trans*-enyne sidechain is obtained in a rather indirect fashion in an eight step process. Alcohol **1444** is converted in three steps to terminal epoxide **1445**. Cyanide opening of the epoxide and DIBAL reduction affords aldehyde **1446**, which is immediately submitted to Wittig olefination with TMS protected alkynylphosphorane **1447**. Deprotection and hydrolysis yields the natural product in 12 steps and 12% overall yield from epoxide **1442**.

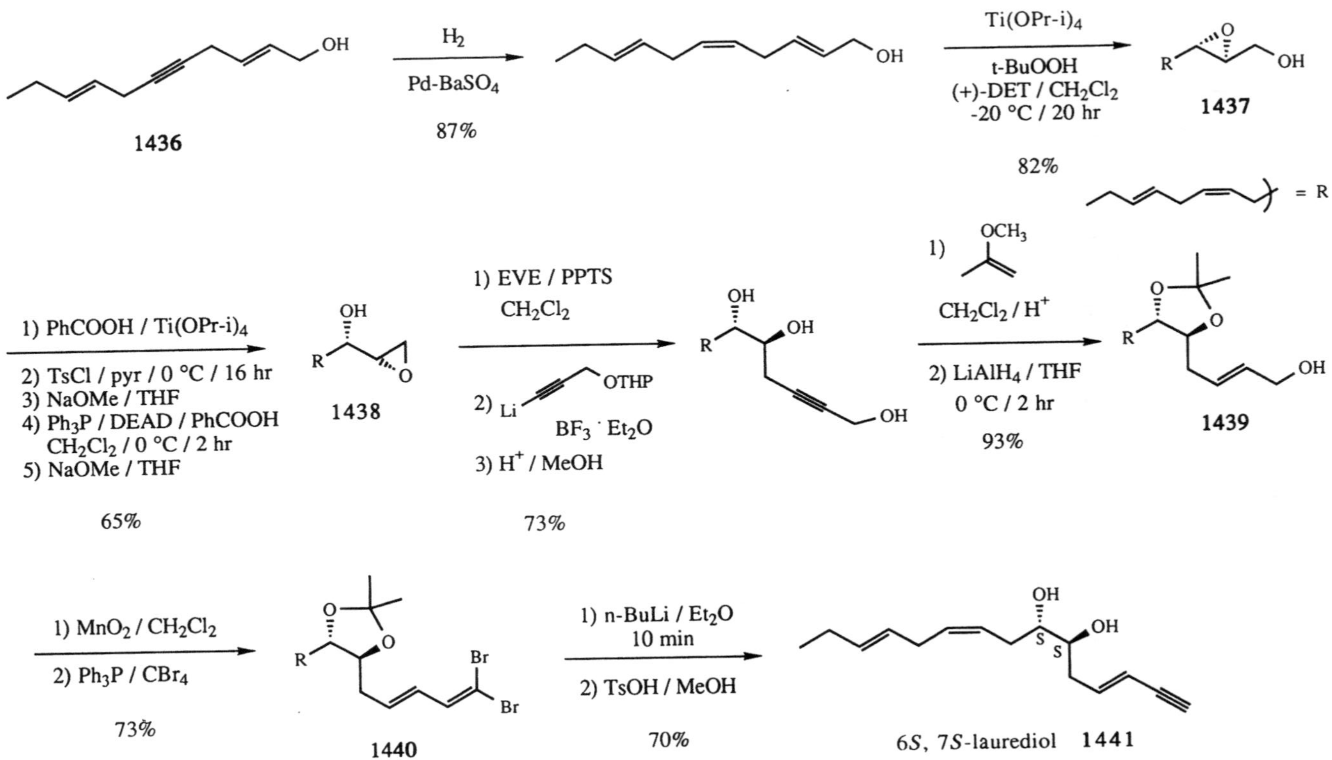

Using similar technology, the Δ3,4 -*cis* isomer, the C6-R-C7-R Δ3,4 *cis* and *trans* isomers were also made.

Scheme 248. Martin Syntheses of *trans* and *cis* 6R, 7r and 6S, 7S-Laurediols

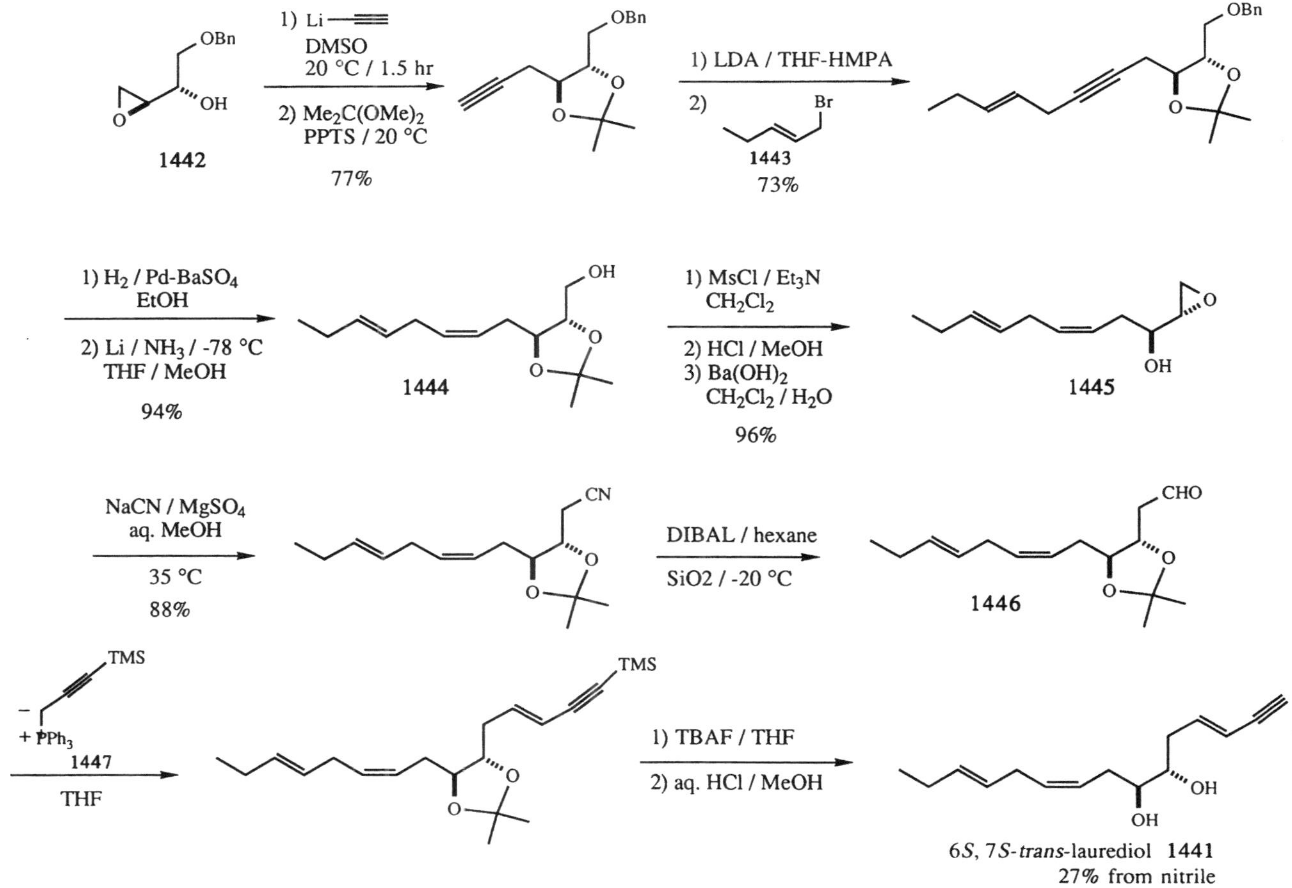

Scheme 249. Masamune Synthesis of 6S, 7S-*trans*-Laurediol

4.1.3 Laurediol Derivatives

An enantioselective total synthesis of trienyne **1459** and its 3-*trans*-isomer has been performed by Martin [393] as shown in Scheme 250. Allylic alcohol **1449** is converted to epoxide **1450** which is then converted to a key intermediate epoxide **1451**. This epoxide **1451** is homologated by addition of THP protected acetylide **1452** and manipulated via a series of four protection-deprotection steps to give **1453**. An interesting use of disposable chirality provides aldehyde **1454** in three steps. For reasons unknown, non-chiral methods of epoxidation in this transformation result in much lower yields. Peterson olefination of **1454** with propargylic lithium reagent **1455** gives **1456** with *cis* stereochemistry. Desilylation and treatment of **1456** with Mukaiyama's reagent [394] gives the natural product **1457** in 16 steps. The *trans*-enyne **1459** is also a natural product. Peterson olefination of aldehyde **1454** with propargylic lithium reagent without the presence of added Lewis acid provides the *trans*-enyne geometry of compound **1458**. Enyne **1458** is treated as the *cis*-enyne **1456** to provide natural product **1459** in a similar fashion.

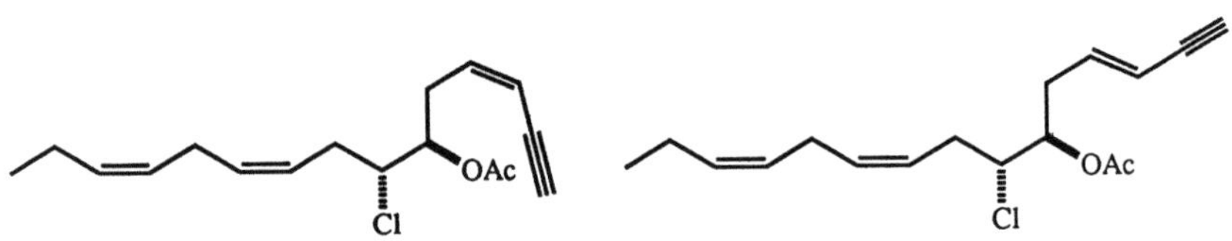

1457 1459

4.1.4 Laurencin

Laurencin has been prepared by Masamune [395] in a rather difficult synthesis (Scheme 251). Dialdehyde **1460** is prepared from 5-ethyl-2-furoic acid in four steps. In a key transformation, the dialdehyde is subjected to the Robinson-Schöpf condensation to give 9-aza-3-oxabicyclo[3.3.1]nonanone, thus setting the relative configuration of two of the final four stereocenters. After acylation, the desired isomer **1461** is obtained in only 2.2% yield, along with 0.6% of the other 2,8-diastereomers. Ketone **1461** is converted to a mixture of β,γ-unsaturated ketones **1462** and **1463** in four steps via standard transformations.

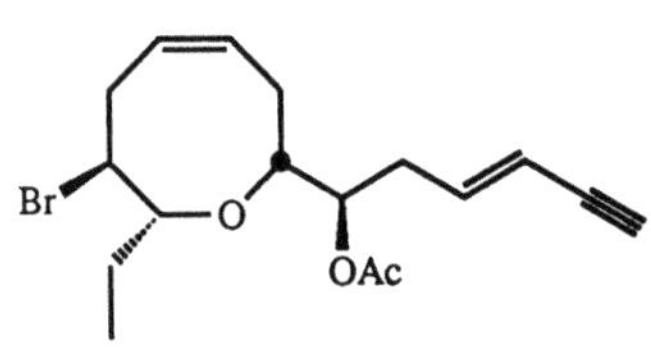

(±)-Laurencin **1468**

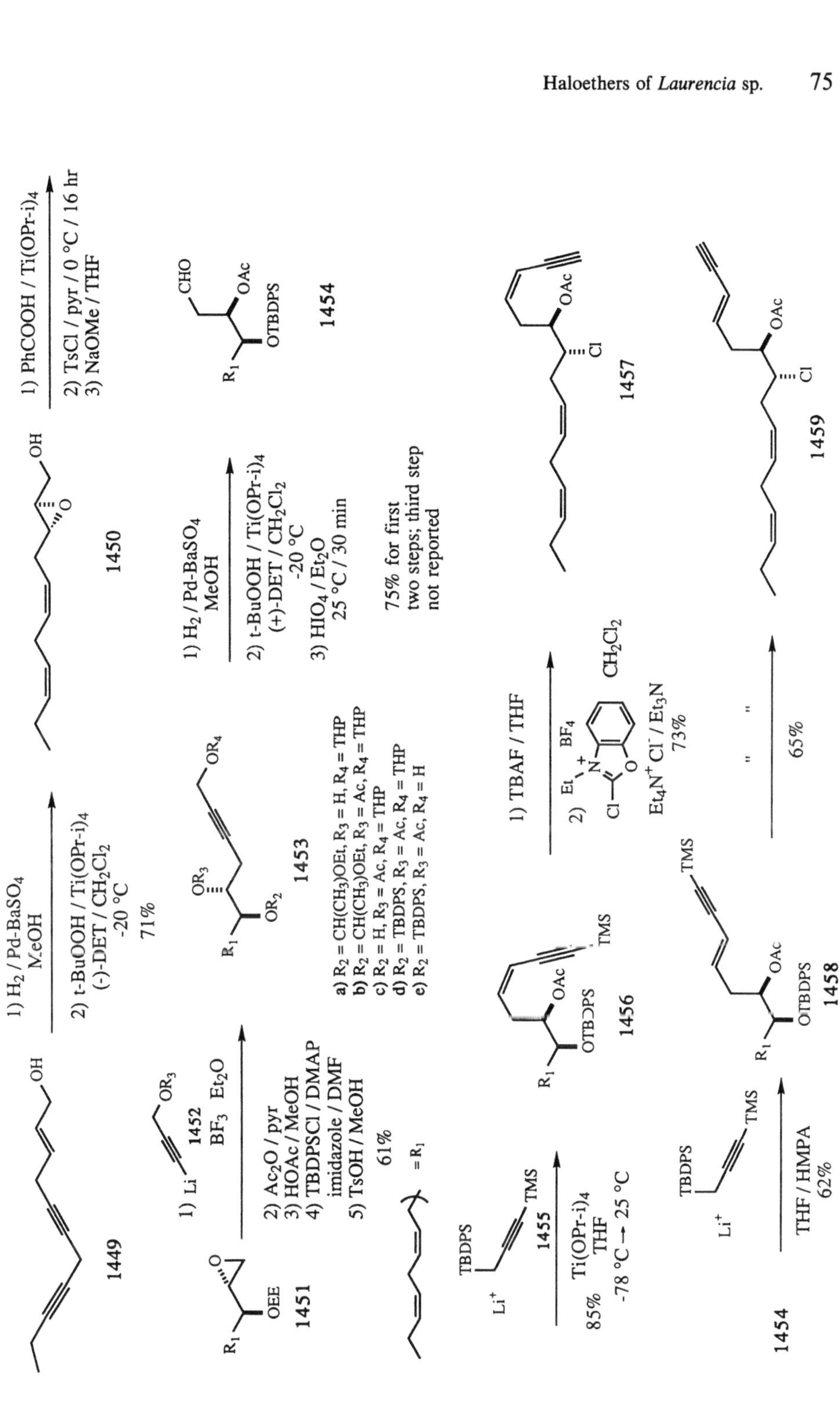

Scheme 250. Martin Synthesis of a Triene-yne from *Laurencia*

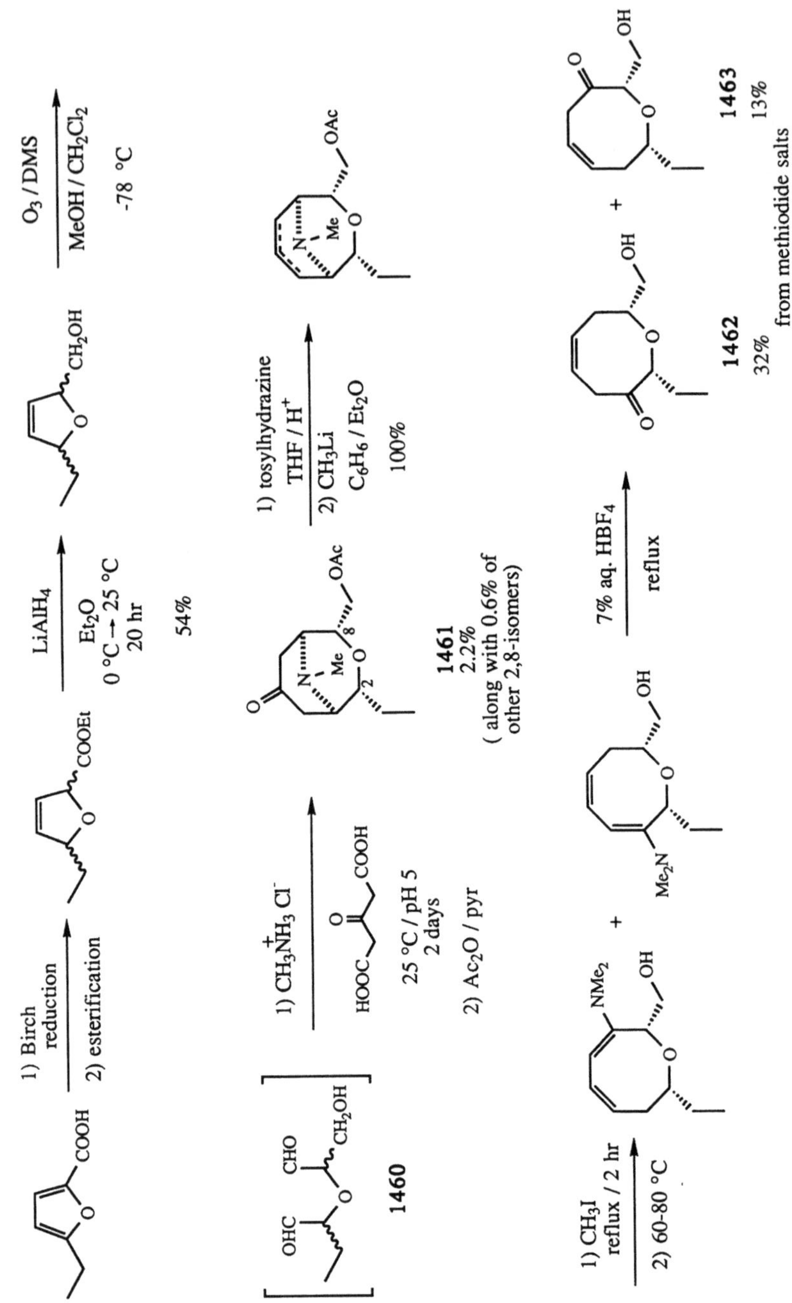
COOH
1) Birch reduction
2) esterification
COOEt
LiAlH4
Et2O
0 °C → 25 °C
20 hr
54%
CH2OH
O3 / DMS
MeOH / CH2Cl2
-78 °C
OHC
CHO
CH2OH
1460
1) CH3NH3 Cl
HOOC
COOH
25 °C / pH 5
2 days
2) Ac2O / pyr
OAc
Me
1461
2.2%
(along with 0.6% of other 2,8-isomers)
1) tosylhydrazine
THF / H+
2) CH3Li
C6H6 / Et2O
100%
1) CH3I
reflux / 2 hr
2) 60-80 °C
NMe2
OH
Me2N
7% aq. HBF4
reflux
1462
32%
1463
13%
from methiodide salts

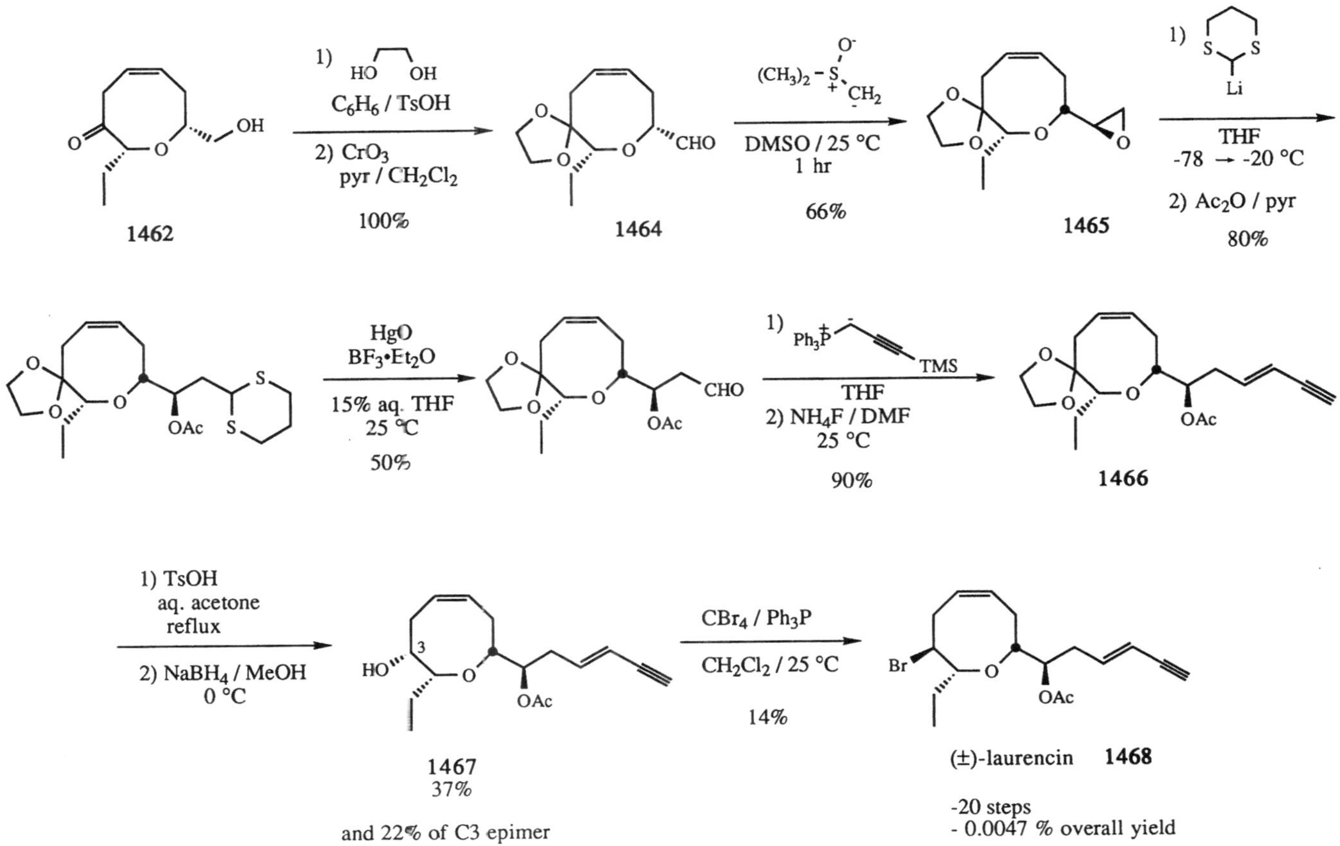

Scheme 251. Masamune Synthesis of (±)-Laurencin

Regioisomer **1462** is protected as the ketal and oxidized to give aldehyde **1464**. Stereospecific conversion of the formyl moiety in **1464** to epoxide **1465** by treatment with dimethyloxosulfonium methylide sets the relative stereochemistry of the third chiral center. Opening of the epoxide with 2-lithio-1,3-dithiane and homologation of the sidechain by Wittig olefination gives *trans*-enyne **1466**. Hydrolysis of **1466** affords a ketone, which upon reduction yields 37% of alcohol **1467** along with 22% of its C3 epimer. Conversion of **1467** to the bromide, with inversion at C3, produces a complex mixture of products from which laurencin **1468** can be isolated in 14% yield. An overall yield of 0.0047% is obtained over 20 steps.

4.1.5 Laurenyne

Overman's [396] synthesis of (−)-laurenyne (**1478**) is illustrated in Scheme 252. The synthetic strategy addresses the problems of oxocane formation and the control of stereochemistry in eight-membered rings. The synthesis is initiated with the $BF_3 \cdot Et_2O$-assisted opening of oxetane with a silicon-stabilized organolithium reagent to afford alcohol **1470**. Conversion of **1470** to allylic alcohol **1471** and Sharpless asymmetric epoxidation provides epoxide **1472** with an enantiomeric excess of 78–81%. Titanium-assisted opening of epoxide **1472** with chloride ion gives chlorohydrin **1473** in 68% yield along with 25% of the regioisomeric product, thus setting the stereochemistry of two of the three stereocenters required in the product. Conversion of **1473** to acetal **1474** sets the stage for the key cyclization reaction. Treatment of **1471** with two equivalents of tin (IV) chloride results in cyclization to give oxocane **1475** in 37% yield after desilylation. Unfortunately, the use of more appropriately substituted side-chains resulted in lower yields of cyclized product. Once the oxocane ring system was obtained (along with the third stereocenter) all that was needed to complete the synthesis was minor refunctionalization and further homologation of the right hand sidechain. Application of standard methodology converts compound **1475** to diene **1476**. Transformation of **1476** to aldehyde **1477** followed by Peterson olefination and desilyation gives (−)-laurenyne with the proper *cis*-enyne geometry. Laurenyne (**1478**) is obtained in approximately 0.67% yield over 20 steps.

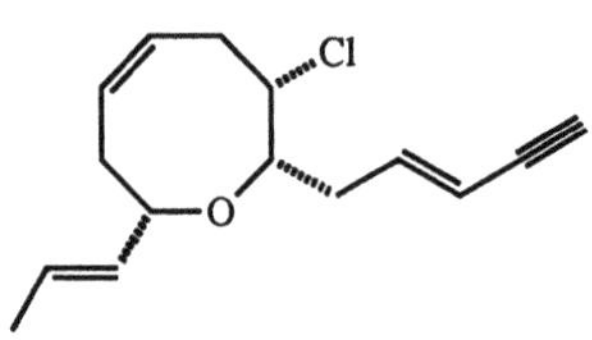

Laurenyne **1478**

4.1.6 *Cis*-Maneonenes A and B

The preparation of *cis*-maneonenes A and B is shown in Scheme 253. The synthesis by Holmes [397] begins with Grignard addition to the readily available lactone **1479** to afford lactol **1480**. This generates five of the six chiral centers needed in the products. A standard sequence of reactions yields aldehyde **1481**, which is converted to the diyne **1482** by addition of 4-lithio-1-trimethylsilylbutadiyne. A mixture of four diastereomers is obtained. Lindlar reduction of **1482** gives *cis*-enyne **1483**, as would be expected due to the steric and electronic influence of the trimethylsilyl group, along with 50% recovered starting material. Pyrolysis of the enyne bromoacetal gives a mixture of *E* and *Z* isomers **1484** and **1485**. Separation of the isomers and treatment of each with the tetramethyl-α-chloro-enamine **1486** forms the chlorides with inversion and leads to *cis*-maneonenes A and B in 24% yield. An overall yield of approximately 1.5% is obtained for each isomer over the course of the 9 step synthesis.

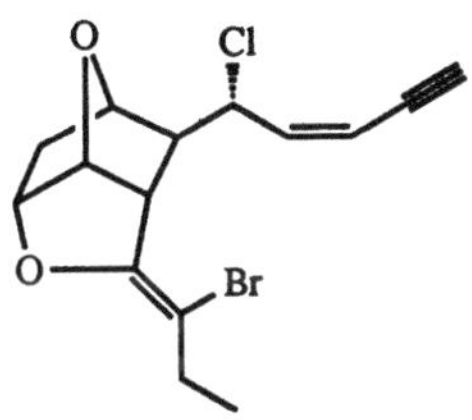

cis-Maneonene A

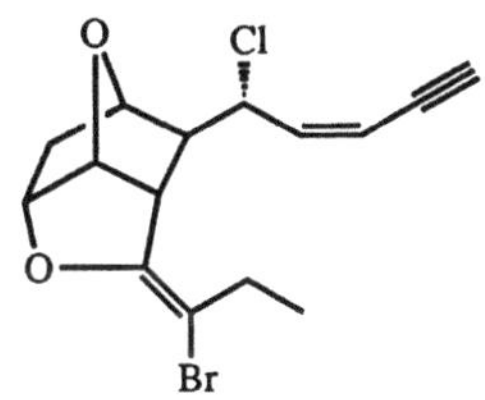

cis-Maneonene B

4.1.7 *Trans*-Maneonene B

Holmes has also prepared [398] *trans*-maneonene B (**1489**) as shown in Scheme 254. The mixture of minor isomers obtained from acetylide addition to aldehyde **1481** in Holmes's synthesis of *cis*-maneonenes A and B (Scheme 253) serves as the starting material for the preparation of **1489**. Hydroxyl directed LAH reduction of the propargylic diyne system produces two *trans*-enynes **1488**. Separation of the isomers and treatment as in the previous synthesis produces *trans*-maneonene B (**1489**) along with its isomer **1490**.

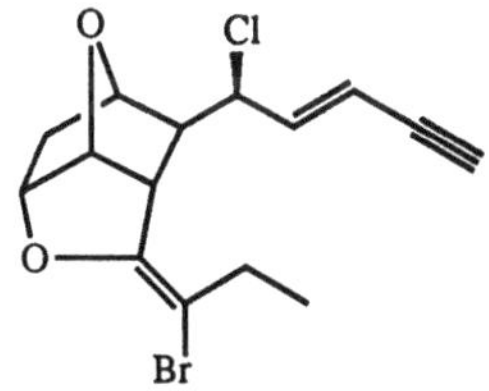

trans-Maneonene B **1489**

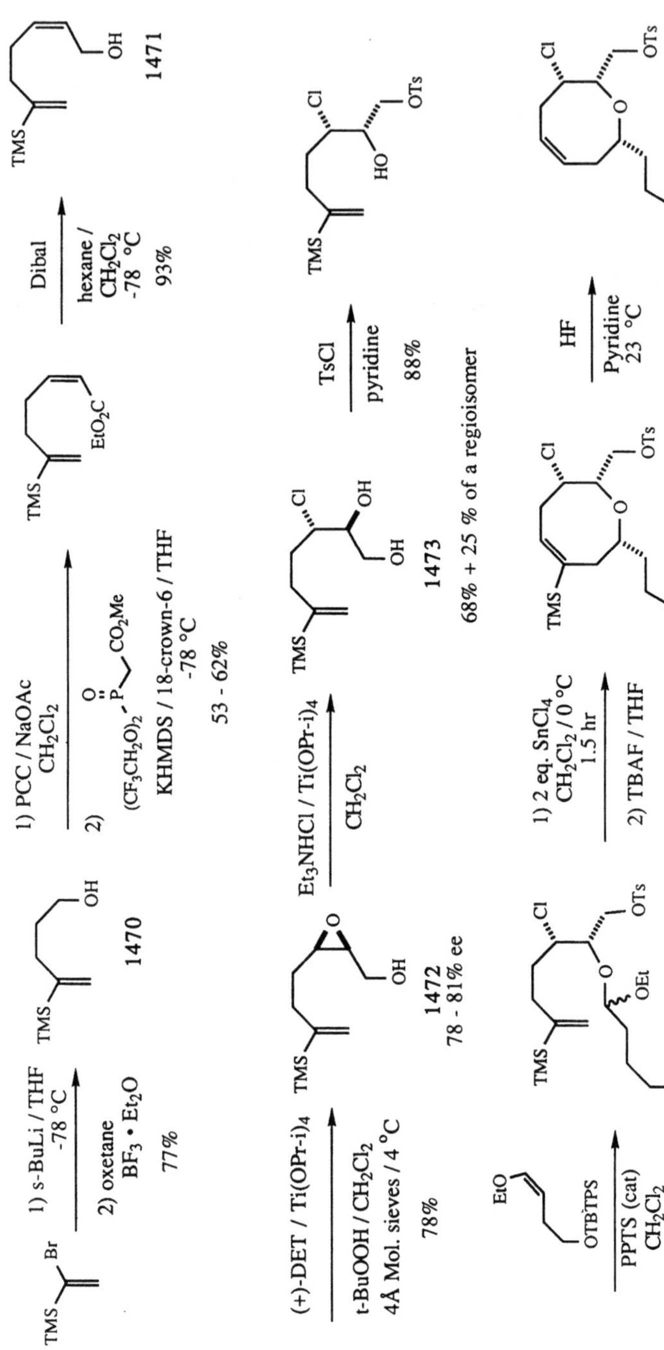
1) s-BuLi / THF
-78 °C
2) oxetane
BF3 • Et2O
77%
1470
1) PCC / NaOAc
CH2Cl2
2) (CF3CH2O)2P(O)CH2CO2Me
KHMDS / 18-crown-6 / THF
-78 °C
53 - 62%
Dibal
hexane / CH2Cl2
-78 °C
93%
1471
(+)-DET / Ti(OPr-i)4
t-BuOOH / CH2Cl2
4Å Mol. sieves / 4 °C
78%
1472
78 - 81% ee
Et3NHCl / Ti(OPr-i)4
CH2Cl2
1473
68% + 25 % of a regioisomer
TsCl
pyridine
88%
PPTS (cat)
CH2Cl2
98%
1474
1) 2 eq. SnCl4
CH2Cl2 / 0 °C
1.5 hr
2) TBAF / THF
37%
1475
HF
Pyridine
23 °C
83%

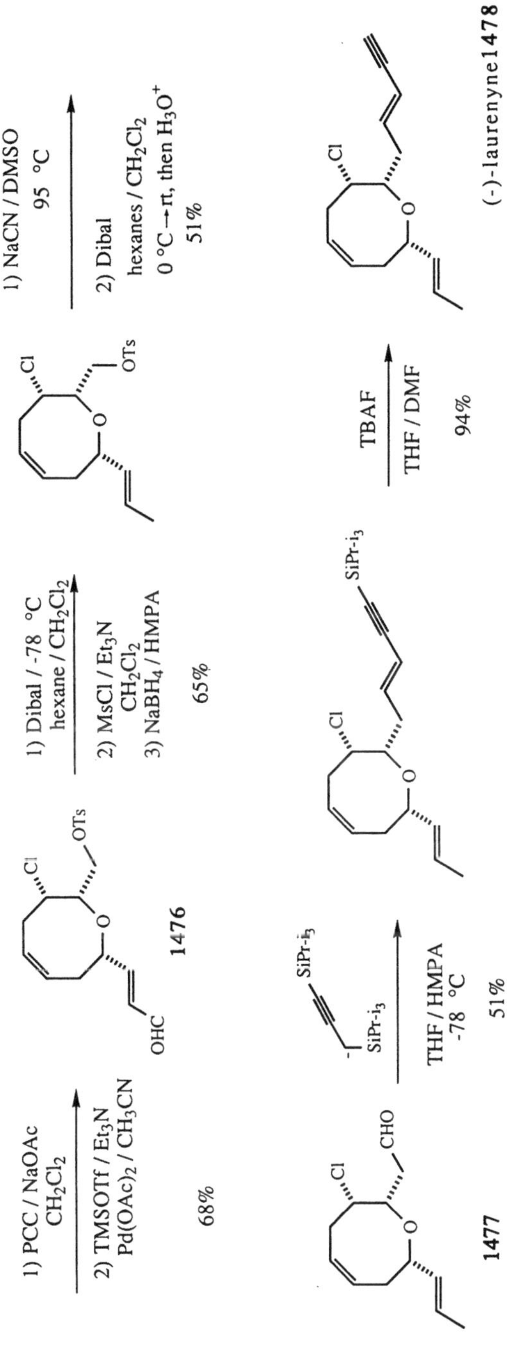

Scheme 252. Overman Synthesis of (−)-Laurenyne

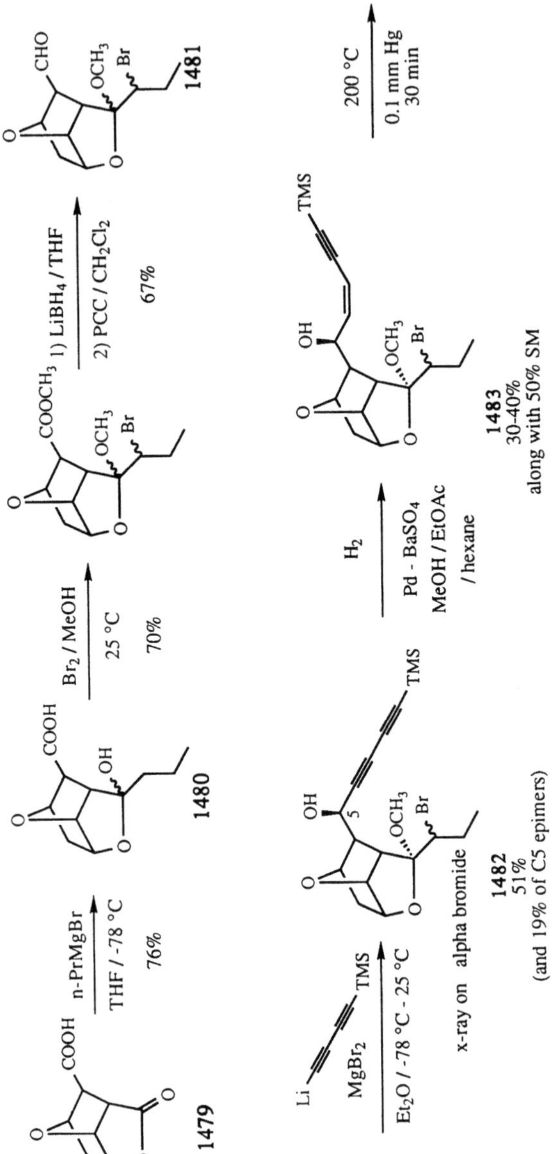
1479
n-PrMgBr
THF / -78 °C
76%
1480
Br2 / MeOH
25 °C
70%
1) LiBH4 / THF
2) PCC / CH2Cl2
67%
1481
MgBr2
Et2O / -78 °C - 25 °C
x-ray on alpha bromide
1482
51%
(and 19% of C5 epimers)
H2
Pd - BaSO4
MeOH / EtOAc
/ hexane
1483
30-40%
along with 50% SM
200 °C
0.1 mm Hg
30 min

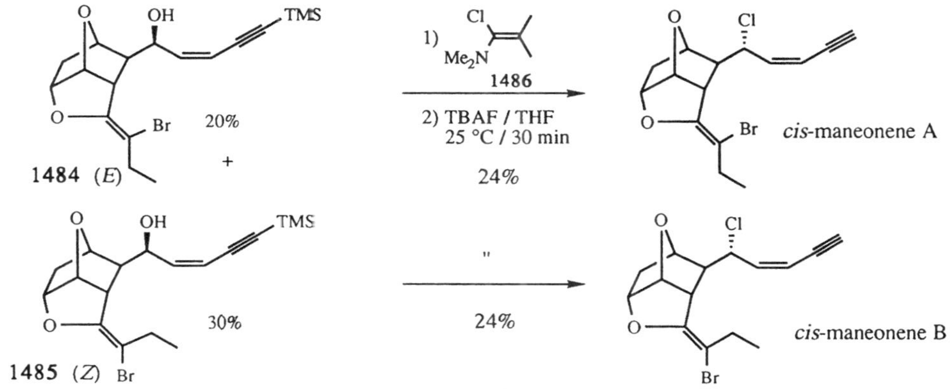

Scheme 253. Holmes Syntheses of *cis*-Maneonenes A and B

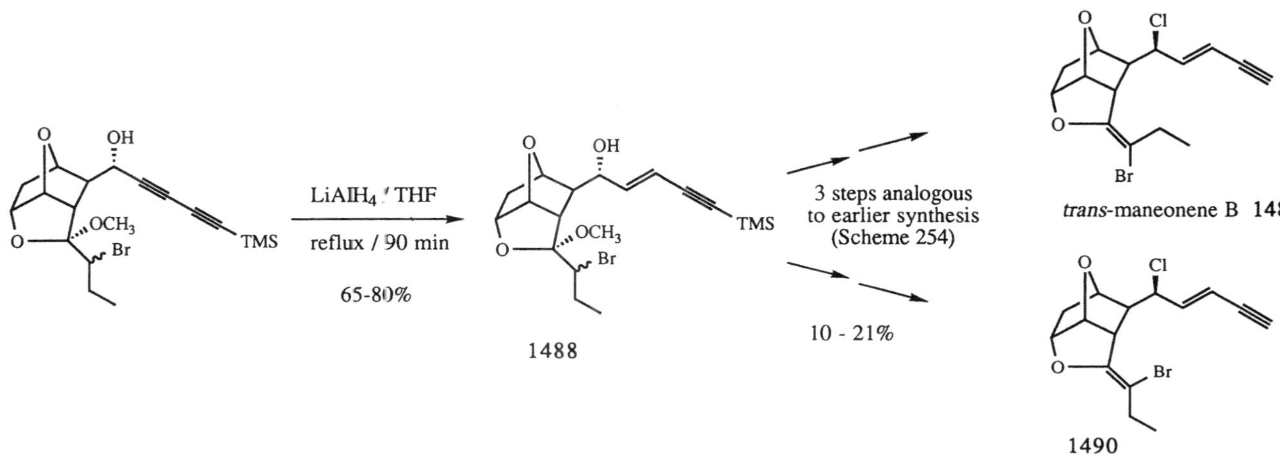

Scheme 254. Holmes Synthesis of *trans*-Maneonene B

4.1.8 Panacene

Panacene (**1497**) is an aromatic diether containing a brominated allene substituent. Its synthesis by Feldman [399] is outlined in Scheme 255. At the onset of Feldman's synthetic planning, the relative stereochemistry of panacene was undetermined. Therefore, a strategy was developed that would allow the stereochemisry of the precursor to be transmitted to the bromoallene unit and would be flexible enough to provide both isomers for comparison purposes. Ethyl 6-ethylsalicylate is converted to benzofuranone **1491** by modification of known methods. Allylation, decarboxylation and reduction produces **1492** which is converted to diether **1493** by bromonium ion initiated cyclization. The tricyclic ring system is obtained as a 3:1 mixture of *exo*:*endo* bromides. Aldehyde **1494** is obtained from **1493** in three steps. Addition of TMS protected acetylide to aldehyde **1494** affords a mixture of alcohols which is separated by formation of the benzoates followed by chromatography and hydrolysis. Mesylation of the α-OH isomer of **1495** and treatment with $LiCuBr_2$ produces 1-epi-panacene **1496**. Similar treatment of the β-OH isomer provides panacene **1497** in 13 steps and a total yield of approximately 5%. The assigned configuration of the allene is based on the assumed stereochemistry of the displacement step as related to a steroidal model of known configuration [400]. Panacene can also be prepared from benzofuranone **1491** in a less stereo-rational, but perhaps biomimetic manner (Scheme 256) [401]. Conversion of **1491** to enyne **1499** occurs in four steps and 14% yield. Bromonium ion initiated cyclization of **1499** provides a 1:1 mixture of 1-epi-panacene and panacene in 8.9% overall yield.

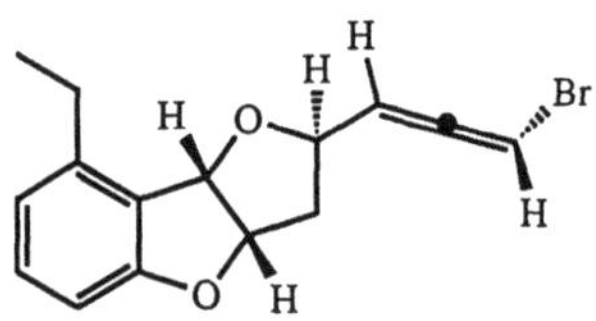

Panacene **1497**

4.2 Prostanoids

4.2.1 Clavulones

The clavulones are a series of marine prostanoids isolated [402] from the Okinawan coral *Clavularia viridis*. They have attracted strong synthetic interest due to their structures and their strong antitumor and antiinflammatory activities [403]. They bear a structural resemblance to a highly cytotoxic metabolite [404] of PGD_2. Biosynthetically, it has been proposed that they are

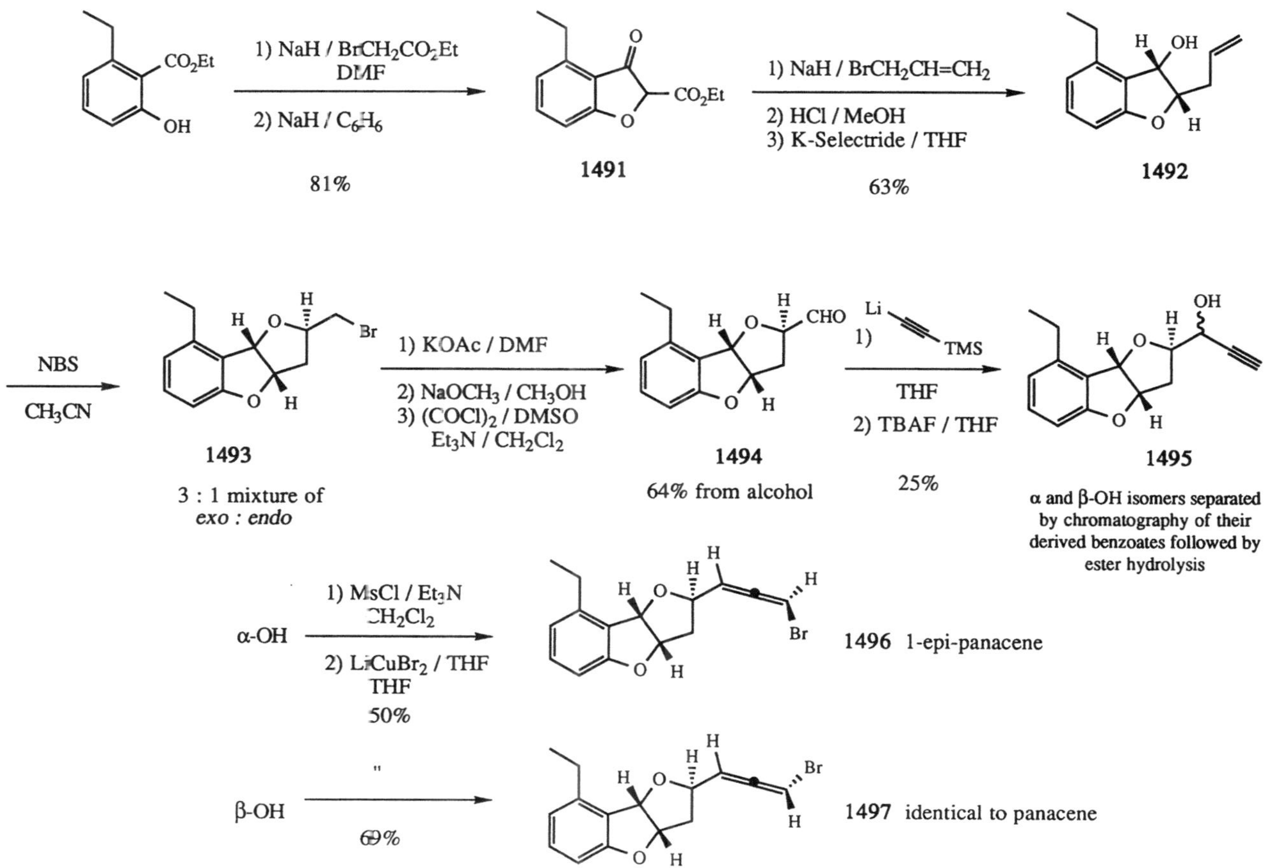

Scheme 255. Feldman Synthesis of (±)-Panacene

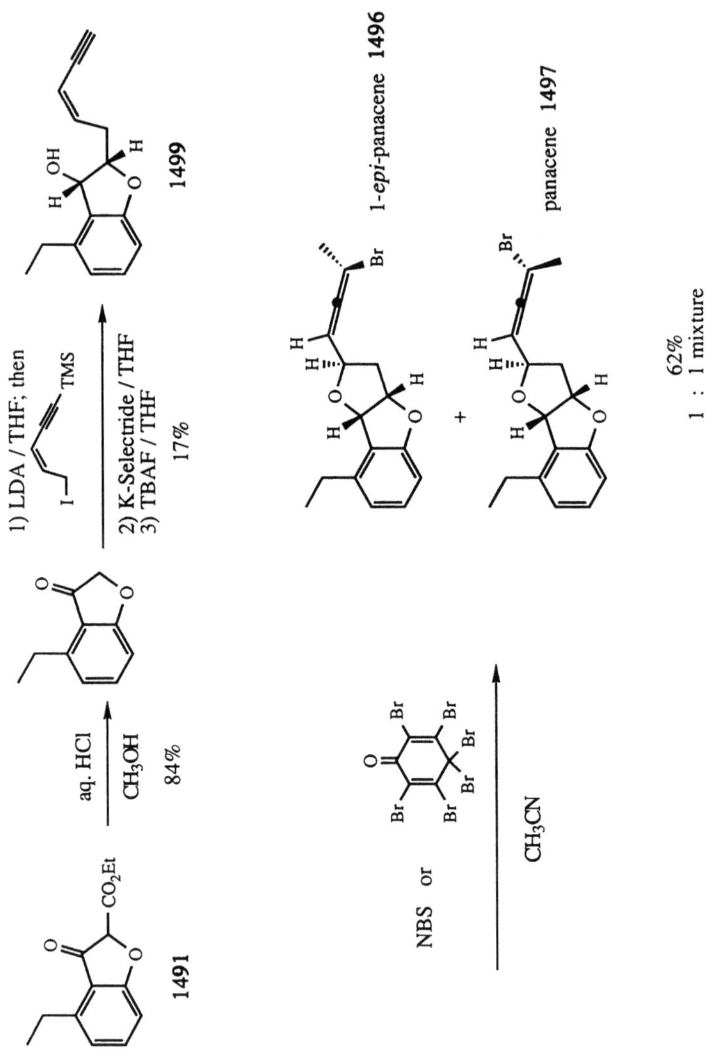

Scheme 256. Alternate Endgame for (±)-Panacene Synthesis

derived from arachidonic acid via a mechanistic pathway involving a series of free radical intermediates [405].

4.2.1.1 Clavulones I and II

Clavulone I has been prepared by Corey [406] in 10 steps and 10% yield (Scheme 257). Sensitized photooxidation of substituted cyclopentadiene **1500** followed by sodium borohydride reduction affords the *cis* diol **1501**. PDC oxidation, Lindlar reduction of the alkyne and silylation of the tertiary alcohol gives the cyclopentenone ring and lower sidechain with proper alkene geometry. Due to the sluggishness of cyclopentadienone forming eliminations, the upper sidechain can be appended by an interesting aldol-coupling process of **1502** and racemic aldehyde **1503** to produce **1504** as a 1:1 mixture of diastereomers. Selective replacement of the tertiary TBS group with an acetyl moiety followed by separation of the diastereomers produces optically pure **1505**. Desilylation and acylation completes the synthesis to provide clavulone I (**1506**).

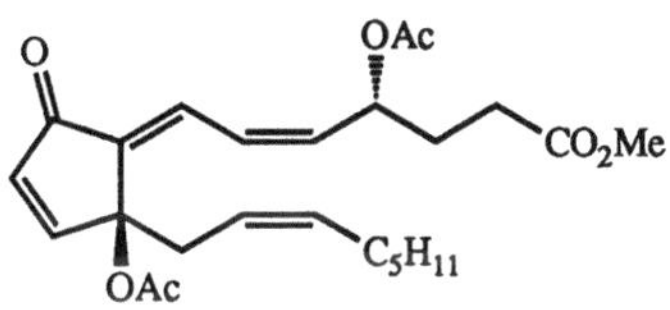

Clavulone I **1506**

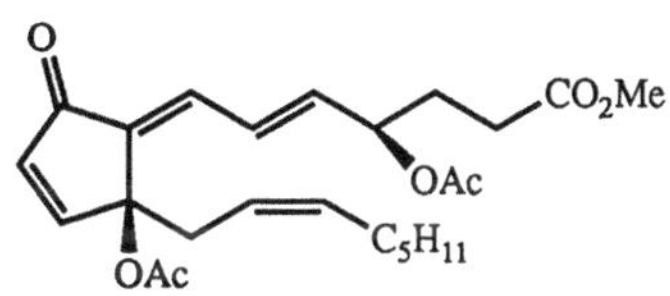

(-)-Clavulone II 1510

The Shibasaki [407] synthesis of clavulone II (**1510**) is shown in Scheme 258. Alkylation of an α-hydroxycyclopentanone dianion with an eight carbon propargylic bromide followed by Lindlar reduction affords the lower side chain with the proper *cis* geometry. Formation of the cyclopentenone via the enol triflate followed by a four-step carbonyl transposition gives the cyclopentenone **1508** with proper regiochemistry. Coupling of **1508** with the enantiomerically pure aldehyde **1509** according to the procedure of Yamada (Scheme 262) and separation of the diastereomers produces (−)-clavulone II (**1510**).

4.2.1.2 Clavulone III

The synthesis by Hamanaka [408] enantiospecifically provides both clavulone II and clavulone III (Scheme 259). Selenylation-elimination and DIBAL reduction of the Corey lactone **1511** provides the multi-functionalized cyclopentane **1512**. Sharpless epoxidation of **1512** and reductive epoxide opening affords triol **1513** which is selectively protected at the secondary alcohol and oxidized to provide aldehyde **1514** in a four-step process. Wittig olefination provides the lower side chain with the proper *cis* unsaturation. Selective deprotection and acylation of the secondary alcohol followed by deprotection and oxidation of the primary alcohol affords aldehyde **1515**. Coupling of vinyl stannane **1516**

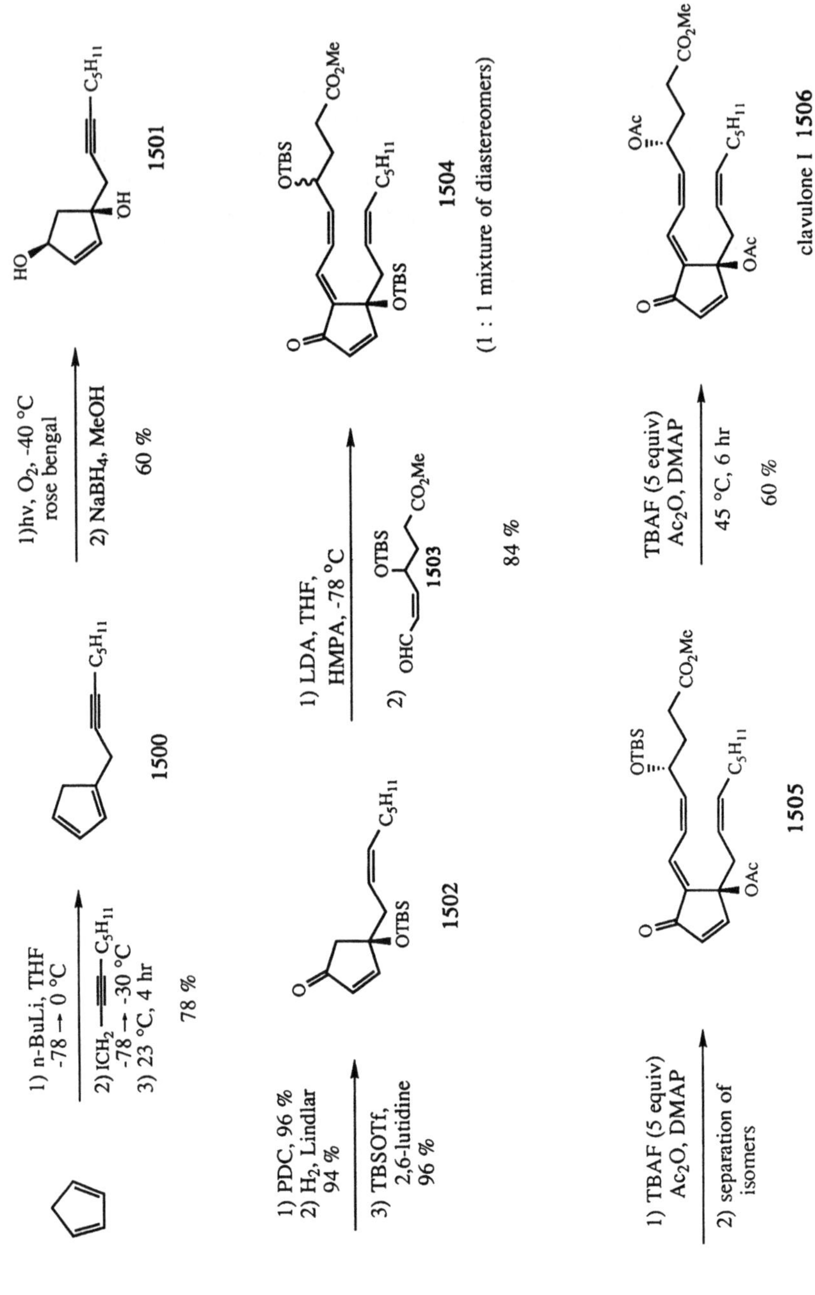

Scheme 257. Corey Synthesis of Clavulone I

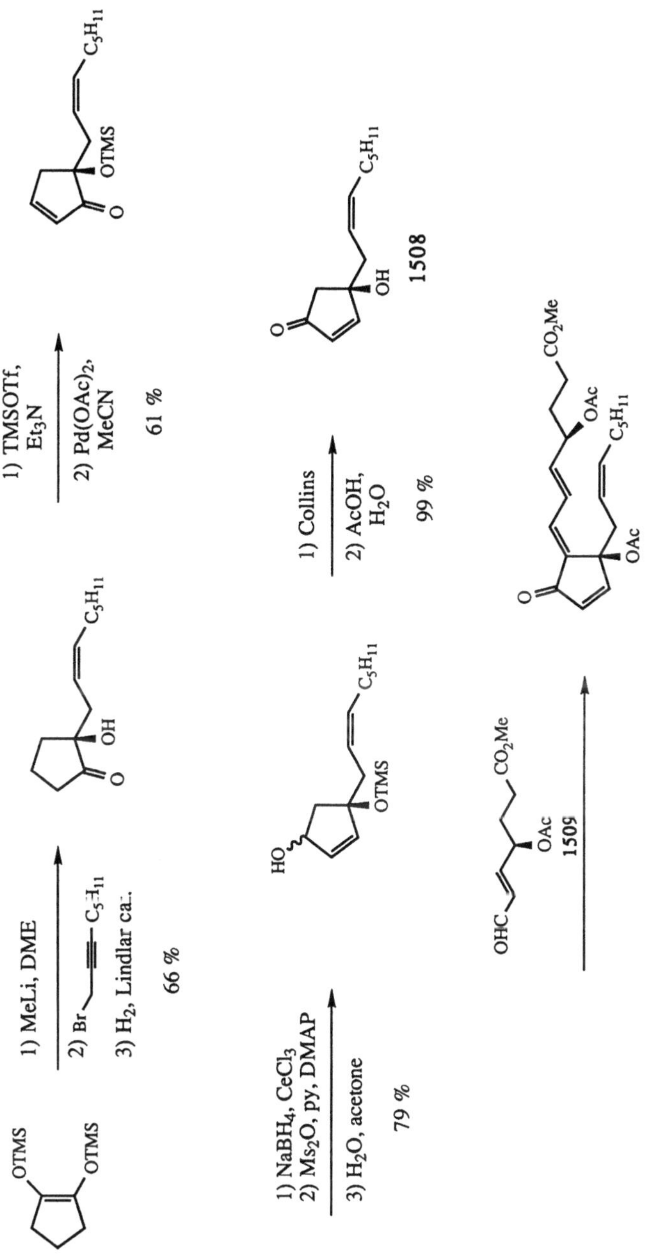

Scheme 258. Shibasaki Synthesis of (−)-Clavulone II

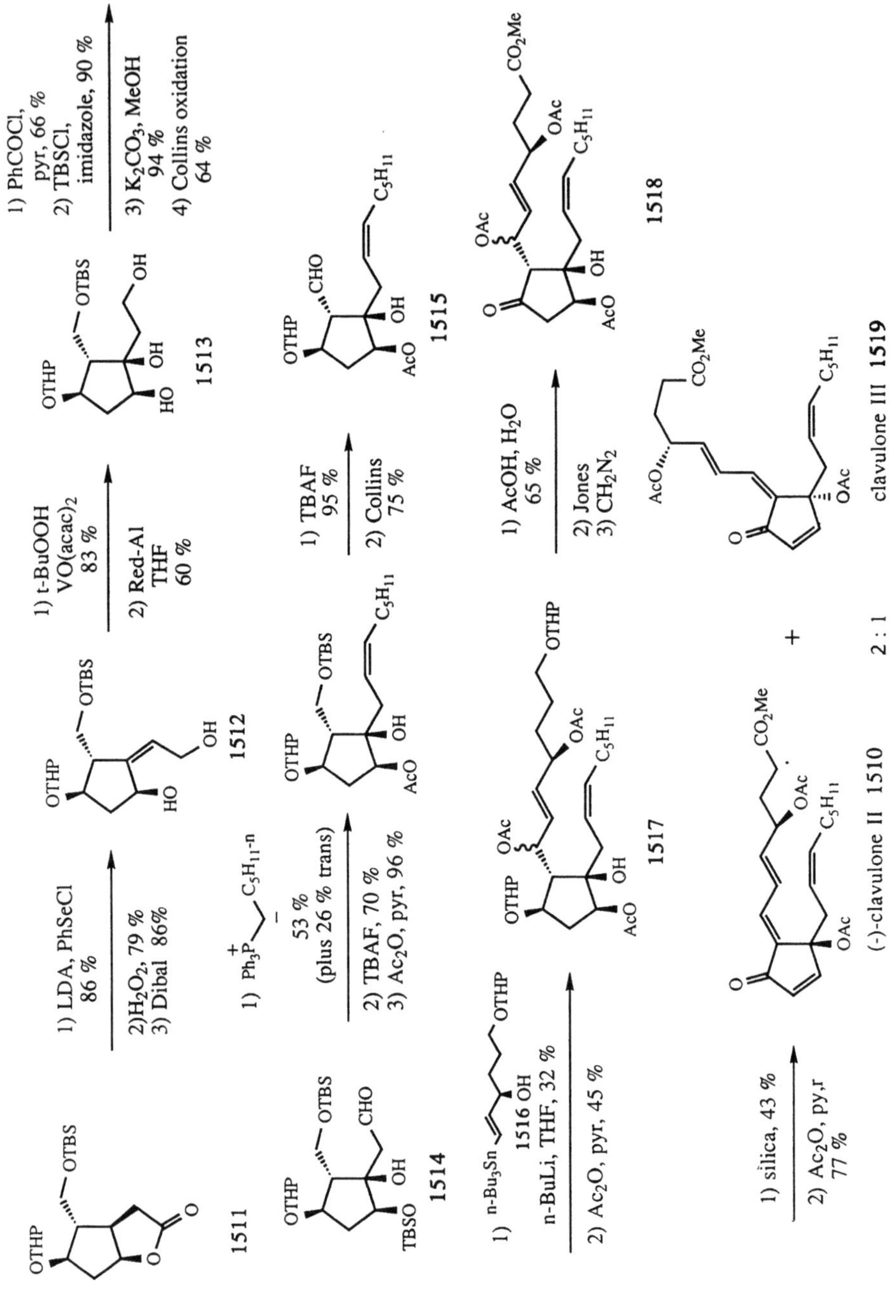

Scheme 259. Hamanaka Synthesis of Clavulone II and Clavulone III

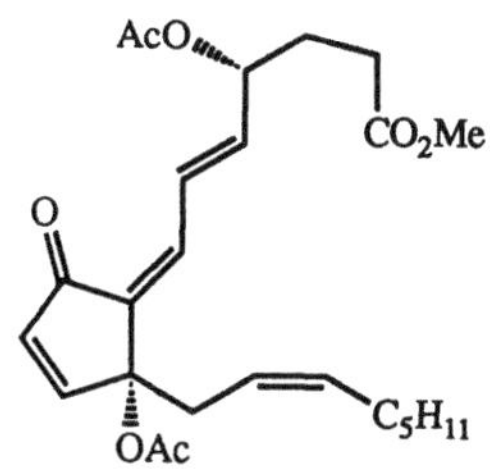

Clavulone III **1519**

(prepared from D-glutamic acid) with **1515** yields **1517** after acylation. Removal of both THP protecting groups, Jones oxidation and treatment with diazomethane produces keto-ester **1518**. Acetate elimination on silica and subsequent acylation of the tertiary alcohol gives clavulone II and clavulone III in a 2:1 ratio and an overall yield of 0.08% over 20 steps.

4.2.1.3 Deacetylclavulone II

An enantioselective and stereoselective synthesis of clavulone II that produces the naturally occurring 12-O-deacetylclavulone II as an intermediate has been demonstrated by Yamada [409] (Scheme 260–262). The basic strategy involves utilizing optically pure **1526** as an acceptor to attach the lower side chain (Scheme 260). A carbonyl transposition allows the cyclopentyl moiety to act as a donor so as to attach the optically pure upper appendage. The upper sidechain is prepared from lactone **1521** which is readily available from D-mannitol (Scheme 261). Reductive opening of the lactone with LAH, followed by acylation of the secondary alcohol in a three-step process yields **1522**. Stepwise oxidation of **1522** and treatment with diazomethane affords ester **1523**, which is converted to aldehyde **1524** by removal of the benzyl protecting group and Swern oxidation. Wittig olefination produces the unsaturated aldehyde **1525** with proper stereochemistry in 10 steps. Reaction of optically pure (*S*)-4-hydroxycyclopentenone **1526** (obtained from (+)-diethyl tartrate) with lithium *t*-butylacetate yields ester **1527** (Scheme 262). Protection of the alcohols, ester reduction-oxidation to the aldehyde and Wittig olefination allows atachment of the lower side chain in 91% yield from the diol **1527** to give **1528**. Removal of the THP groups and PCC oxidation produces cyclopentenone **1529**, which is

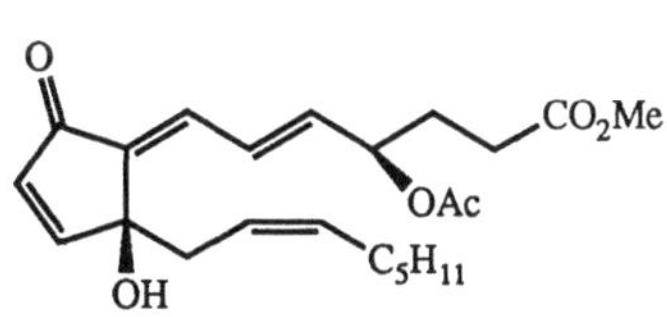

Deacetylclavulone II **1531**

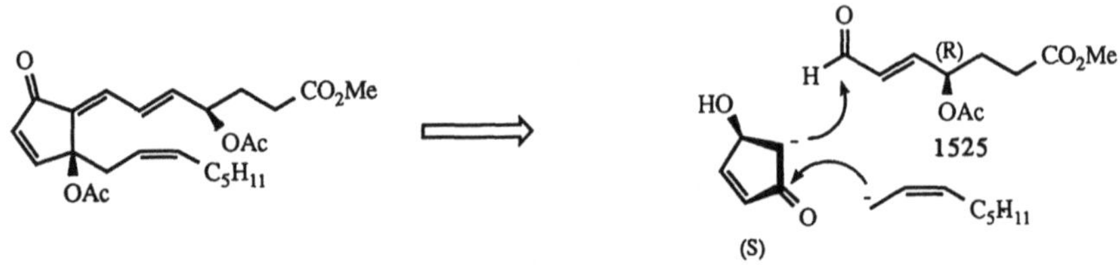

Scheme 260. Yamada Basic Strategy for Synthesis of Clavulone II and 12-*O*-Deacetylclavulone II

coupled through aldol reaction with the optically pure aldehyde **1525** to give **1530**. Mesylation produces desacetylclavulone II **1531**, which upon treatment with acetic anhydride results in the formation of clavulone II. Alternatively, the diol **1530** can be converted directly to clavulone II in 87% yield by treatment with acetic anhydride in pyridine. Clavulone II (**1510**) is produced in 12 steps and 34% yield from lactone **1526**.

4.2.1.4 Chlorovulone II

Optically pure 4-hydroxy-2-cyclopentenone is also utilized by Yamada [410] in an enantioselective synthesis of (−)-chlorovulone II (**1541**) (Scheme 263). The chlorine functionality is introduced early in the synthesis by dichlorination of cyclopentenone **1533**, followed by elimination to afford **1534**. Sodium borohydride reduction of the ketone, TBS protection of the alcohol and LAH reduction of the ester yields the primary alcohol **1535**. The lower appendage is attached by Swern oxidation and Wittig olefination to give **1537**. Desilylation and Jones oxidation produces the ketone **1538**, which is then coupled in aldol fashion with aldehyde **1539** to produce **1540** as a single isomer. Removal of the MOM group gives (−)-chlorovulone II (**1541**) in 13 steps and approximately 20% yield. When compared, the natural product **1541** was found to be of the opposite configuration.

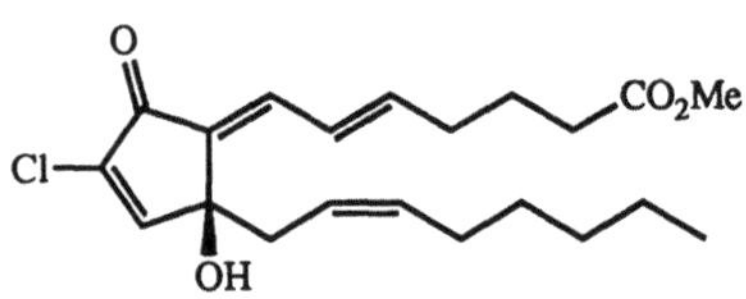

(-)-Chlorovulone II **1541**

4.2.1.5 Preclavulone A

A combination of physico-chemical methods has detected the presence of preclavulone-A as an intermediate in the biosynthetic pathway utilized by certain corals in the production of clavulones [411]. The synthesis of this substance by Corey [412] allowed comparison with the small amount of the

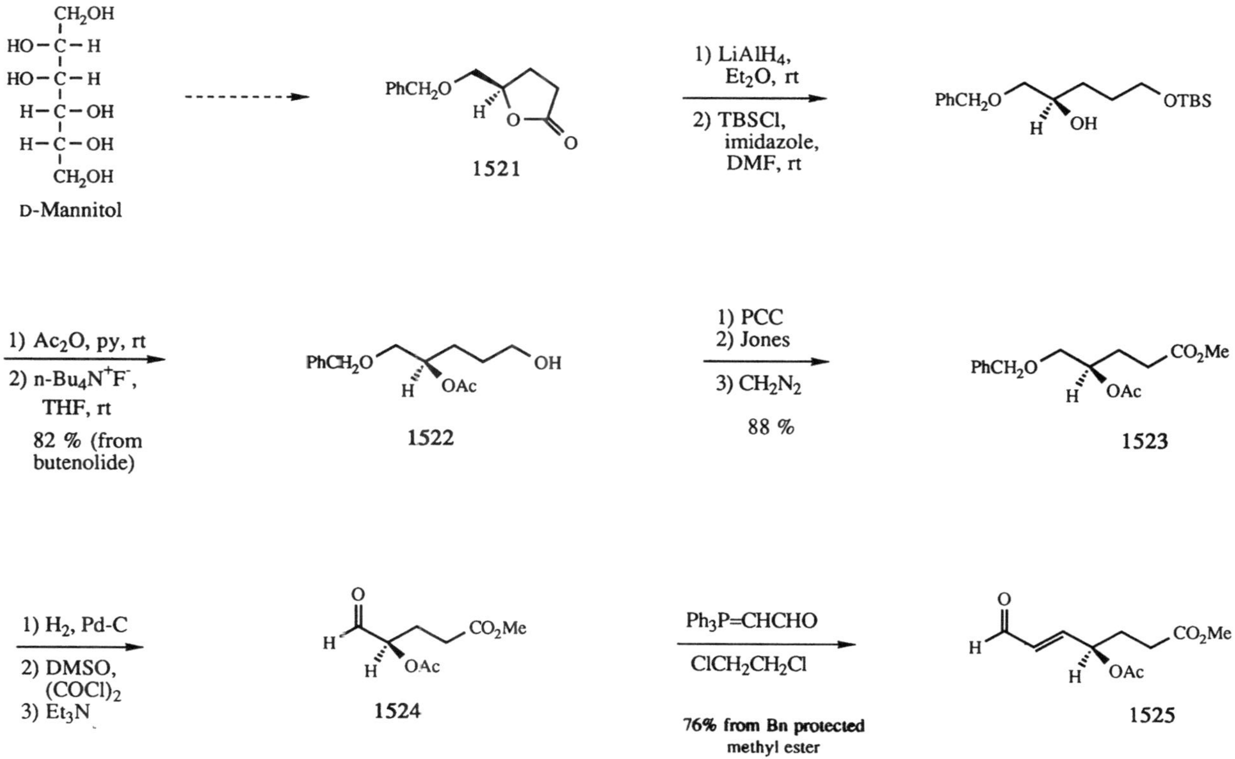

Scheme 261. Yamada Preparation of Aldehyde Intermediate **1525** for Synthesis of Clavulone II and 12-*O*-Deacetylclavulone

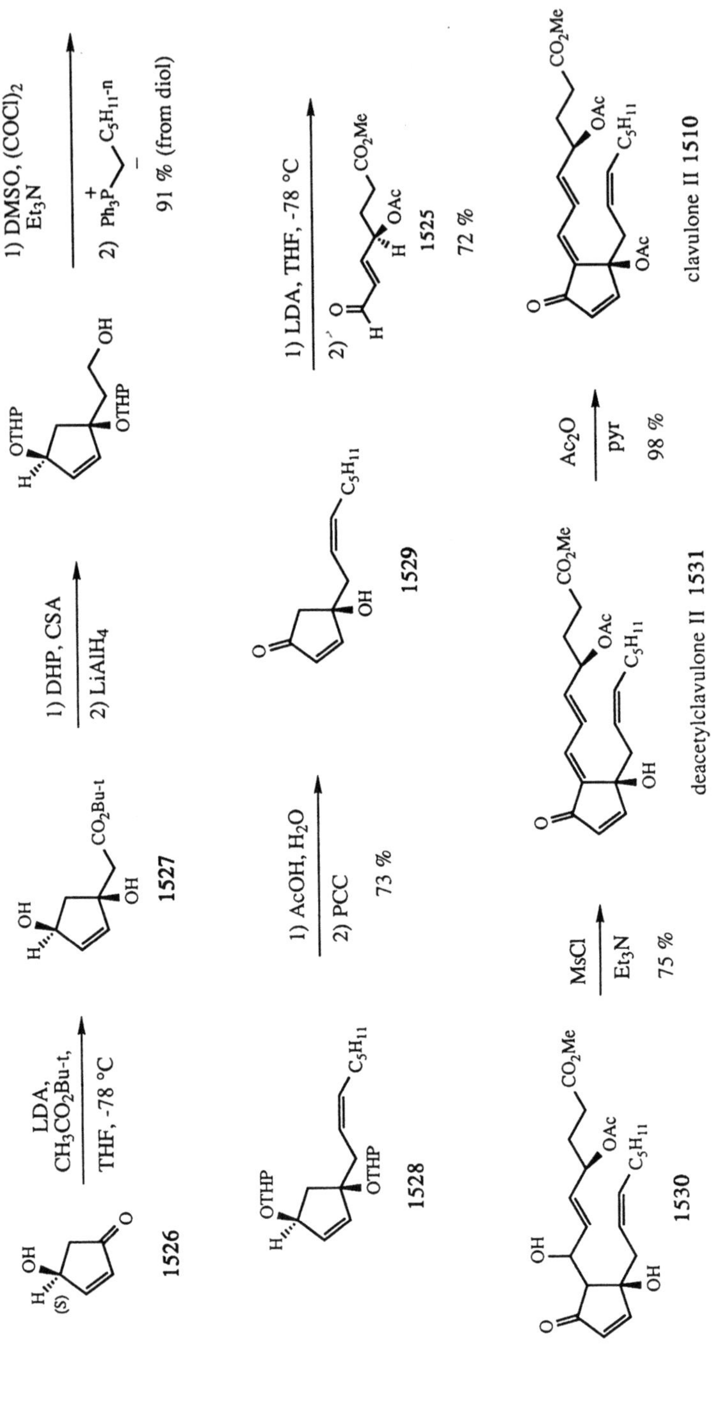

Scheme 262. Yamada Synthesis of Clavulone II and 12-*O*-Deacetylclavulone

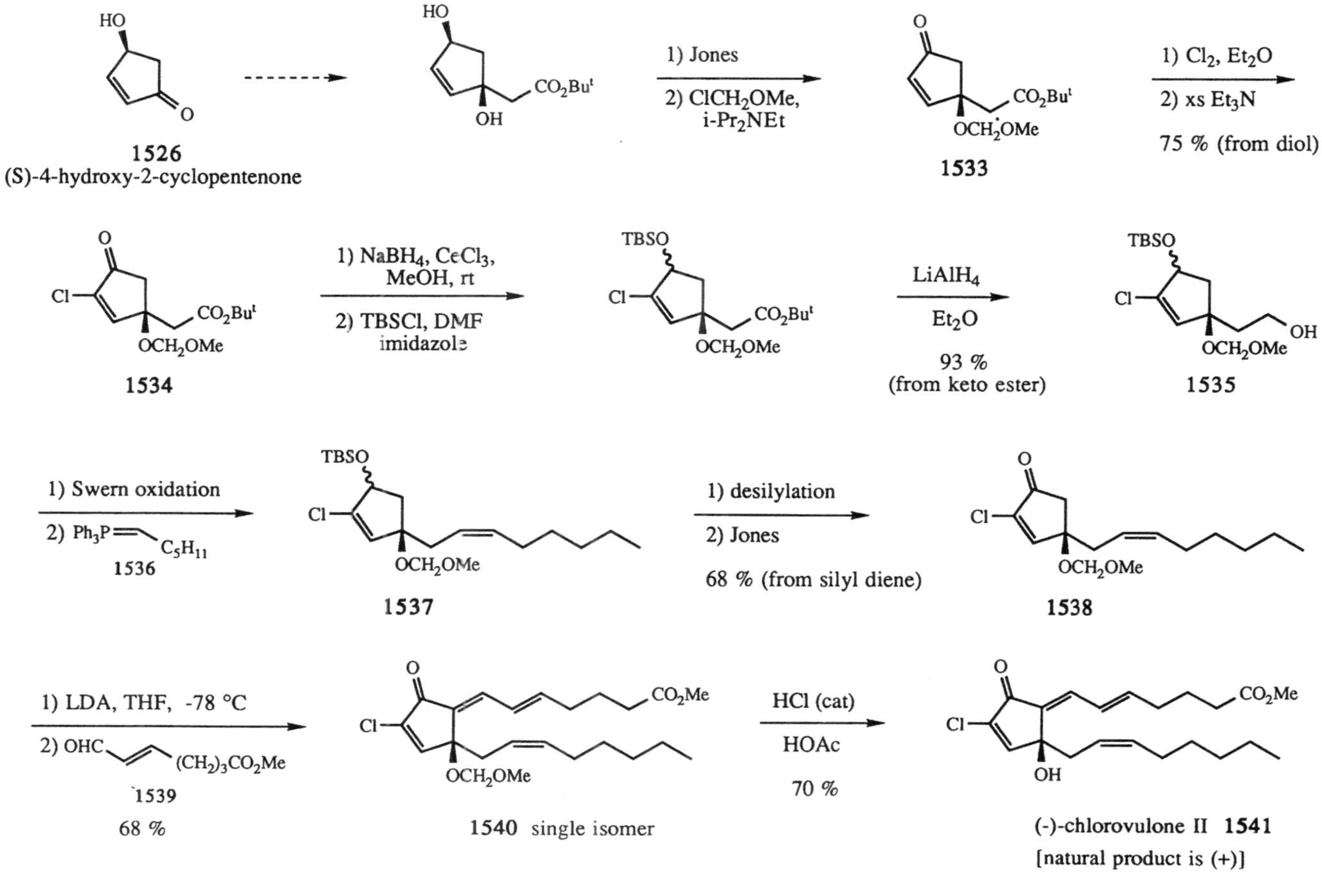

Scheme 263. Yamada Synthesis of (−)-Chlorovulone II

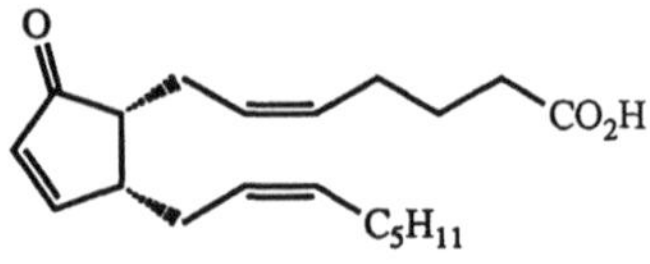

Preclavulone-A

compound obtainable from incubation of the extract from *Clavularia viridis* with arachidonic acid (Scheme 264). The (−)-Diels Alder adduct **1542** was enolized and trapped as the TMS enol ether **1543**. Cope rearrangement led to the *cis* fused bicyclic **1544**. Oxidative cleavage led to the differentially *cis*-substituted cyclopentene **1545**. Wittig reaction was used to append the lower sidechain. A sequence utilizing iodolactonization allowed functionalization of the 5-membered ring to the desired lactone **1548**. DIBAL reduction and a second Wittig olefination established the upper sidechain, with the synthesis being finished off with a Dess-Martin oxidation to give (−)-preclavulone-A (**1549**).

4.2.2 Punaglandins

The punaglandins are a series of C-10 chlorinated marine prostanoids, isolated [413] from the Hawaiian octocoral *Telesto riisei* that show strong antitumor activity. Punaglandin 3 exhibits (*Z*)-7 geometry, whereas punaglandin 4 exhibits both (*Z*)- and (*E*)-7 geometry, but differs in that the 17,18-position is saturated. Syntheses of the punaglandins follow a very similar strategy. The lower sidechain is attached to a cyclopentenone intermediate, followed by coupling with an optically active aldehyde to obtain the upper side chain. In this way the problem of obtaining the correct relative stereochemistry between C-12, of the cyclopentenone ring, and C-5 and C-6, on the upper sidechain, is circumvented.

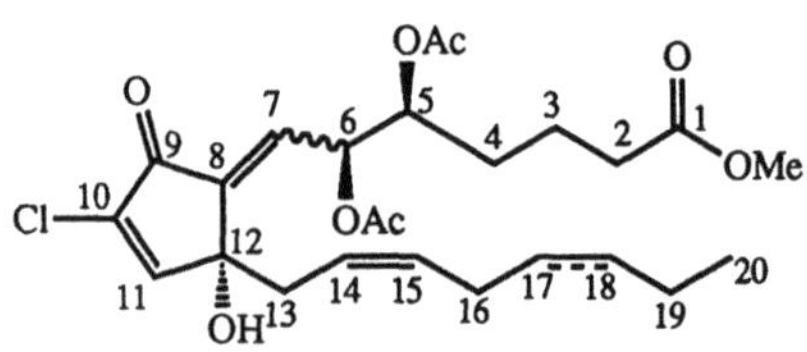

4.2.2.1 *Punaglandin 4*

The synthesis of punaglandin 4 (**1562**) by Mori [414] involves the coupling of two optically active fragments to give **1562** in an enantiospecific manner (Schemes 265 and 266). The chiral precursor **1551**, prepared from (+)-tartaric acid, is a common intermediate in two other syntheses of **1562**. Conversion of **1551** to the iodide **1552**, followed by a photo-induced radical addition to methyl acrylate affords **1553**. Removal of the benzyl protecting group and Swern

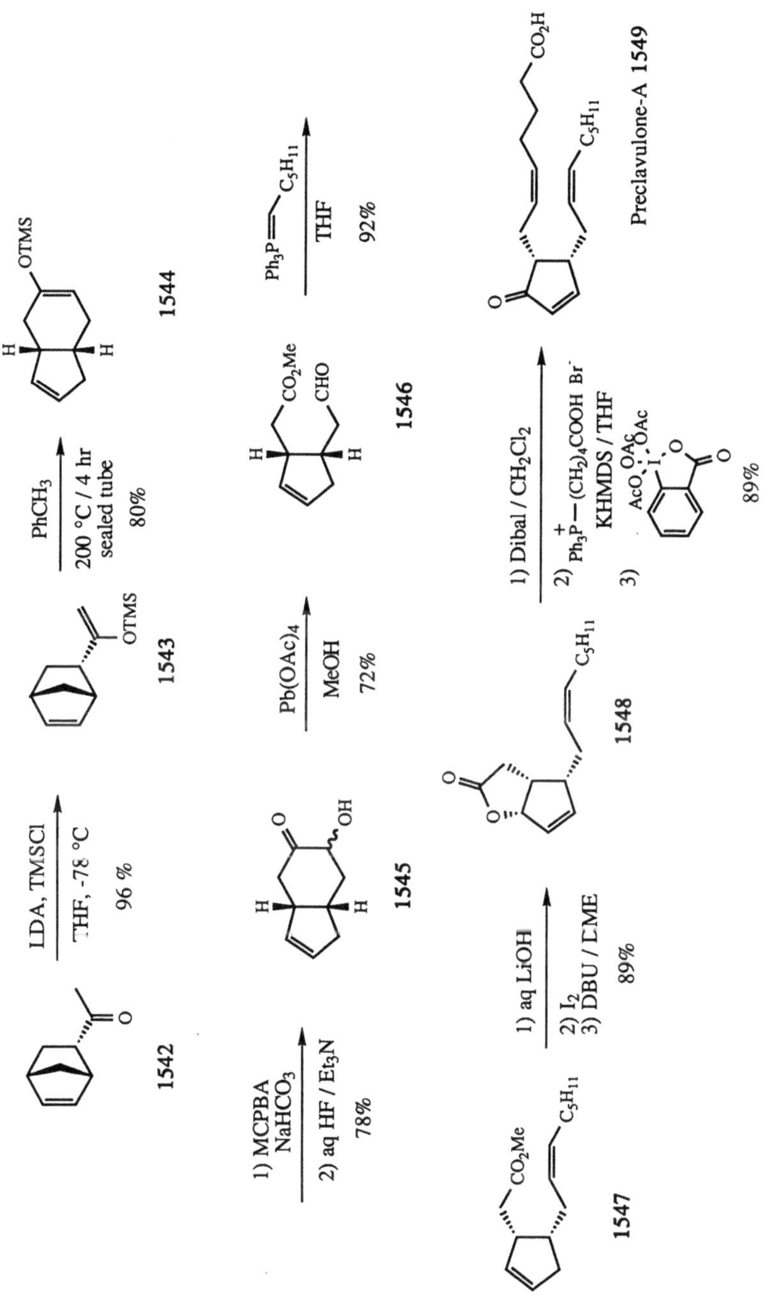

Scheme 264. Corey Synthesis of (−)-Preclavulone A

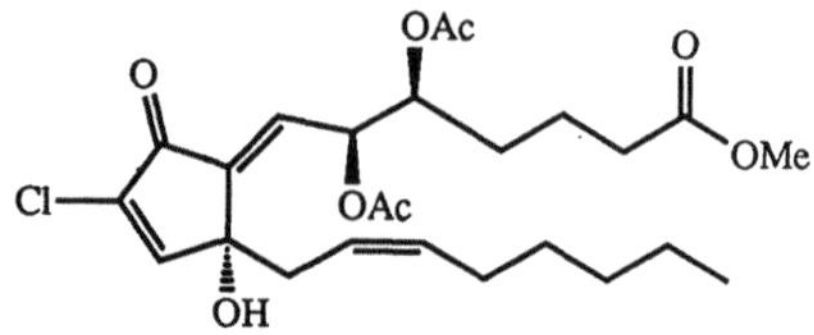

Punaglandin 4 1562

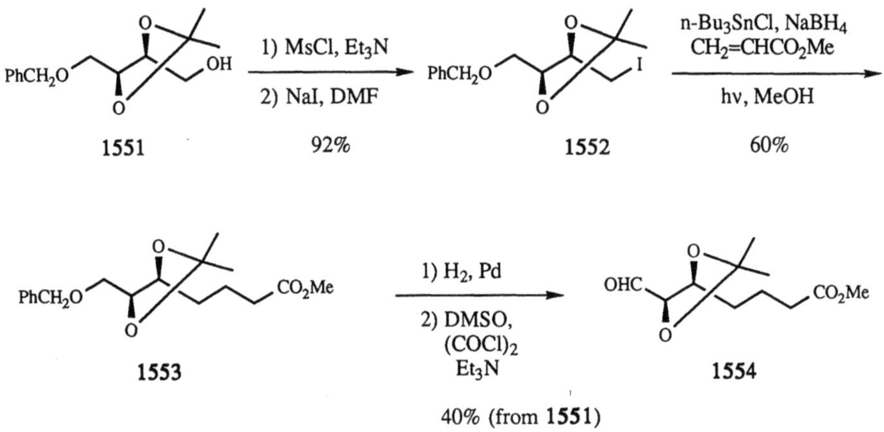

Scheme 265. Preparation of Aldehyde Fragment **1554** for Mori Synthesis of Punaglandin 4

oxidation gives the aldehyde fragment **1554**. The cyclopentenone fragment **1560** is obtained from hydroxycyclopentenone **1555**. Acetylation and chlorination-elimination of **1555** followed by reduction and silylation affords the acetate **1556** as a mixture of diastereomers. Resolution of **1556** by enzymatic hydrolysis gives enantiomerically pure **1557**. Oxidation of **1557** with PDC, addition of the lower side chain and Lindlar reduction of the alkyne **1558** yields **1559** with the proper olefin geometry. PCC oxidation of **1559** and protection of the tertiary alcohol forms the second optically pure fragment **1560**. Coupling of **1560** and **1554** in an aldol condensation with elimination gives **1561** in 25% yield, along with 37% of the (*Z*)-isomer. Deketalization of **1561** and acylation produces punaglandin 4 (**1562**) in 16 steps and 0.03% yield.

The preparation of **1562** by Shibasaki [415] begins with attachment of the lower side chain during the initial steps of the synthesis to give **1568** (Scheme 268). Bis-chlorination of **1568** followed by lithium chloride-assisted thermal elimination yields the unsaturated ketone **1569**. A 1,3-carbonyl migration in four steps, followed by protection of the alcohol as its MOM ether gives the fragment **1570**, ready to be coupled to the optically pure aldehyde. The aldehyde fragment is prepared from (+)-diethyl tartrate via the similar intermediate **1563**

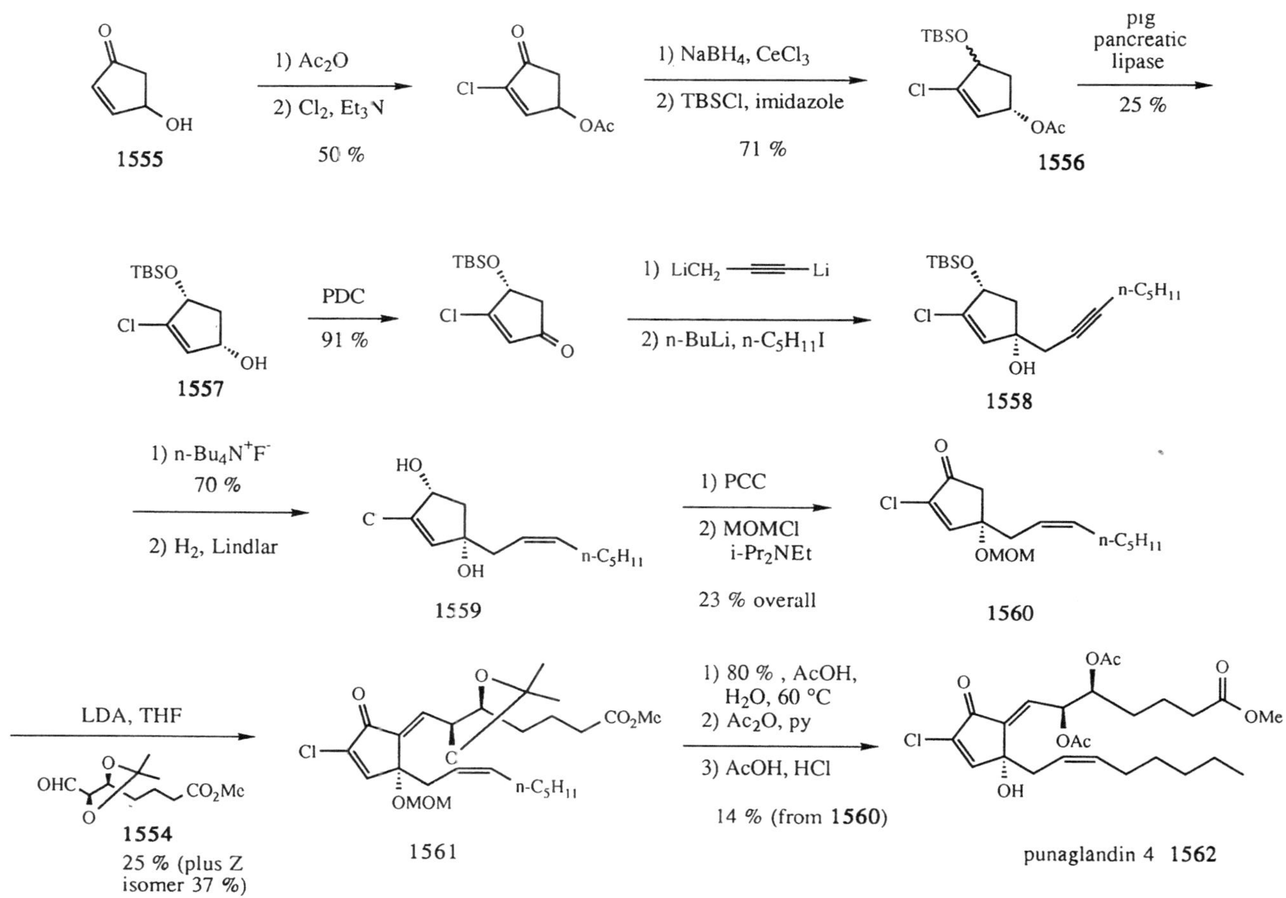

Scheme 266. Mori Synthesis of Punaglandin 4

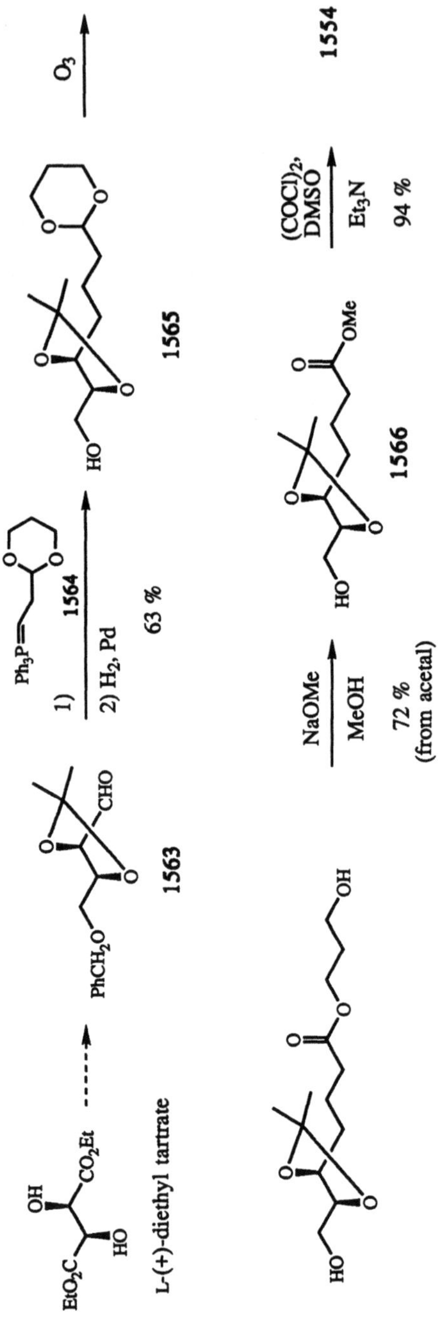

Scheme 267. Preparation of Chiral Aldehyde Fragment **1554** for Shibasaki Synthesis of Punaglandin 4

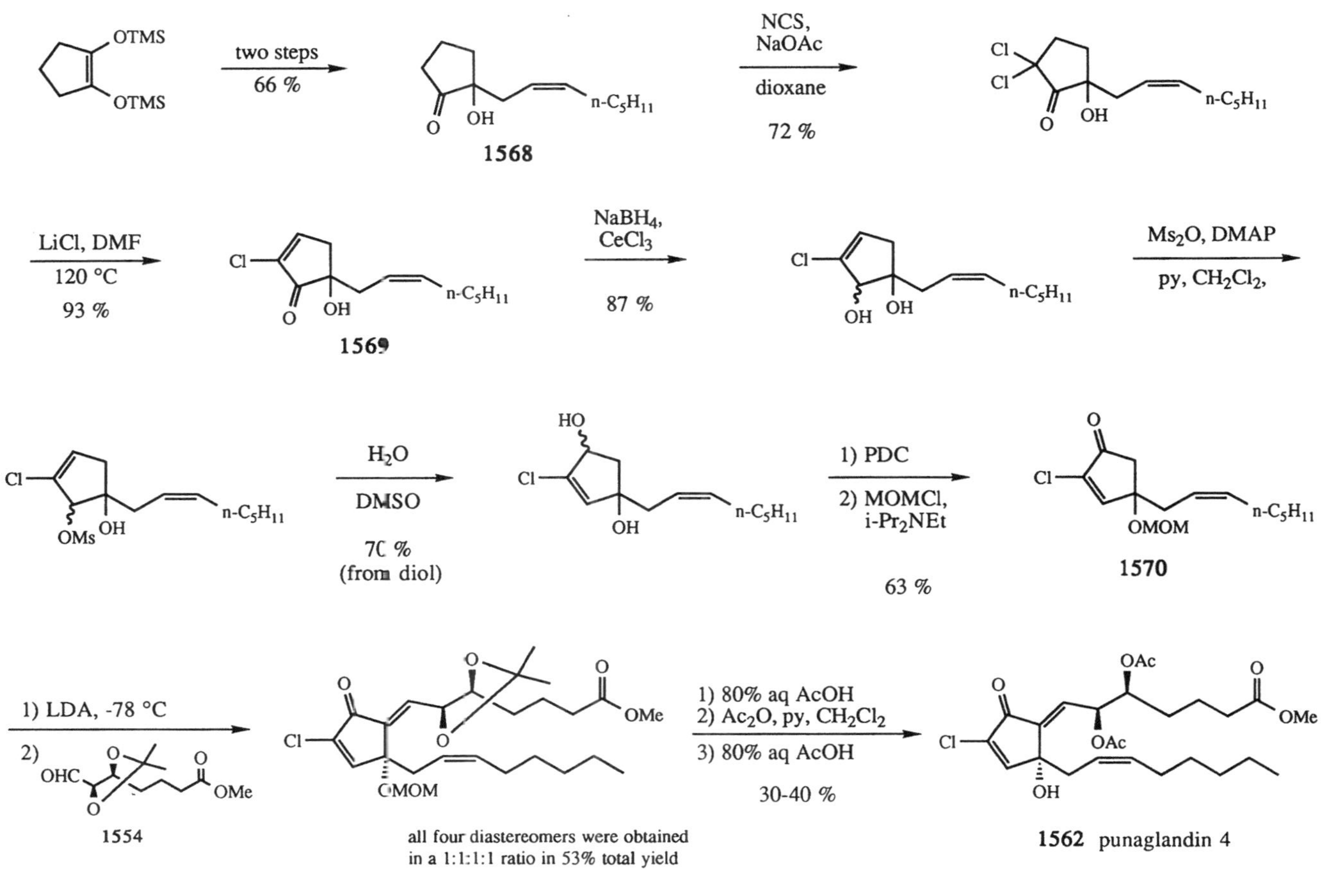

Scheme 268. Shibasaki Synthesis of Punaglandin 4

(Scheme 267). Chain extension of **1563** by Wittig olefination with **1564** and hydrogenation gives **1565**. Ozonolysis and transesterification followed by Swern oxidation gives the needed chiral aldehyde fragment **1554**. Aldol coupling of **1570** and **1554** with elimination gives an equal mixture of all four diastereomers at the 7- and 12-positions. Deketalization and acylation of the proper diastereomer gives punaglandin 4 (**1562**) in 13 steps and 0.8% yield.

The Noyori [416] preparation of punaglandin 4 provides both the (7*E*)- and (7*Z*)-isomers (Scheme 270) in an enantiospecific synthesis. The chiral aldehyde fragment is prepared from the allylic alcohol **1571** via Sharpless asymmetric epoxidation to afford the chiral epoxide **1572** (Scheme 269). Protection of the alcohol, basic opening of the epoxide, esterification with diazomethane and acetylation gives acetate **1573** in good yield. Deprotection of the primary alcohol and Swern oxidation provides the chiral aldehyde in 7 steps. Chirality at the 12-position of **1578** is obtained through the chiral cyclopentenone **1576**, which is obtained in four steps from 2,4,6-trichlorophenol using known chemistry. The lower appendage is attached via the equilibrating mixture of organolithiums obtained by lithiation of allenylstanne **1577** to give a 42% yield of **1578** (along with 22% of the cyclopentyl allene). Lindlar reduction affords the desired *cis* geometry in the sidechain. The complete punaglandin skeleton is obtained by aldol condensation of the enolate from **1579** with aldehyde **1574**. Acylation and desilylation yields a mixture of (7*E*)- and (7*Z*)-punaglandin 4 in a 2:5 ratio. This ratio can be reversed to 7:3 by irradiation with light.

4.2.2.2 Punaglandin-3 and (7E)Punaglandin 4

Yamada's [417] synthesis of (7*E*)-punaglandin 4 (Scheme 271) involves the coupling of the usual optically pure aldehyde **1582** with the enantiomer of an intermediate utilized in his synthesis of chlorovulone II (Scheme 264). Coupling of **1582** with **1538** in an aldol reaction followed by acylation and elimination gives the ketal **1583** as a mixture of 7,8-double bond isomers. Deketalization of the (7*E*)-isomer, acylation, and removal of the MOM protecting group affords (7*E*)-punaglandin 4 (**1562**) in what might be called an "off the shelf" synthesis.

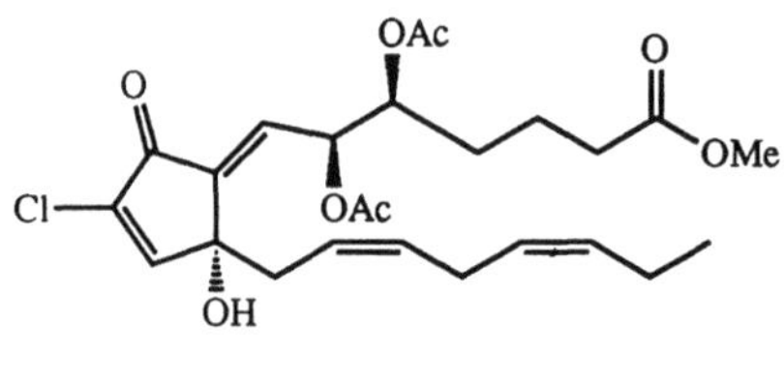

Punaglandin 3

Punaglandin-3 was assembled in an entirely analogous manner (Scheme 272). Wittig reaction of the unsaturated phosphorane with aldehyde **1585** established the lower sidechain. Desilylation and Jones oxidation led to enone **1587**. Attachment of the upper sidechain as in the punaglandin-4 synthesis eventually yielded punaglandin-3 (**1588**).

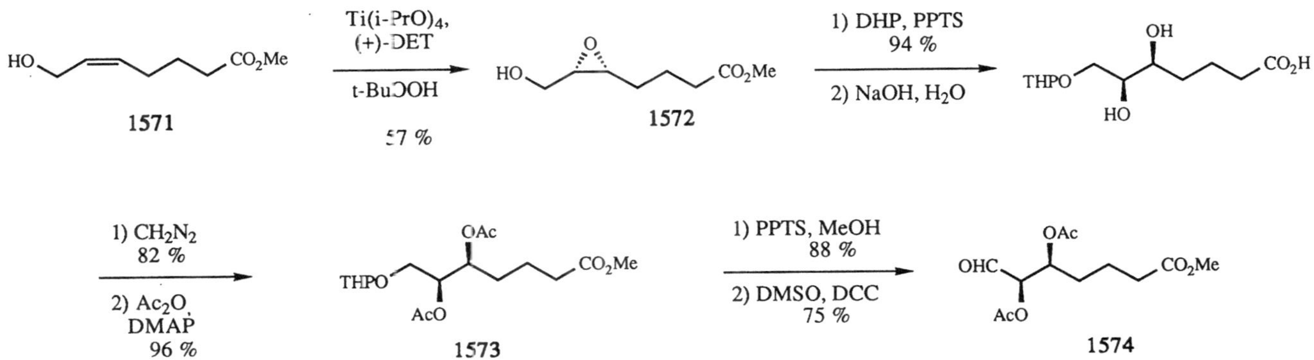

Scheme 269. Preparation of Chiral Aldehyde Fragment **1574** for Noyori Synthesis of (7*Z*)- and (7*E*)-Punaglandin 4

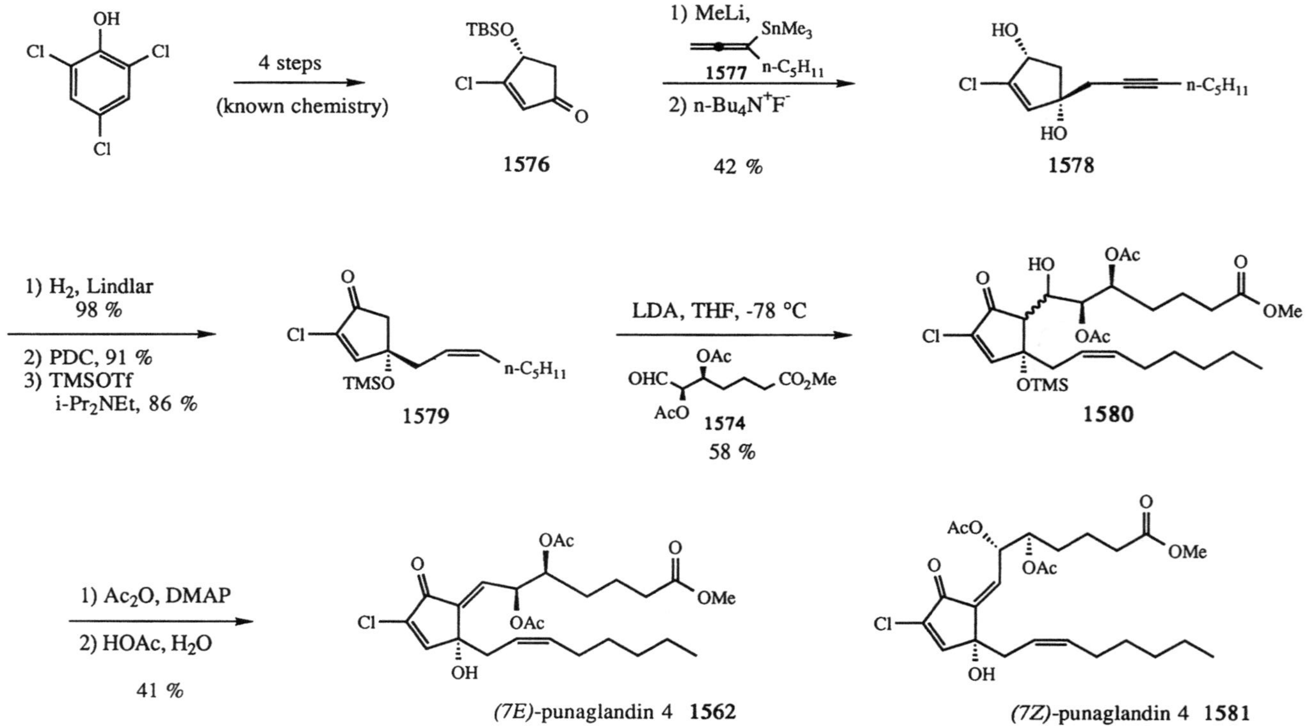

2 : 5 $7E : 7Z$ ratio; irradiation with light gives 7 : 3 $7E : 7Z$ ratio

Scheme 270. Noyori Synthesis of (7*Z*)- and (7E)-Punaglandin 4

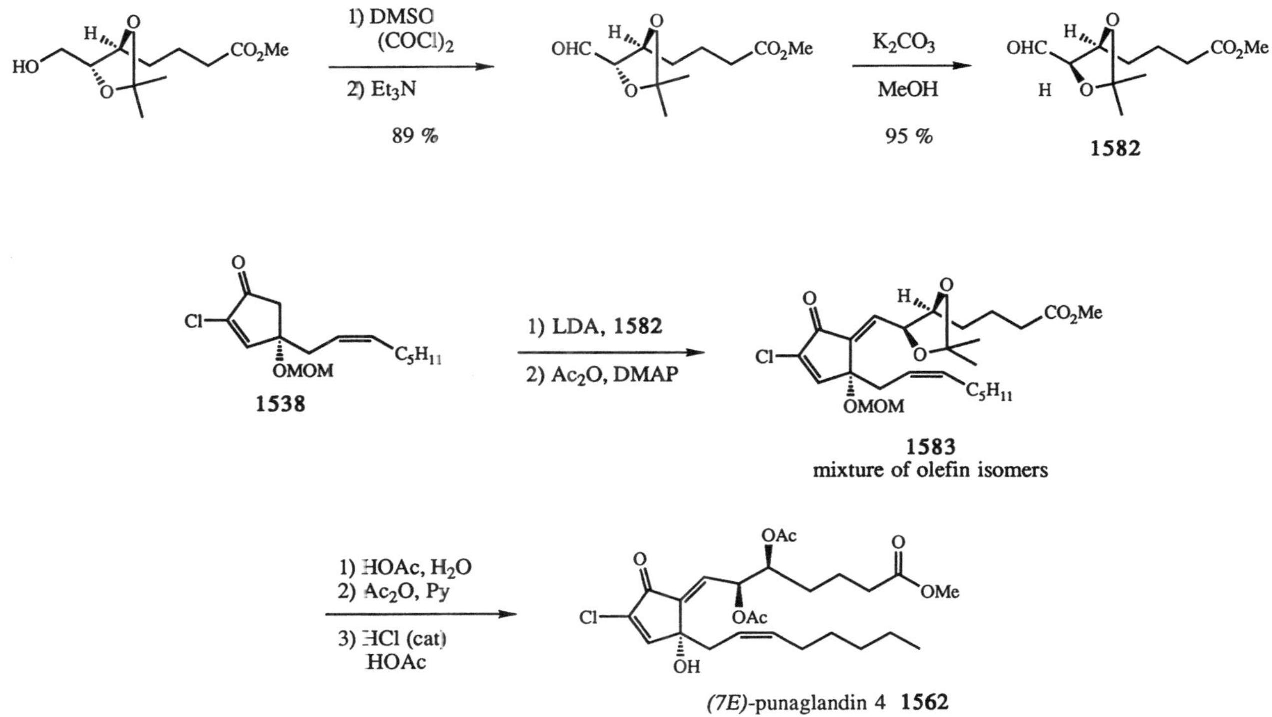

Scheme 271. Yamada Synthesis of (7*E*)-Punaglandin 4

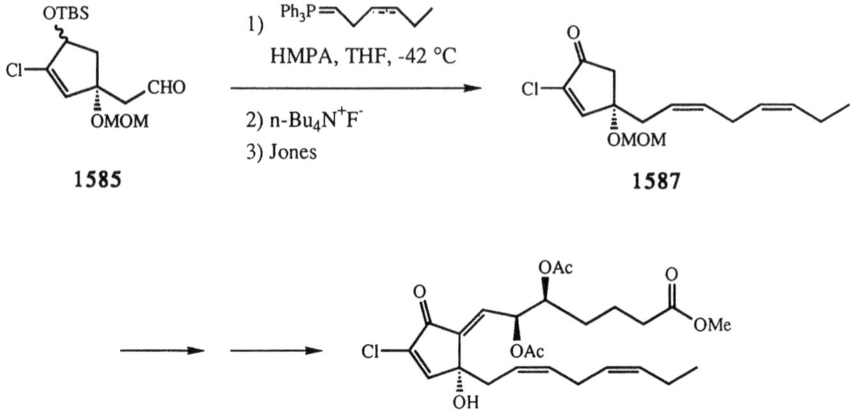

punaglandin 3 **1588**

Scheme 272. Yamada Synthesis of of Punaglandin 3

4.2.3 Hybridalactone

Hybridalactone (**1593**) is a macrocyclic lactone isolated from the marine alga *Laurencia hybrida* [418]. An enantiospecific synthesis of **1593** by Corey [419] is illustrated in Scheme 273. Problems that need to be addressed in the synthesis of **1593** include control of stereochemistry at the seven chiral centers and the problem of macrolactonization. These difficulties are mitigated somewhat by the fact that all seven chiral centers are contiguous and the number of degrees of rotational freedom in the ring-opened lactone are restricted due to the two *cis* double bonds and the fused five-membered ring. Four of the seven chiral centers are obtained by the coupling of tosylate **1587** and cyclopropylstannane **1588**, both of which are optically pure. Fluoride-catalyzed fragmentation of **1589**, followed by reduction with L-Selectride provides the fifth chiral center (as a 6:1 mixture of the epimeric carbinol) along with the necessary functionality to further elaborate the upper sidechain. The final two stereocenters are obtained via Sharpless epoxidation to afford **1590**. Lithiation of acetylene **1590**, conversion to the Gilman reagent and 1,3-addition to the iodoallene **1591** yields the diyne **1592** which upon Lindlar reduction affords the *cis* stereochemistry in the

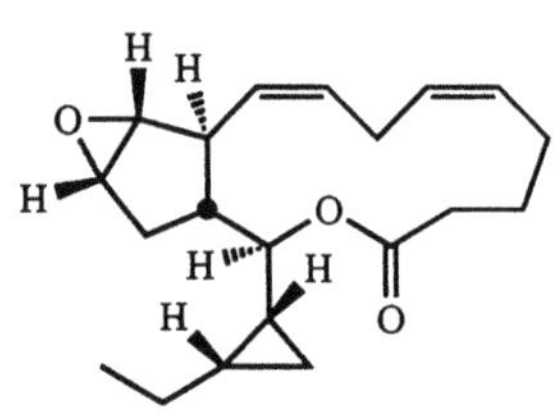

Hybridalactone 1593

upper appendage. The carbinol stereochemistry is corrected by an oxidation-reduction sequence followed by deprotection of the latent carboxyl functionality. Macrolactonization is achieved by the double activation method in 83% yield to give (−)-hybridalactone (**1593**) in 13 steps and 10% overall yield.

4.3 C8–C11 Algal Pheromones

A variety of C11 and C8 hydrocarbons have been isolated from brown algae of the genera *Dictyopteris*, *Ectocarpus*, and *Cutleria*. These metabolites are often important in algal reproduction, acting as pheromones and chemotactic agents. They possess linear unbranched structures, either cyclic or acyclic, characterized by varying degrees of unsaturation.

4.3.1 Dictyoprolene

Dictyoprolene (**1598**) is the acetate of one of the undec-1-en-3-ols that have been implicated as possible key biosynthetic intermediates leading to many of the C11 hydrocarbons [420]. It has been isolated from the brown alga *Dictyopteris proliferra* by Yamada, who determined its absolute stereochemistry through synthesis (Scheme 274) [421]. Condensation of acrolein with 1-bromooct-2-yne in the presence of activated zinc yields allylic alcohol **1595**. Esterification with optically pure steroidal acid chloride **1596** and separation of the diastereomers on silver nitrate impregnated silica provides optically pure adduct **1597**. Lindlar reduction, reductive cleavage of the ester and acylation provides (+)-dictyoprolene in 5 steps and 5.0% overall yield.

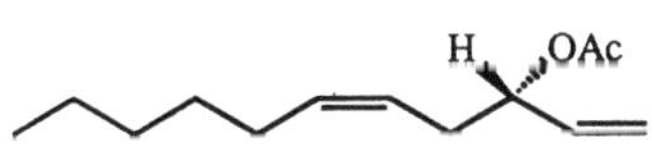

(+)-Dictyoprolene **1598**

4.3.2 Dictyopterenes

Dictyopterene A (**1601**) is a cyclopropane-containing diene first isolated [422] in 1968 from a mixture of the brown algae *Dictyopteris plagiogramma* and *D. australis*. The related triene dictyopterene B (**1628**) has been isolated from the essential oils of an unidentified *Dictyopteris sp.* by Moore in 1970 [423]. It was later found in a variety of other brown algae by Jaenicke and co-workers and named hormosirene. Dictyopterene B has an intense "ocean smell" and has been

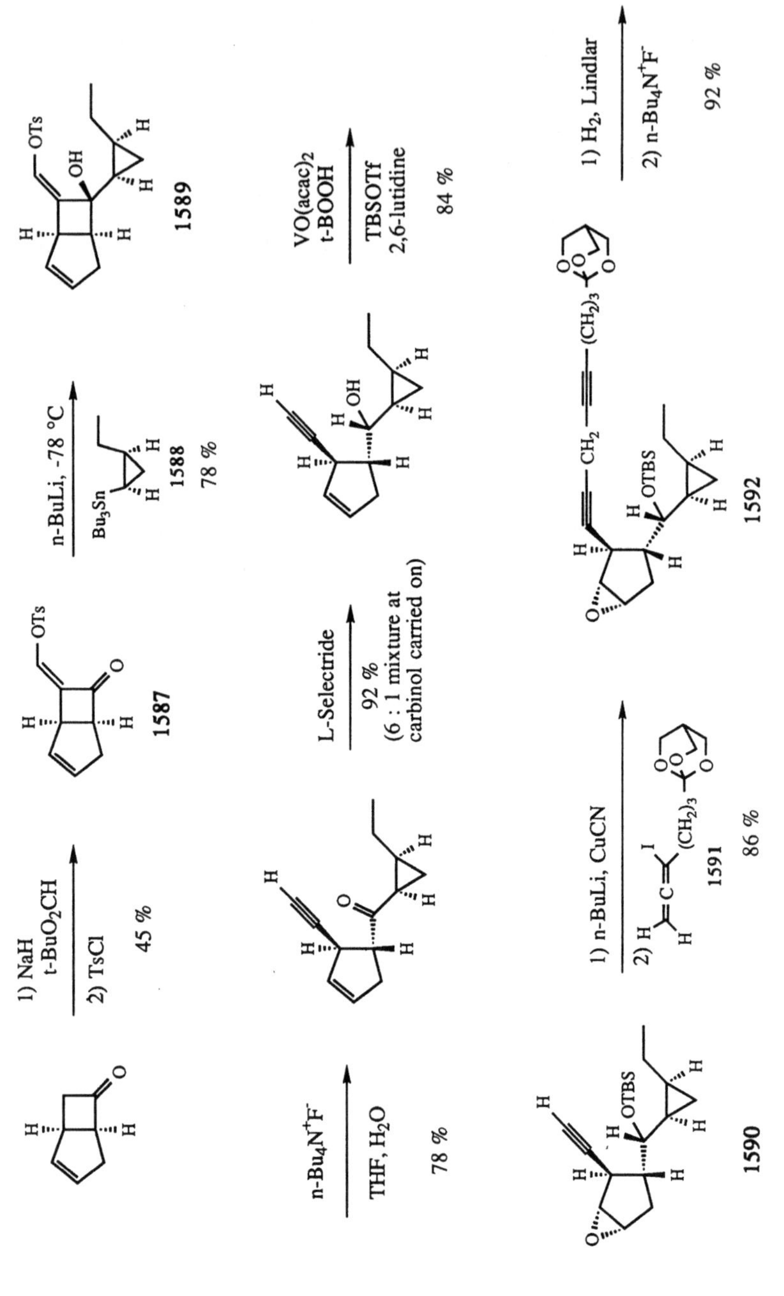
1) NaH
t-BuO2CH
2) TsCl
45 %
1587
n-BuLi, -78 °C
Bu3Sn
1588
78 %
1589
n-Bu4N+F-
THF, H2O
78 %
L-Selectride
92 %
(6 : 1 mixture at carbinol carried on)
VO(acac)2
t-BOOH
TBSOTf
2,6-lutidine
84 %
1590
1) n-BuLi, CuCN
2)
1591
86 %
1592
1) H2, Lindlar
2) n-Bu4N+F-
92 %

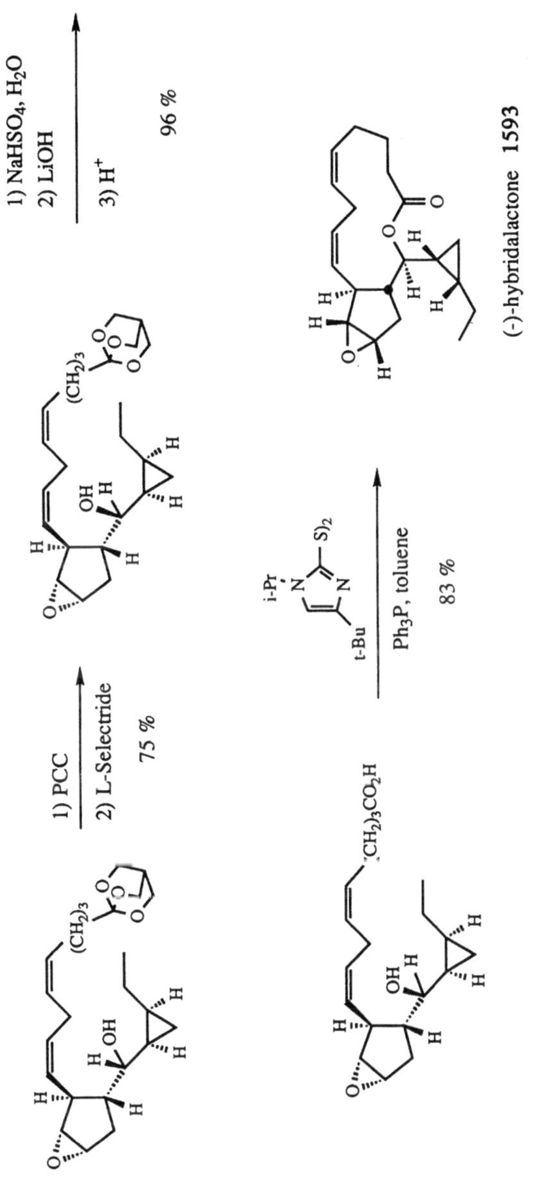

Scheme 273. Corey Synthesis of (−)-Hybridalactone

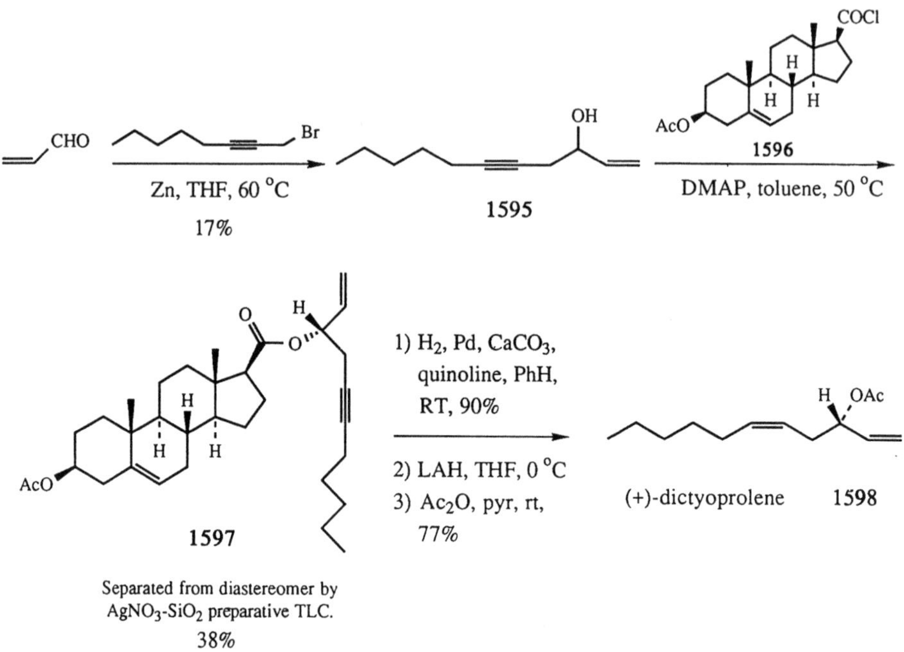

Scheme 274. Yamada Synthesis of (+)-dictyoprolene

shown by Jaenicke to act as a sperm attractant [424]. Dictyopterene C (**1606**) is a common constituent of many marine brown algae. It was first isolated from the Pacific seaweeds *Dictyopteris plagiogramma* and *D. australis* collected near Hawaii [425]. It has also been found to occur in several North Pacific representatives of the same genus [426]. The closely related dictyopterene D′ (**1629**) (variously known as sirenin and ectocarpene) has been obtained from the brown alga *Ectocarpus siliculosus* and shown to be a sperm attractant of the female gametes [427].

4.3.2.1 Dictyopterene A

The first synthesis of dictyopterene A was that of Ohloff [428] in 1969 (Scheme 275). Separation of **1599** from a mixture of the *cis* and *trans* isomers, followed by oxidation with activated manganese dioxide produces aldehyde **1600** with the needed *trans* stereochemistry at the cyclopropane ring. A non-stereoselective Wittig reaction yields dictyopterene A (**1601**) along with its (Z)-isomer in a 2:3 ratio, favoring the unwanted (Z)-isomer.

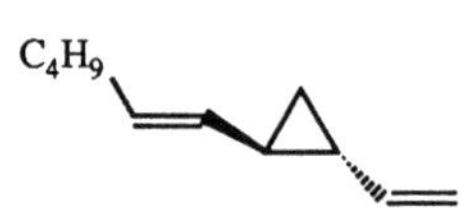

Dictyopterene A 1601

Later in the same year a second synthesis of dictyopterene A was described by Weinstein [429]. Addition of ethyl diazoacetate to butadiene, followed by hydrolysis and treatment with thionyl chloride produces the acid chloride **1603** as a mixture of *cis* and *trans* isomers (Scheme 276). Reduction of **1603** to the aldehyde and *cis*-selective Wittig olefination affords **1601** as a 1:1 mixture of the *cis* and *trans* cyclopropane isomers. Dictyopterene A is secured by preparative GC in 5 steps.

Dictyopterene A has been prepared in low yield, along with dictyopterene C′ as the major product by Billups [430]. Treatment of **1604**, prepared by reaction of *trans*-dec-3-ene and dichlorocarbene, with potassium *t*-butoxide in DMSO affords a mixture of all possible isomers of **1605** in 80–90% yield (Scheme 277). Pyrolysis of the mixture at 80 °C gives dictyopterene A (**1601**) in 4% overall yield along with dictyopterene C′ (**1606**) produced via Cope rearrangement of the cyclopropyl *cis*-isomers. Pyrolysis of **1601** at 175 °C gives **1606** as the sole product in 30% overall yield.

Yamada [431] has prepared dictyopterene A in a biomimetic fashion (Scheme 278). Beginning with the alcohol portion of the ester dictyoprolene, thiol **1608** can be prepared in three steps. Solvolysis of the mesylate of **1608** in aqueous acetone and potassium acetate affords cyclopropane **1609** via a biogenetically patterned homoallyl-cyclopropylcarbinyl rearrangement. Due to the lability of cyclopropylcarbinols under acidic conditions, the elimination of the hydroxyl was performed with sodium hydride and $MeO_2CN^-SO_2^+NEt_3$, giving a 1:1 mixture of the *cis* and *trans* olefins **1610**. Elimination of the sulfide and separation of the diastereomers affords dictyopterene A in 8 steps and > 6% overall yield.

An enantiospecific (but not diastereospecific) approach (Scheme 279) that produces (+)-dictyopterene A and unnatural (+)-dictyopterene C′ has been developed by Genet [432]. Lithiation, transmetallation and acylation of commercially available silylacetylene **1611** affords ketone **1612**. Introduction of chirality is achieved by enantioselective reduction of **1612** with (*S*)-Alpine borane producing the (*S*)-propargylic alcohol **1613** in 96% yield with 85% ee after desilylation. Desilylation of **1613**, silylation of the hydroxyl and hydroxymethylation to give **1614** is achieved by application of standard methodology. Acylation, desilylation and Lindlar reduction of **1614** affords the *cis*-alkene **1615**. Treatment of **1615** with sulfone in the presence of DBU and a palladium catalyst results in alkylation and concomitant isomerization of the alkene to the (*E*)-isomer. Benzoylation affords **1618**, which is the critical intermediate needed for transfer of chirality from C–O to C–C via cyclopropanation. Palladium-catalyzed cyclopropanation of allylic benzoate **1618** gives **1619** as a mixture of diastereomers with exclusive (*E*)-geometry at the alkene. Desulfonylation, ester reduction and PCC oxidation gives the aldehydes **1620** and **1621** as a 3:2 mixture of *cis*/*trans* diastereomers at the cyclopropane ring. Wittig olefination of the mixture followed by Cope rearrangement of the *cis*-isomer produces (+)-dictyopterene A (**1601**) (85% ee) and unnatural (+)-dictyopterene C′ (**1606**) via a 15 step synthesis.

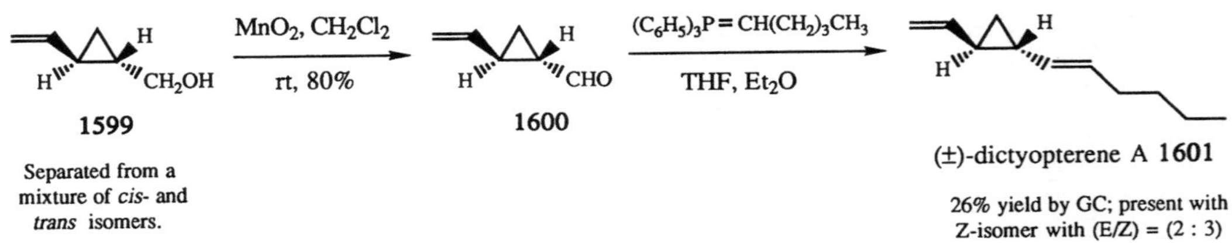

Scheme 275. Ohloff Synthesis of (±)-Dictyopterene A

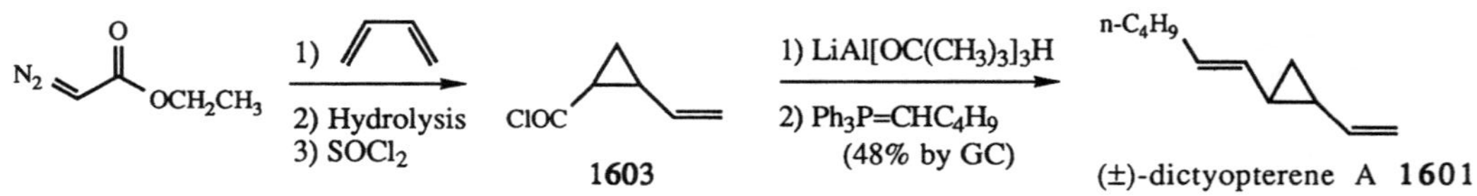

Scheme 276. Weinstein Synthesis of (±)-Dictyopterene A

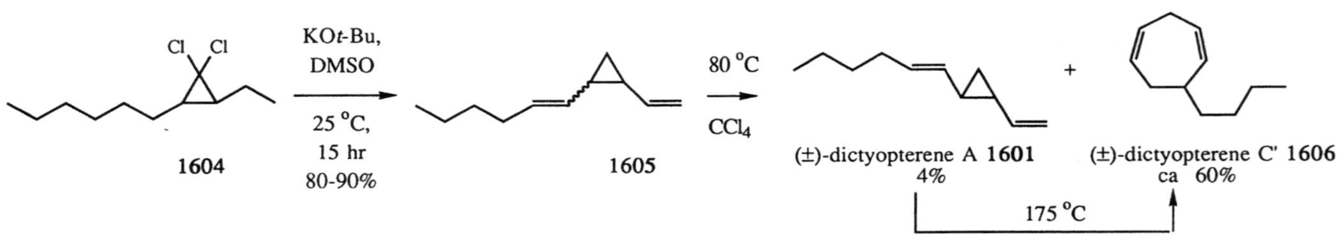

Scheme 277. Billups Syntheses of (±)-Dictyopterene A and (±)-Dictyopterene C′

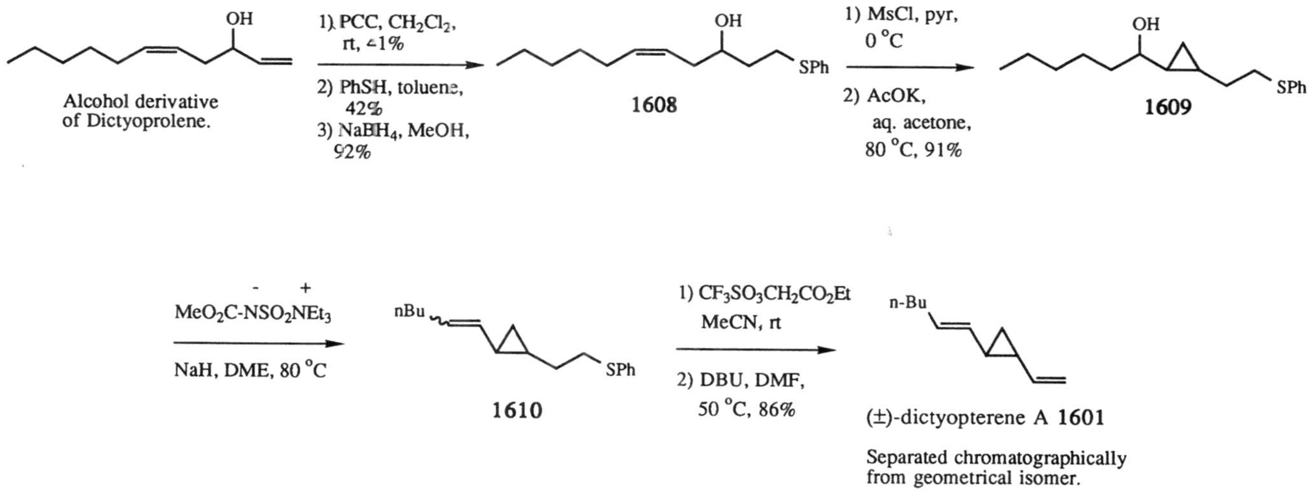

Scheme 278. Yamada Synthesis of (±)-Dictypterene A

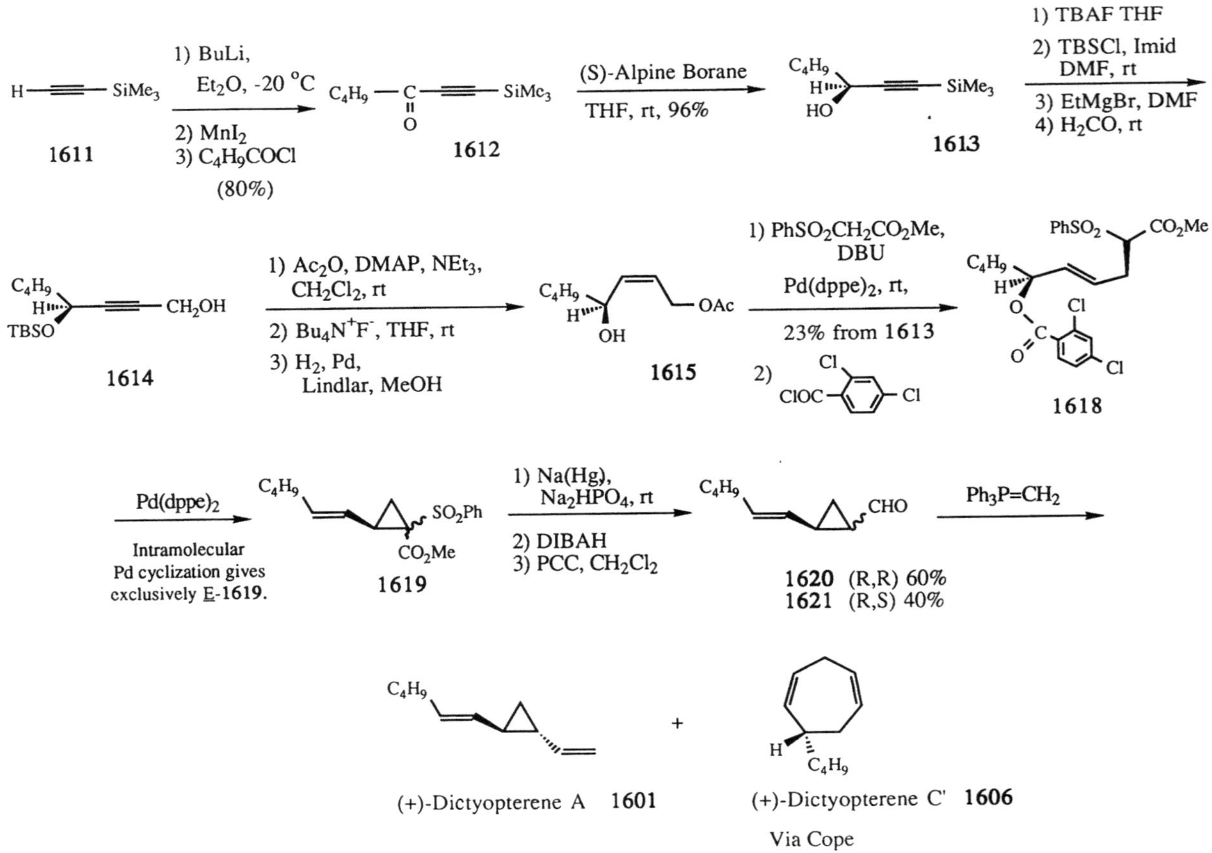

Scheme 279. Genet Syntheses of (±)-Dictyopterene A and (±)-Dictyopteren C'

A shorter enantiospecific synthesis of dictyopterene A (**1601**) (Scheme 280) has been developed by Jaenicke [433]. The optically pure ester **1623** is obtained by resolution of the acid and esterification. Condensation of **1623** with pentylidenediphenylphosphine oxide affords the β-ketophosphine oxides **1624**. Reduction of **1624** to the alcohol and treatment with sodium hydride gives an 85:15 (*E*/*Z*) mixture of isomers. Separation on silver nitrate-impregnated silica produces (+)-dictyopterene A (**1601**) in 19% overall yield.

Dictyopterene A has also been obtained enantiospecifically by Jaenicke [434] as an intermediate in the preparation of dictyopterene C (Scheme 284).

4.3.2.2 Dictyopterene B (Hormosirene)

The first synthesis of dictyopterene B (**1628**) was developed by Weinstein [435] (Scheme 281). Hydrogenation of **1625** over Lindlar catalyst, followed by treatment with phosphorus tribromide and then triphenylphosphine dibromide produces the phosphonium bromide **1626**. Wittig olefination of **1626** with the isomeric mixture of aldehydes **1627** gives dictyopterene B as a mixture of isomers. Upon heating, the *cis*-isomer is transformed by Cope rearrangement into **1628** and dictyopterene D′ (**1629**) in 6 steps.

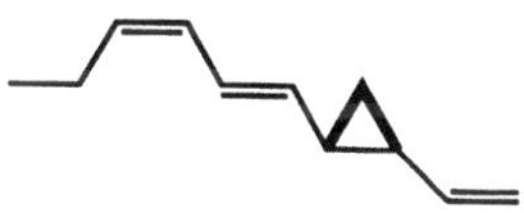

Dictyopterene B 1628

Dictyopterene B (**1628**) has been prepared enantiospecifically by Jaenicke (Scheme 282) [436]. The optically pure ester **1623** (a common intermediate in the synthesis of dictyopterene A, Scheme 280) is obtained by resolution of the acid and esterification. Reduction of the ester to the alcohol **1631**, oxidation to the aldehyde and olefination with (formylmethylidene)triphenylphosphorane yields the α,β-unsaturated aldehyde **1632** with a 93:7 *E*/*Z* ratio. A second Wittig olefination unfortunately gives a 1:1 *cis*/*trans* ratio. Selective Diels-Alder cycloaddition of the *trans* isomer with 4-phenyl-1,2,4,-triazoline-3,5-dione allows separation of the two isomers to yield (−)-dictyopterene B (**1628**) optically pure in 4 steps, after resolution of the starting material.

Dictyopterene B (**1628**) has also been prepared by Schneider [437] from fucoserratene (Scheme 302, see p. 134).

Helmchen [438] has developed the highly enantio- and diastereoselective synthesis of dictyopterene B (**1628**) shown in Scheme 283. The key transformation in this synthesis involves a novel diastereoface-selective intramolecular alkylation to construct the cyclopropane ring of **1628**. Alkylation of the acetate of camphor derivative **1633** with (*E*)-1,4-dibromo-2-butene yields **1634**. Treatment of **1634** with potassium *t*-butoxide produces cyclopropane **1635** in 85% yield and a diastereomeric purity of 96.5%. Reduction of **1635** with LAH, PCC oxidation and reaction with the formylphosphorane **1636** affords aldehyde

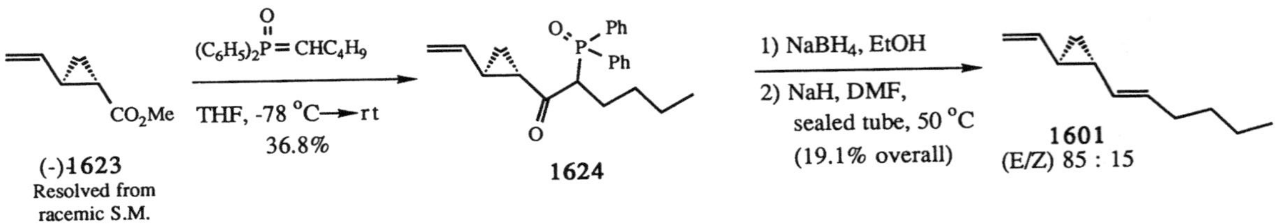

Scheme 280. Jaenicke Synthesis of Dictyopterene A

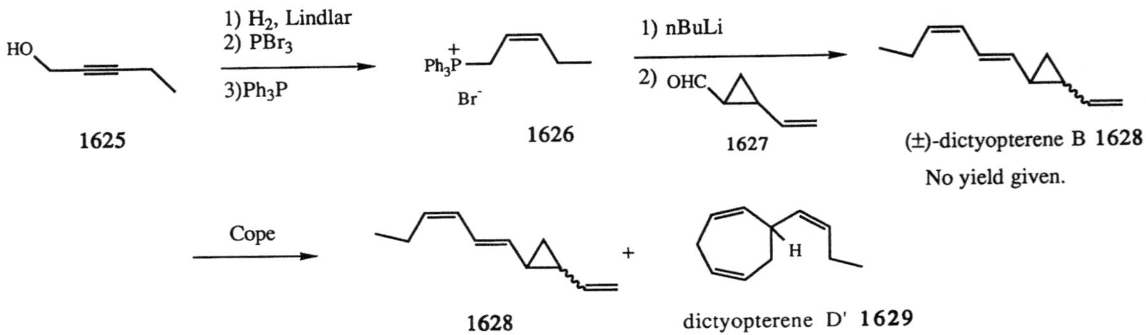

Scheme 281. Weinstein Synthesis of (±)-Dictyopterenes B and D′

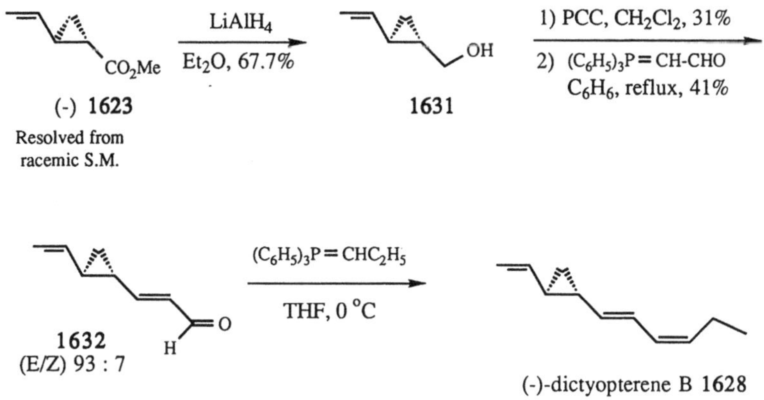

Scheme 282. Jaenicke Synthesis of (−)-Dictyopterene B (Hormosirene)

1637 with an E/Z ratio of 93:7. Salt-free Wittig olefination and removal of the unwanted 3-E isomer by reaction with 4-phenyl-1,2,4-triazolidine-3,5-dione provides pure dictyopterene B (**1628**) in 7 steps and 19% overall yield.

4.3.2.3 Dictyopterene C and Dictyopterene C′

Dictyopterene C′ has been prepared as the major product in a dictyopterene A synthesis by Billups [439] (Scheme 277).

(-)-Dictyopterene C **1641** (+)-Dictyopterene C′ **1606** (the unnatural isomer)

Dictyopterene C′ (**1606**) (the unnatural isomer) has been prepared as a mixture with dictyopterene A in an enantiospecific synthesis by Genet [440] (Scheme 279).

An enantioselective synthesis of dictyopterene C, by Jaenicke [441], is shown in Scheme 284. Lactone **1638** is obtained optically pure by separation of the diastereomeric amides obtained by reaction with (S)-phenylethylamine. DIBAL reduction of **1638** and immediate addition of the salt-free Wittig reagent **1639** provides **1640** with a Z/E ratio of $\geq 97{:}3$. PCC oxidation of the alcohol **1640** and Wittig olefination provides dictyopterene A (**1628**) in 24% yield. Cope rearrangement of **1628** provides dictyopterene C (**1641**) quantitatively and with 97% ee.

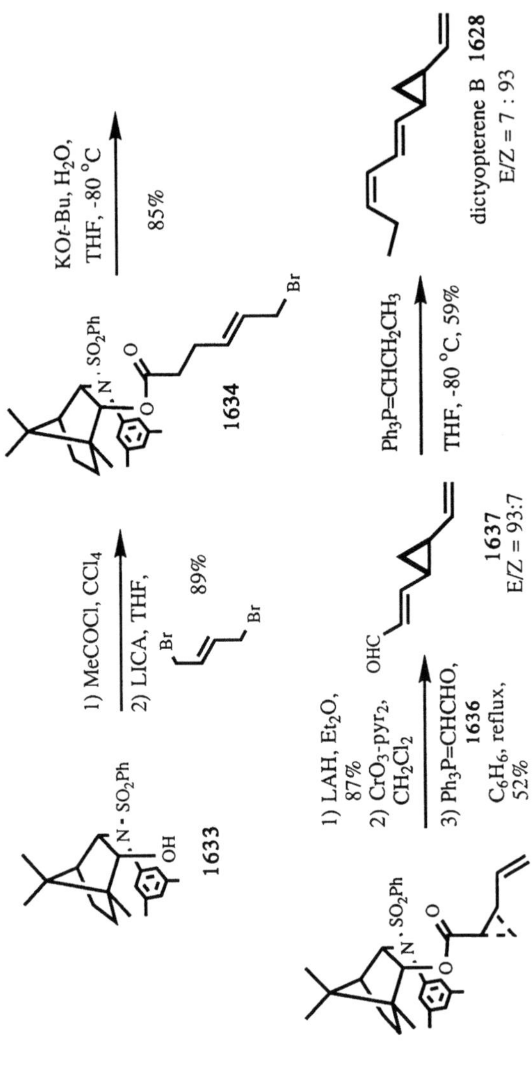

Scheme 283. Helmchen Synthesis of Dictyopterene B (Hormosirene)

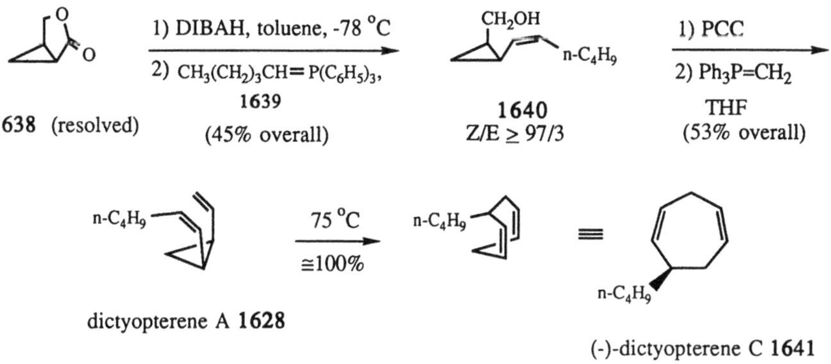

Scheme 284. Jaenicke Synthesis of *R*-(−)-Dictyopterene C

4.3.2.4 Dictyopterene D′ (Serenin or Ectocarpene)

The first synthesis of dictyopterene D′ was that of Mueller [442] (Scheme 285). Wittig olefination of a mixture of *cis*- and *trans*-2-vinylcyclopropane-carboxaldehyde **1643** gives a mixture of all four possible geometric isomers **1644**. Cope rearrangement followed by Lindlar reduction affords **1629** in 30% yield after separation by preparative GC.

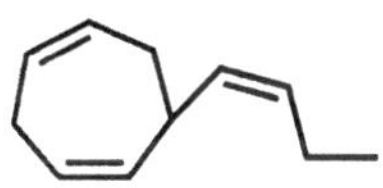

Dictyopterene D' **1629**

Another synthesis of dictyopterene D′ has been performed by Jaenicke [443] (Scheme 286). Reaction of butadiene with dibromocarbene and subsequent monodehalogenation of the adduct affords the bromovinylcyclopropane as a

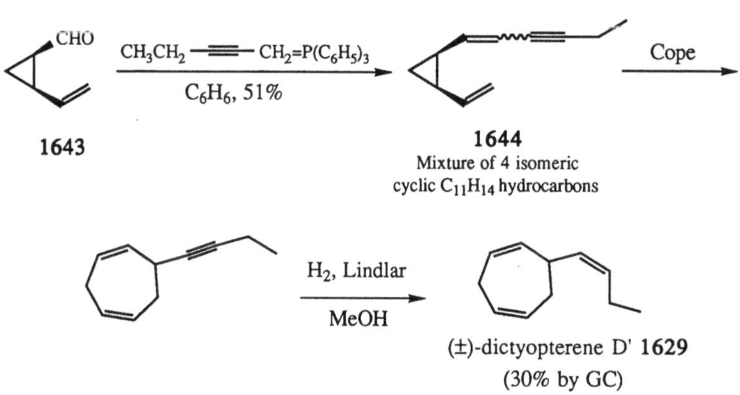

Scheme 285. Mueller Synthesis of (±)-Dictyopterene D′ (Serenin)

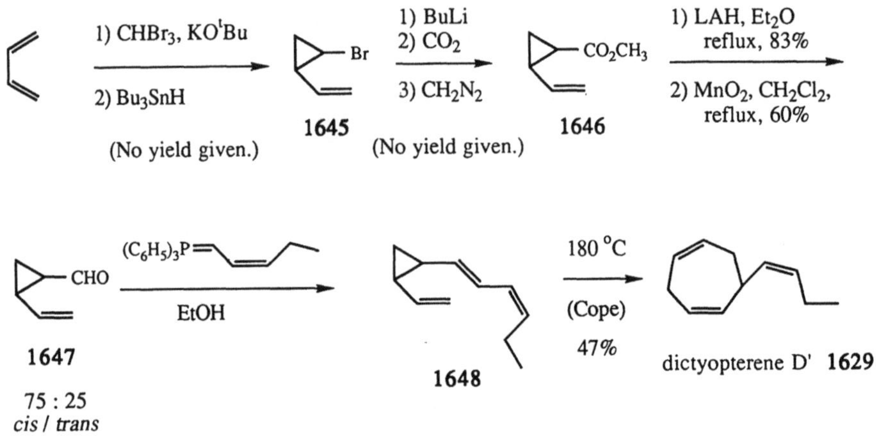

Scheme 286. Jaenicke Synthesis of Dictyopterene D′ (Ectocarpene)

mixture of isomers **1645**. Metalation of **1645** followed by carbonation and esterification with diazomethane produces **1646**. Reduction of the ester with LAH, oxidation to the aldehyde **1647**, and Wittig olefination produces the divinyl cyclopropane system **1648**. Cope rearrangement by heating to 180 °C yields dictyopterene D′ in 8 steps.

Dictyopterene D′ has been prepared by Weinstein [444] as a mixture with dictyopterene B (Scheme 281).

Dictyopterene D′ (**1629**) has also been prepared by Schneider [445] from fucoserratene (Scheme 302, see p. 134).

4.3.2.5 4-n-Butyl-2,6-Cycloheptadienone

A compound related to the C11 hydrocarbons is 4-*n*-butyl-2,6-cycloheptadienone (**1653**). It has been isolated from the essential oil of Hawaiian *Dictyopteris* and shown to have a dihydrotropone structure. It is structurally related to dictyopterene C (**1641**), and has been prepared by derivatization of **1606** [446]. The only complete synthesis of **1653** has been by Asaoka [447] (Scheme 287). Copper-catalyzed 1,4-addition of butylmagnesium bromide to **1650** in the presence of TMSCl, followed by desilylation of the silyl enol ether

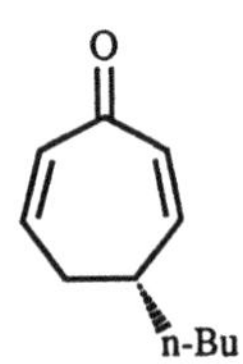

(+)-4-n-Butyl-2,6-cycloheptadienone
1653

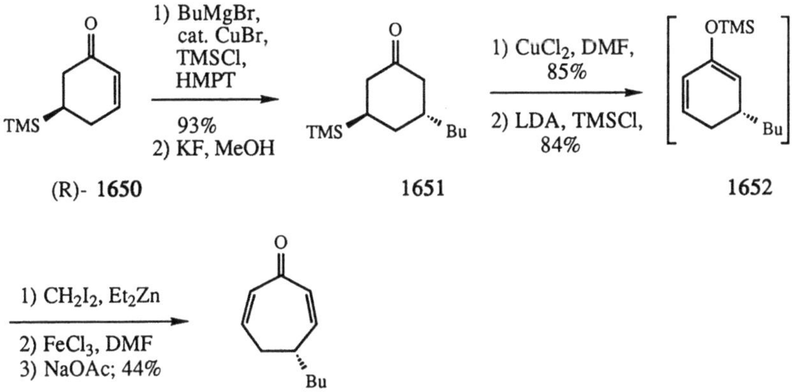

1653
(+)-4-n-Butyl-2,6-cycloheptadienone
19% overall yield

Scheme 287. Asaoka Synthesis of (*R*)-(±)-4-n-Butyl-2,6-cycloheptadienone

yields ketone **1651** stereospecifically. Oxidative removal of the remaining TMS group to afford the enone, followed by treatment with LDA and TMSCl gives the TMS dienol ether **1652**. Cyclopropanation of **1652** followed by oxidative ring opening with ferric chloride and dehydrochlorination furnishes 4-*n*-butyl-2,6-cycloheptadienone **1653** optically pure in 7 steps and 19% overall yield.

4.3.2.6 Multifidene

Multifidene (**1659**) has been isolated from the anisogamous brown alga *Cutleria multifida*, and identified as the male-attracting sex attractant [448]. The major structural aspects of multifidene that must be considered when designing a synthesis are the 1,2-*cis* stereochemistry of the two substituents on the cyclopentene ring and the (*Z*)-olefin geometry.

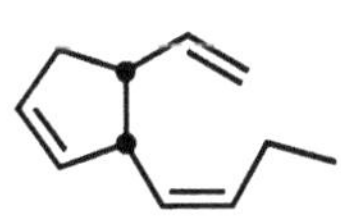

Multifidene **1659**

The first synthesis of multifidene (**1659**) was that of Jaenicke [449] in 1978 (Scheme 288). The synthetic strategy involves using the norbornene ring system of cyclopentadiene dimer **1654** to control the relative stereochemistry of the two *cis* substituents on the cyclopentene ring. After elaboration of the sidechains, the norbornene system is removed via a retro Diels-Alder reaction. Thermal [2+2] cycloaddition of dichloroketene affords **1655** as the major regioisomer in a 4:1 ratio. Treatment of ketene adduct **1655** with base, esterification and reduction produces dichloroalcohol **1657**. Silver-assisted hydrolysis of **1657** and Wittig

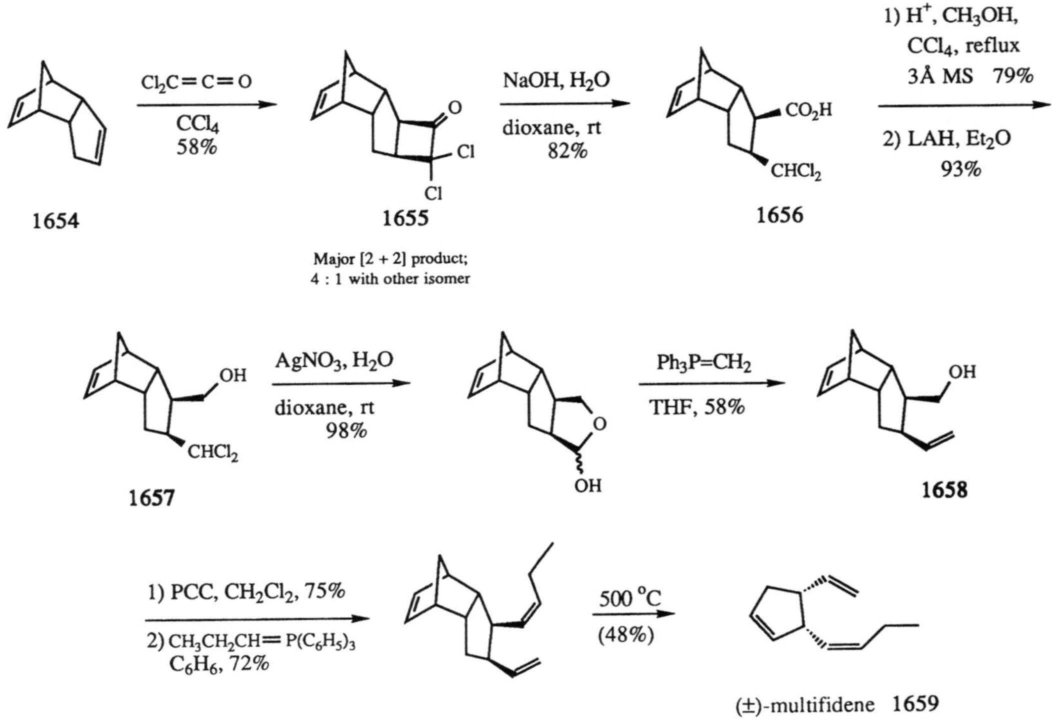

Scheme 288. Jaenicke Synthesis of (±)-Multifidene

olefination of the resulting lactol gives alkene **1658**. A second Wittig olefination and thermally induced retro Diels-Alder reaction completes the synthesis to give multifidene (**1659**) in 9 steps and approximately 8% overall yield.

The second synthesis of multifidene (**1659**) byJaenicke [450] begins with the isomerically pure acid **1656**, available from dicyclopentadiene and dichloroacetyl chloride in two steps (Scheme 289). Esterification of **1656** and monodechlorination with tri-*n*-butytin hydride affords **1661**. Treatment with silver nitrate gives the lactone **1662** which is converted to diene **1663** by DIBAL reduction and Wittig olefination under salt free conditions. Silylation of **1663** followed by a thermally induced retro Diels-Alder reaction and desilylation yields cyclopentene **1664** with the required 1,2-*cis* substitution pattern. Conversion of **1664** to **1659** is achieved by PCC oxidation and Wittig olefination to give multifidene (**1659**) in 10 steps and 26% yield overall. It was subsequently found that intermediate alcohol **1664** could be resolved via carbamate formation with (+)- or (−)-1-phenylethylisocyanate and chromatographic separation of the diastereomeric mixture [451].

Jaenicke has also prepared multifidene via an enantiospecific route beginning with *meso*-diol **1665** (Scheme 290) [452]. Enzymatic oxidation of the diol **1665** with horse liver alcohol dehydrogenase (HLADH) gives lactone **1666** in 90% yield with an enantiomeric excess of 97.5%. Reduction of the lactone with DIBAL and Wittig olefination provides the lower sidechain with the required *cis*-stereochemistry **1667**. Further oxidation and methylenation produces (+)-multifidene in five steps and an overall yield of 25%. The unnatural enantiomer was also synthesized in an analogous manner.

The strategy followed by Paquette [453] takes advantage of the stereocontrolled anionic oxy-Cope ring contraction of all *cis*-2,4,7,-cyclononatrienol (**1668**) to provide the required 1,2-*cis* disubstitution pattern (Scheme 291). Treatment of alcohol **1668** with potassium hydride followed by trapping with TMSCl affords the rearranged product **1669**. Selenylation of **1669** produces the selenoaldehyde **1670**, which upon treatment with ethylmagnesium bromide gives β-hydroxy selenide **1671** in a stereoselective fashion. As β-hydroxy selenides are not known to undergo *cis* elimination, which is required in this case to obtain the *cis* olefin, a double displacement process is utilized to obtain the correct olefin geometry. Formation of epoxide **1672** by internal displacement of phenylselenide by oxygen, followed by opening of the epoxide by diphenyl phosphide and elimination of methyldiphenylphosphine oxide yields isomerically pure multifidene (**1659**) in seven steps and 18% overall yield.

A preparation of multifidene (**1659**) by Crandall [454] in which there are several problems in controlling stereochemistry throughout the synthesis, is shown in Scheme 292. Dichloroketene adduct **1674** is transformed in three steps to cyclobutanol **1675**. MCPBA oxidation of **1675** to the sulfone and opening of the cyclobutanol to the aldehyde occurs with concomitant trapping of the aldehyde with methylidene triphenylphosphorane to give a mixture of *cis* and *trans* isomers **1677**. Epimerization of the aldehyde occurring before olefination jeopardizes the needed *cis* stereochemistry. A 9:1 *cis*/*trans* ratio of **1677** can be

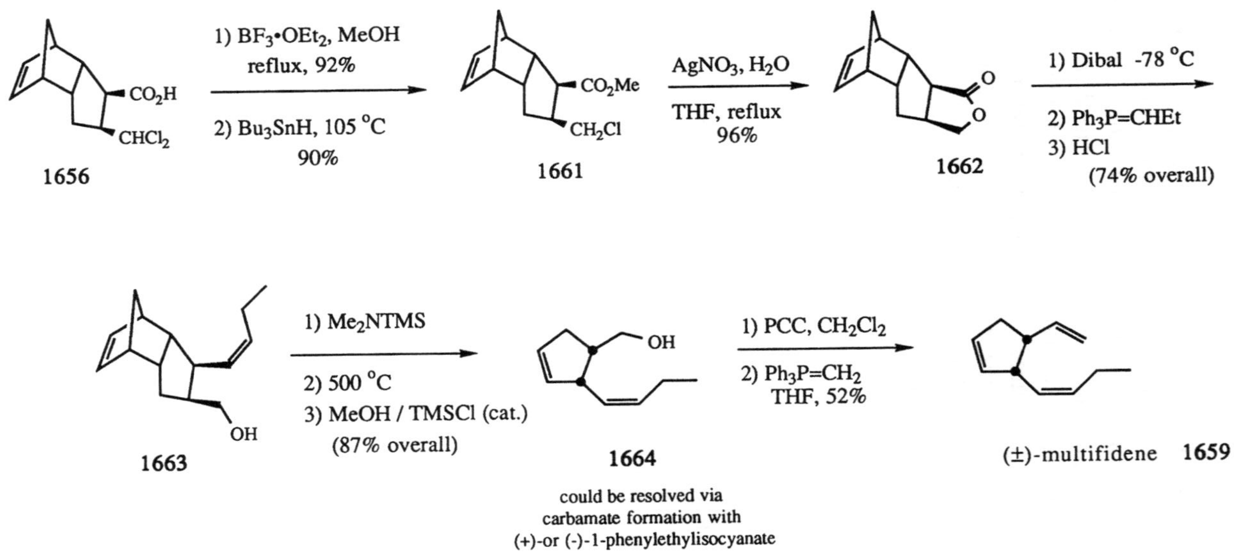

Scheme 289. Jaenicke Synthesis of (±)-Multifidene

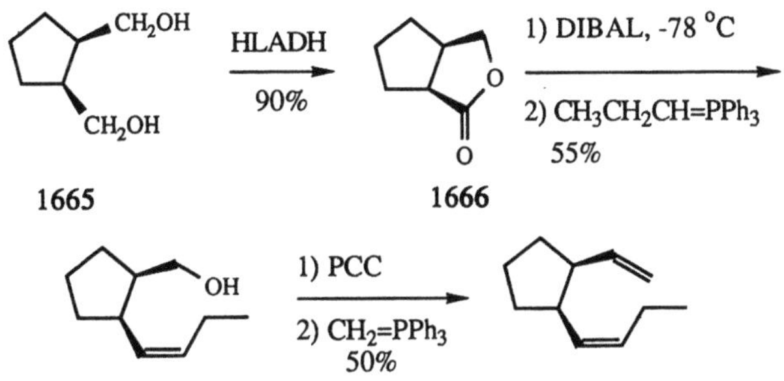

1665 1666 1667

(+)-multifidene 1659

Unnatural (-)-Multifidene was also synthesized in an analogous manner.

Scheme 290. Jaenicke Synthesis of Natural (±)-Multifidene

obtained by ring opening of the sulfide followed by oxidation. However, the synthesis is continued with a 4:1 *cis/trans* mixture of **1677**. Treatment of this mixture with *n*-butyllithium and then propionaldehyde gives **1678**, which is converted to a 3.5:1 *cis/trans* mixture of ketones **1679** by Jones oxidation. Reductive elimination of the phosphonate derivative **1680** yields an alkyne which can be reduced under Lindlar conditions to afford multifidene **1659** with the proper *cis*-alkene stereochemistry. After separation from the 4:1 *cis/trans* mixture, **1659** is obtained in 2.6% overall yield.

4.3.2.7 Desmarestene and Viridiene

The chemical messengers desmarestene (**1682**) and viridiene (**1688**) are discharged [455] by the mature eggs of the Northern Atlantic seaweeds *Desmarestia aculeata* and *D. viridis*. Both pheromones have been prepared by Boland [456]. Desmarestene (**1682**) (Scheme 293) is prepared in two steps from the readily available ester **1681**. DIBAL reduction of **1681** followed by *cis*-selective Wittig olefination of the resulting aldehyde gives **1682** in 26% yield. Viridiene (**1688**) is prepared via two different routes from dibromoester **1684** (Scheme 294). The first pathway involves monodebromination of **1684** followed by silver promoted lactonization to afford **1685**. DIBAL reduction of the lactone **1685** and Wittig olefination of the resulting aldehyde gives alcohol **1686**, which is converted to **1687** by PCC oxidation and Grignard addition of the lower side chain. LAH reduction of the alkyne gives a mixture of geometric isomers. The

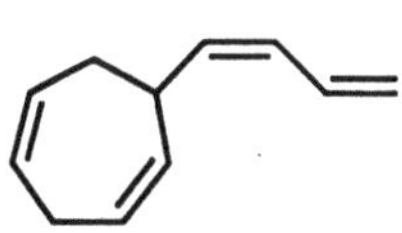

Desmarestene **1682**

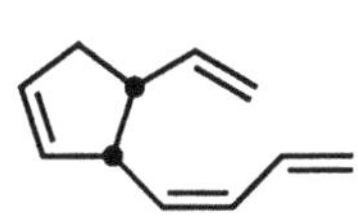

Viridicnc **1688**

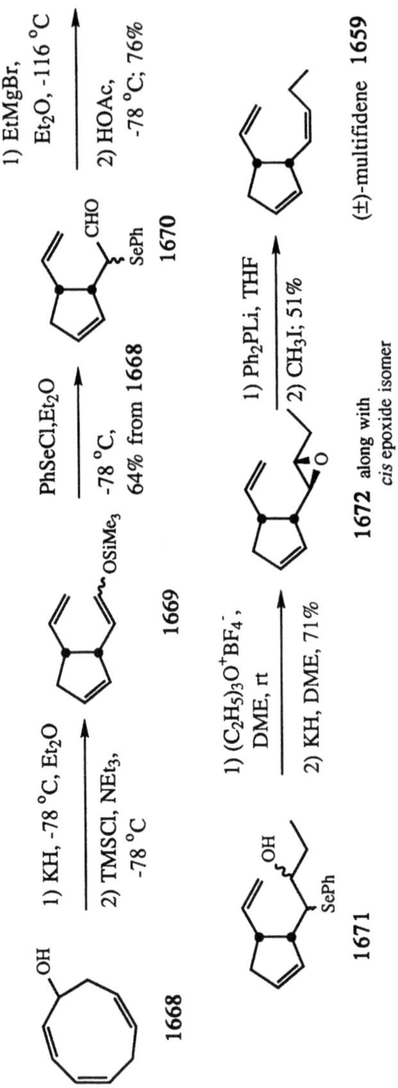

Scheme 291. Paquette Synthesis of (±)-Multifidene

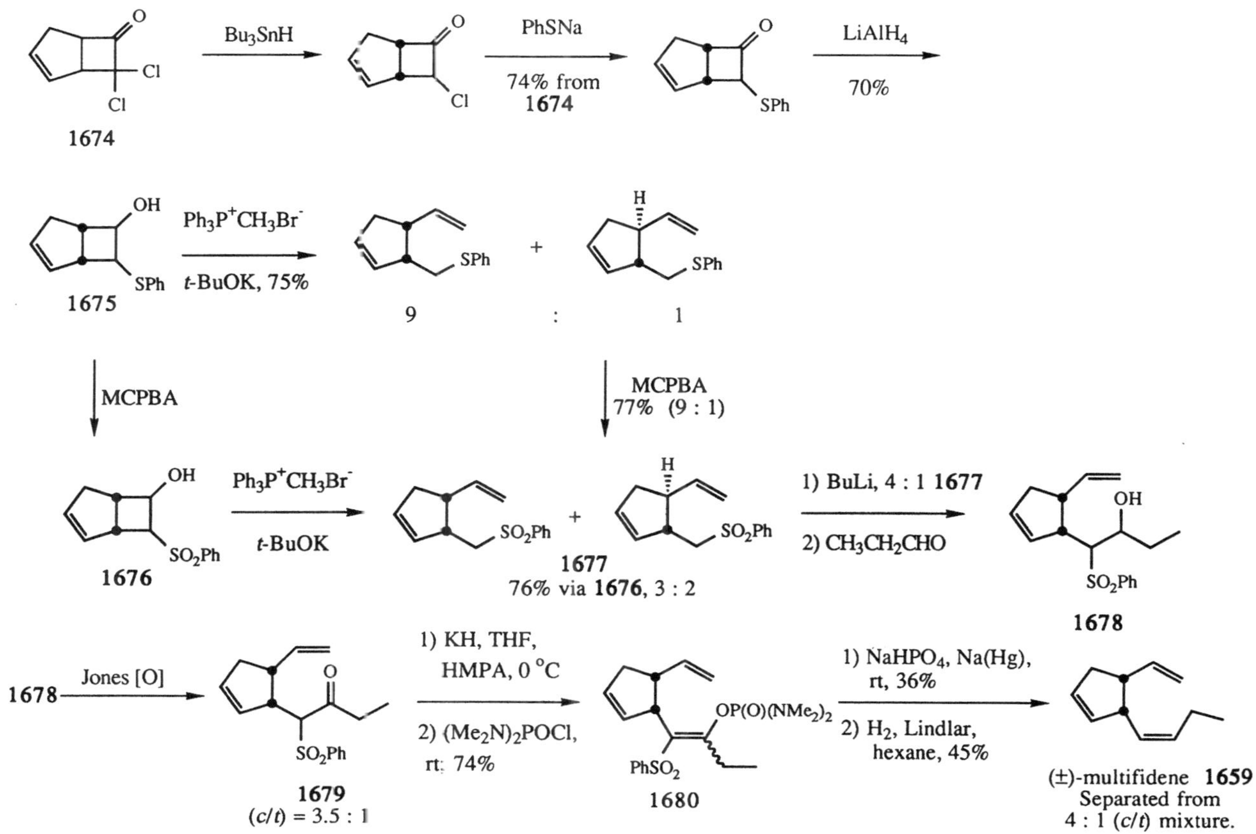

Scheme 292. Crandall Synthesis of (±)-Multifidene

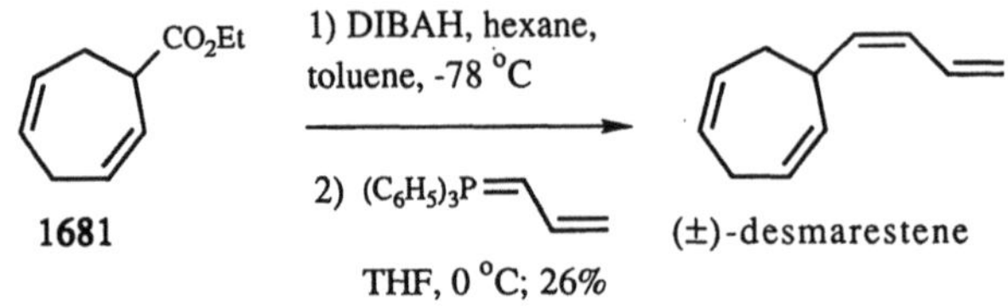

Scheme 293. Boland Synthesis of (±)-Desmarestene

unwanted (*E*)-isomer is removed via selective Diels-Alder reaction by treatment with 4-phenyl-1,2,4-triazoline-3,5-dione to give viridiene (**1688**) in seven steps. A second pathway to **1688** (Scheme 294) involves initial reduction of the dibromoester **1684** followed by silver promoted solvolysis and ring closure to form the lactol **1689**. Addition of the alkynyl Grignard **1690** and LAH reduction yields alcohol **1691**. Conversion of **1691** to viridiene (**1688**) is achieved by PCC oxidation and Wittig olefination to give a mixture of isomers which are separated as in the first route to give **1688** in six steps and 10% overall yield.

Boland [457] has also developed a stereospecific synthesis of (+)-viridiene (Scheme 295). Enzymatic oxidation of **1692** with horse liver alcohol dehydrogenase yields enantiomerically pure **1693**. Reduction of lactone **1693**, addition of Grignard reagent **1690** and LAH reduction of the alkyne affords a 3:2 ratio of the *E*/*Z* isomers. Removal of the unwanted isomer by selective Diels-Alder reaction with 4-phenyl-1,2,4-triazoline-3,5-dione gives pure **1694**. Conversion of **1694** to the final product is achieved by oxidation of the aldehyde to the alcohol and Wittig olefination to afford **1688** in 6 steps from the racemic diol **1692**.

4.3.2.8 Lamoxirene

Lamoxirene has been prepared by Jaenicke [458] (Scheme 296) as an extension of the desmarestene synthesis of Boland and Jaenicke (Scheme 293). Epoxidation of desmarestene (**1682**) with MCPBA gives a mixture of four diastereomers. Lamoxirene can be separated from this mixture by column chromatography. No yield or stereochemical configuration information is provided.

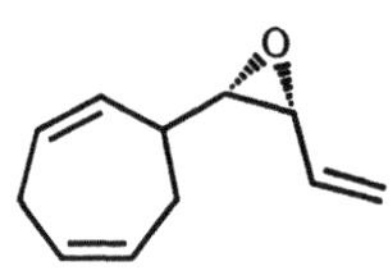

Lamoxirene **1696**

4.3.2.9 Aucantene

Another chemical signal compound isolated [459] from the brown alga *Cutleria multifida* is the C11 triene aucantene (**1698**). The first synthesis of (±)-aucantene

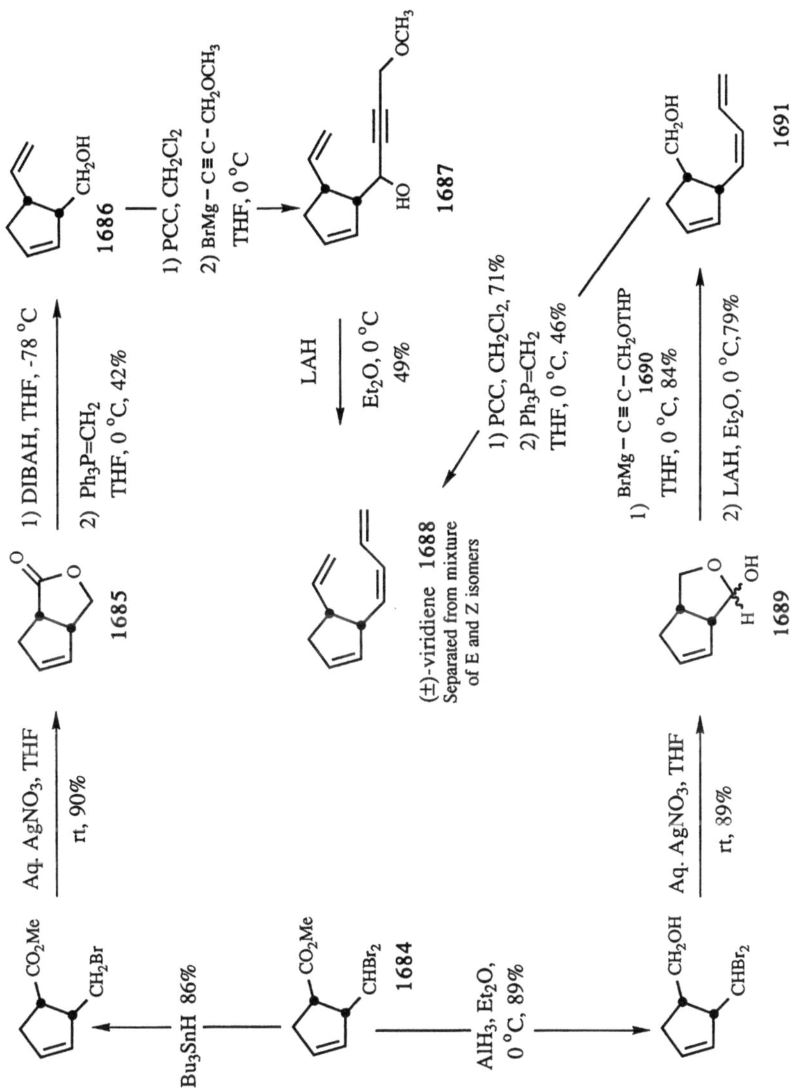

Scheme 294. Boland Synthesis of (±)-Viridiene

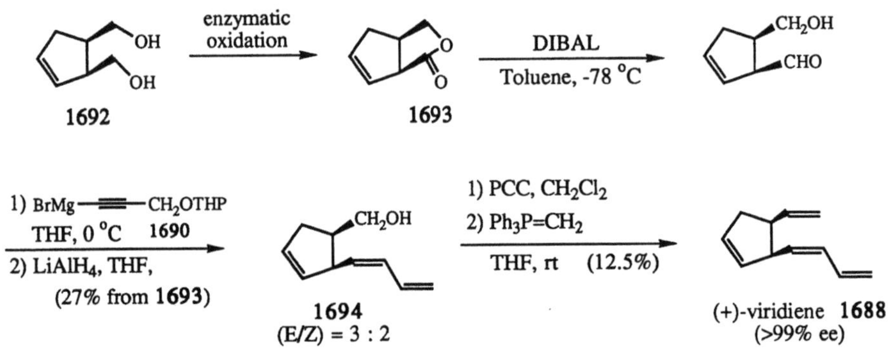

Scheme 295. Boalnd Synthesis of (+)-Viridiene

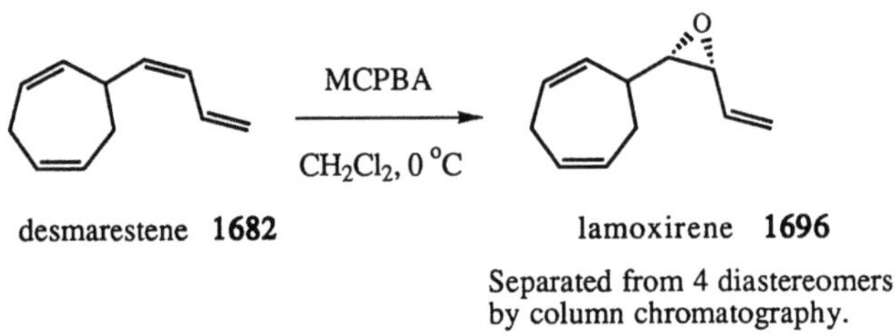

No yields or stereochemical configuration given.

Scheme 296. Jaenicke Synthesis of Lamoxirene

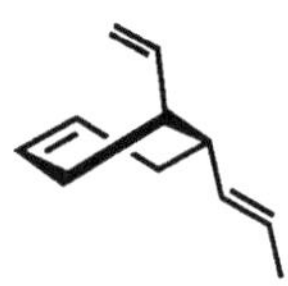

(+)-Aucantene **1698**

is that of Jaenicke [460] in 1975 (Scheme 297). Diels-Alder reaction of methyl 2*E*,4*E*-hexadienoate with butadiene gives the desired cycloaddition product **1697** in 5.2% yield, along with a mixture of three other isomers. Reduction and oxidation of **1697** to the aldehyde followed Wittig olefination gives **1698** in 1.2% yield over four steps.

An enantiospecific synthesis (Scheme 298) of aucantene (**1698**) was later developed by Boland and Jaenicke [461], which utilized a strategy similar to their synthesis of (+)-viridiene (Scheme 295). Enzymatic oxidation of diol **1700** with horse liver alcohol dehydrogenase yields lactone **1701** with an enantiomeric excess of 96%. Reduction to the lactol and Wittig olefination followed by PCC oxidation affords aldehyde **1702**. Epimerization with base and treatment of the aldehyde **1702** with carbon tetrabromide and triphenylphosphine yields

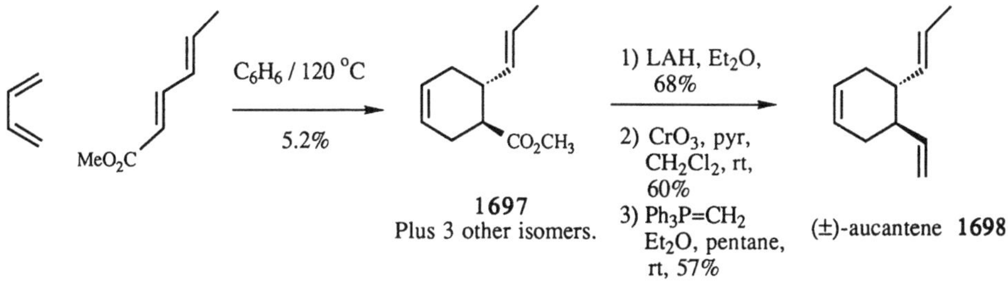

Scheme 297. Jaenicke Synthesis of (±)-Aucantene

1703. Treatment of **1703** with *n*-butyllithium results in elimination to the bromoalkyne and metallation to give the alkynyllithium. Addition of methyl iodide yields alkyne **1704** as a 3:2 mixture of *cis*/*trans* isomers. Birch reduction with sodium in ammonia gives (+)-aucantene **1698** in 8 steps and 1.8% overall yield.

A short and elegant synthesis of (±)-aucantene is that by Schneider [462] shown in Scheme 299. Epoxide **1705** is readily available through singlet oxygenation of cyclopentadiene followed by thermal rearrangement and base catalyzed isomerization. Diels-Alder reaction of **1705** with butadiene gives aldehyde **1706** in good yield with correct stereochemistry and suitable functionalization for further elaboration. Wittig olefination, followed by trimethylsilyl iodide mediated deoxygenation provides **1698** isomerically pure in 3 steps and 62% overall yield.

4.3.2.10 Fucoserratene

Fucoserratene (**1712**), a simple conjugated triene, has been isolated [463] from the mature eggs of the marine brown alga *Fucus serratus*. It exhibits enormous chemotactic activity on the mobile spermatozoa of this alga. Similar activity is displayed by ectocarpene **1629**, isolated [464] from the female gametes of *Ectocarpus siliculosus*. The first synthesis of fucoserratene (**1712**) is that Jaenicke [465] in 1975 (Scheme 300). Control of stereochemistry of the 3,5-diene system is obtained by *cis* hydrogenation and *E*-selective Wittig olefination. Propargyl alcohol is protected as the THP ether, alkylated with ethyl bromide and deprotected to give **1708**. Lindlar reduction followed by allylic oxidation affords the aldehyde **1709**. Olefination with the stabilized ylide gives aldehyde **1711** which can be converted to fucoserratene **1712** by reduction, allylic oxidation and Wittig olefination. The ten step synthesis yields **1712** in 4.4% overall yield.

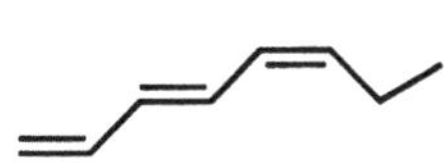

Fucoserratene 1712

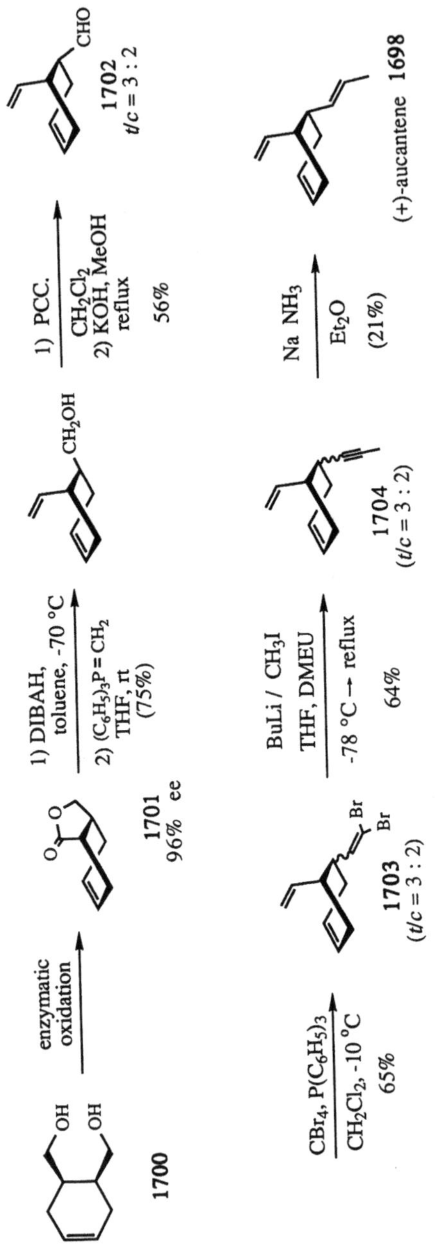

Scheme 298. Boland Synthesis of (+)-Aucantene

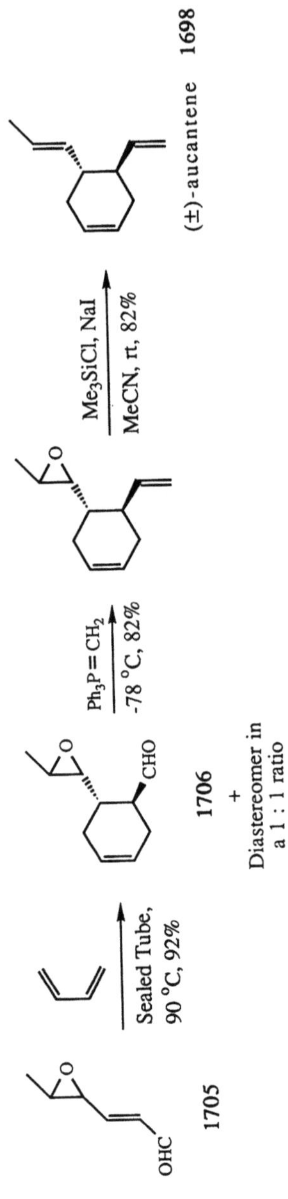

Scheme 299. Schneider Synthesis of (±)-Aucantene

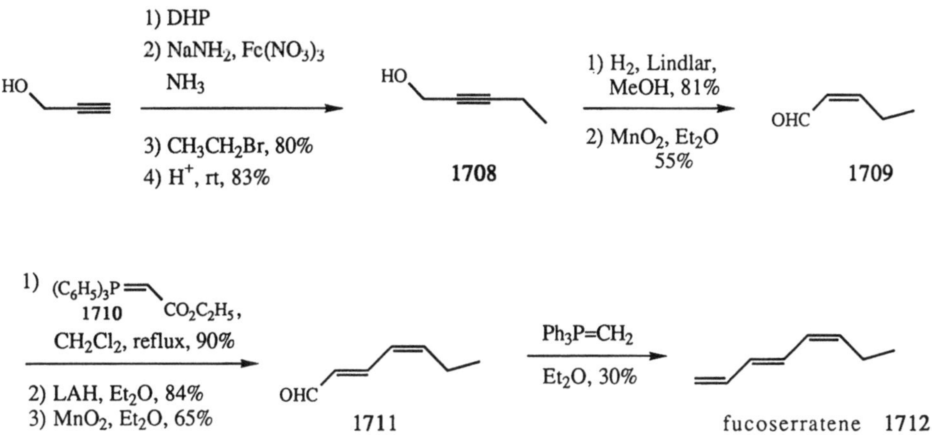

Scheme 300. Jaenicke Synthesis of Fucoserratene

Hopf [466] has prepared fucoserratene (**1712**) in a short three step synthesis as shown in Scheme 301. Diyne **1713**, readily available from 1,5-hexadiyne, is isomerized to **1714** by treatment with potassium *t*-butoxide and dicyclohexyl 18-crown-6. Lindlar reduction of **1714** affords **1712**.

Fucoserratene (**1712**) has also been prepared by Schneider [467] in three steps from (*E*)-2,4-pentadienoic acid (**1716**) in a stereospecific manner (Scheme 302). Reduction of **1716** with LAH, followed by PCC oxidation gives the aldehyde **1717** in 50% yield. A *Z*-selective Wittig reaction, employing the silazide method, affords fucoserratene (**1712**) in three steps and 39% yield overall. The synthesis can be extended to give dictyopterene B (**1628**) and dictyopterene D′ (ectocarpene) (**1629**) by 1,3-dipolar cycloaddition of 3-diazo-1-propene to give **1718**. Heating **1718** in refluxing hexane produces dictyopterene B (**1628**) and dictyopterene D′ as a mixture in a 1:1.2 ratio. Nitrogen extrusion produces **1628**, which can be converted completely to **1629** via Cope rearrangement by heating to >125 °C.

The most recent synthesis of fucoserratene (**1712**) is a second approach by Schneider [468], shown in Scheme 303. *Z*-selective Wittig olefination, via the silazide method, of (*E*)-3-oxiranylprop-2-enal **1719** followed by reaction with thiourea yields thiirane **1720**. Desulfuration of **1720** by treatment with triphenylphosphine results in the formation of **1712** in three steps and 69% overall yield.

4.3.2.11 Giffordene

Giffordene (2*Z*,4*Z*,6*Z*,8*Z*)-2,4,6,8-undecatetraene) has been isolated from laboratory cultures the brown alga *Giffordia mitchellae*. Confirmation of its structure was obtained through its synthesis by Boland [469] (Scheme 304) in a very simple manner. THP-protected enyne alcohol **1722** is transformed into diendiyne **1723** in five steps. Reduction of **1723** with Zn(Cu/Ag), a reagent that is

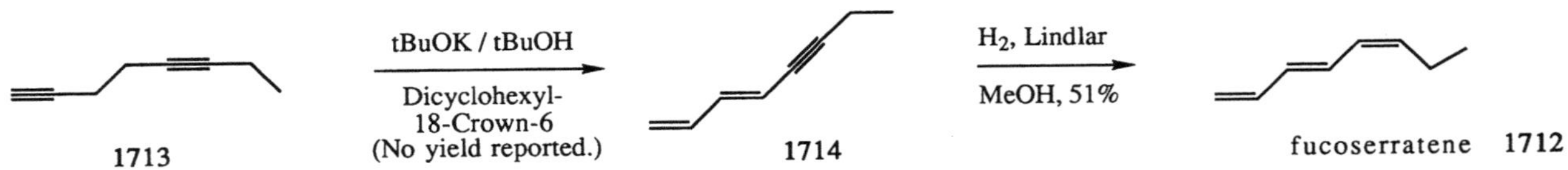

Scheme 301. Hopf Synthesis of Fucoserratene

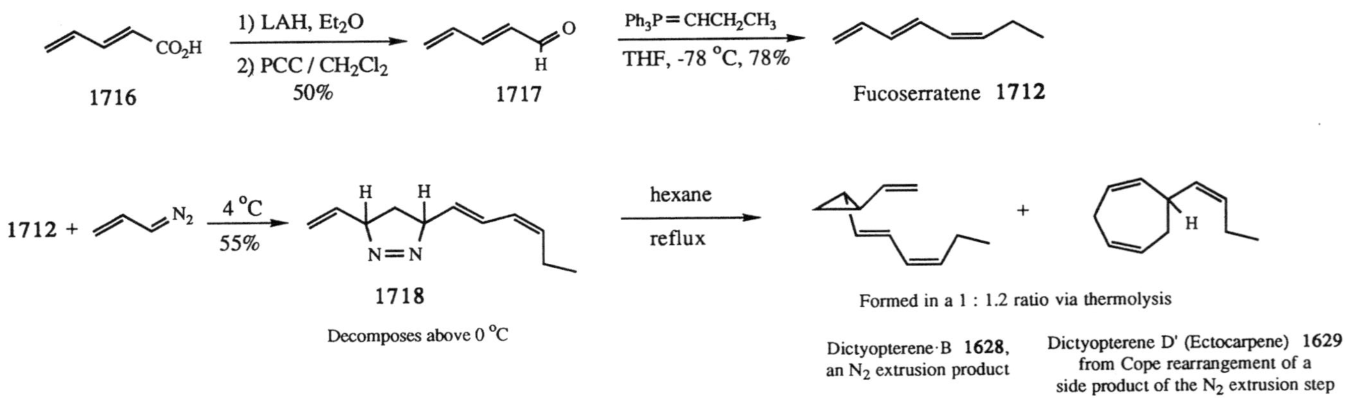

Scheme 302. Schneider Synthesis of Fucoserratene, (±)-Dictyopterenes B and D′ (Ectocarpene)

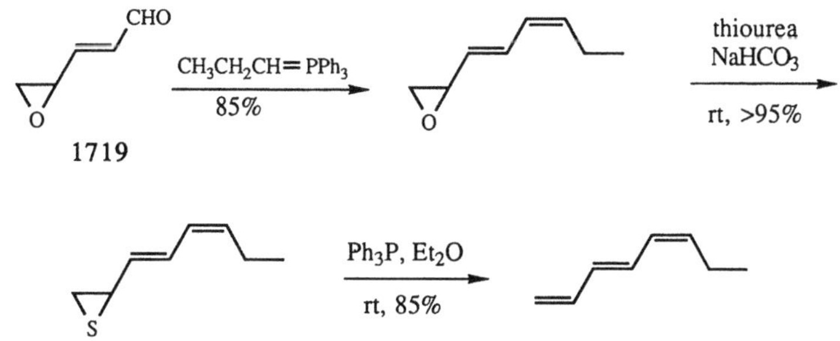

1719 1720 fucoserratene 1712

Scheme 303. Schneider Synthesis of Fucoserratene

Giffordene **1724**

specific for the reduction of internal alkynes, gives giffordene **1724** in five steps and 2.8% overall yield.

Related to giffordene are the four metabolites **1734–1737** isolated from *D. plagiogramma* [470] and *Spermatochnus paradoxus.* Naf [471] has prepared these metabolites from the sulfolene **1725** by making use of the Ramberg-Backlund reaction (Scheme 305). Treatment of **1725** with KOtBu proceeded with ring-opening to the anion **1726**, which was alkylated with allyl bromide to give sulfones **1727** and **1728**. 1,6-Addition of either di-*n*-butylcuprate or the l-butenylcuprate **1729** to this geometric mixture led to the corresponding addition product mixtures **1730–1733** in moderate yield due to competing polymerization processes. These were among the first cuprate additions to unsaturated sulfones. Classical Ramberg-Backlund reaction of each of these mixtures led to mixtures of the natural products which were not further purified.

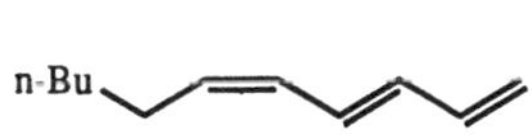

1734

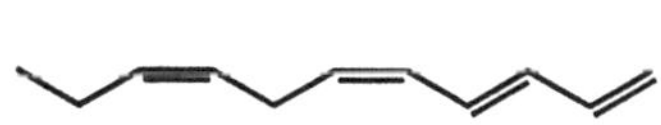

1736

1735

1737

4.3.2.12 Clavularins A and B

Clavularins A and B were originally isolated from the soft coral *Clavularia koellikeri* by Endo [472] and, although not algal pheromones, bear a structural resemblance to members of this class. Subsequently, clavularin A was isolated

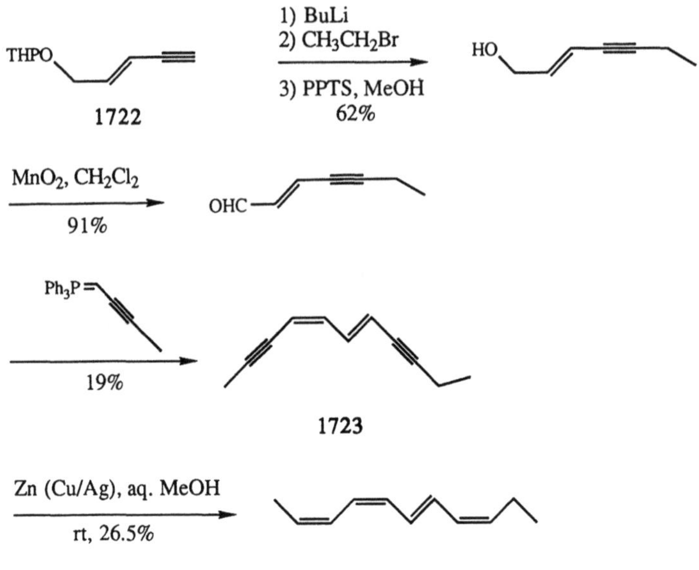

Scheme 304. Boland Synthesis of Giffordene

Initially proposed structures for Clavularins A and B

Clavularin A (*cis*) **1739**
Clavularin B (*trans*) **1740**

Revised Clavularin A and B Structures

Clavularin A (*cis*) **1748**
Clavularin B (*trans*) **1745**

[473] from a second soft coral, *C. viridis.* Both compounds show significant cytotoxicity. On the basis of extensive NMR studies the cyclononanone structures A and B were proposed for the epimeric clavularins A and B. Subsequently, the initially proposed structures were revised [474]. The new structure of clavularin B (**1745**) was confirmed through its synthesis by Urech [475] (Scheme 306). Conjugate addition of $LiCuMe_2$ to cycloheptadienone **1741** and trapping with TMSCl affords dienyl ether **1742**. Regeneration of the enolate with methyllithium and treatment with silyl substituted MVK **1743** provides the diastereomeric Michael adducts in good yield. Use of the silyl substituted Michael acceptor retards further Michael additions of the adduct under the

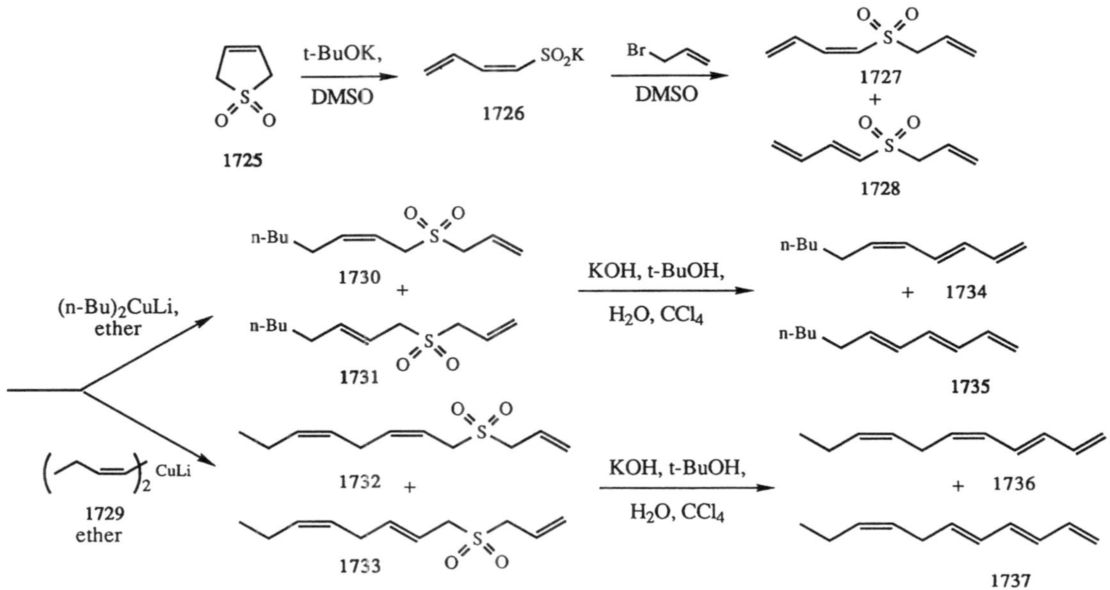

Scheme 305. Naf Synthesis of Brown Algal Metabolites

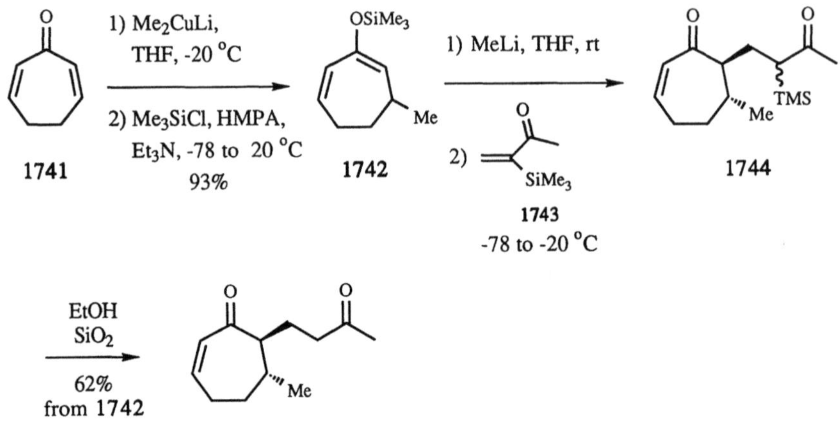

(±)-clavularin B **1745**

Scheme 306. Urech Synthesis of (±)-Clavularin B

basic aprotic conditions. Ethanolysis of **1744** gives clavularin B in three steps and 58% overall yield.

The epimeric clavularin A (**1748**) has been synthesized by Still [476] as outlined in Scheme 307. Cycloheptenone (**1746**) is converted to unsaturated ketone **1747** by a standard series of transformations. Hydrogenation, introduction of the endocyclic double bond and deprotection of the ketal provides clavularin A (**1748**) in seven steps and an overall yield of 19%.

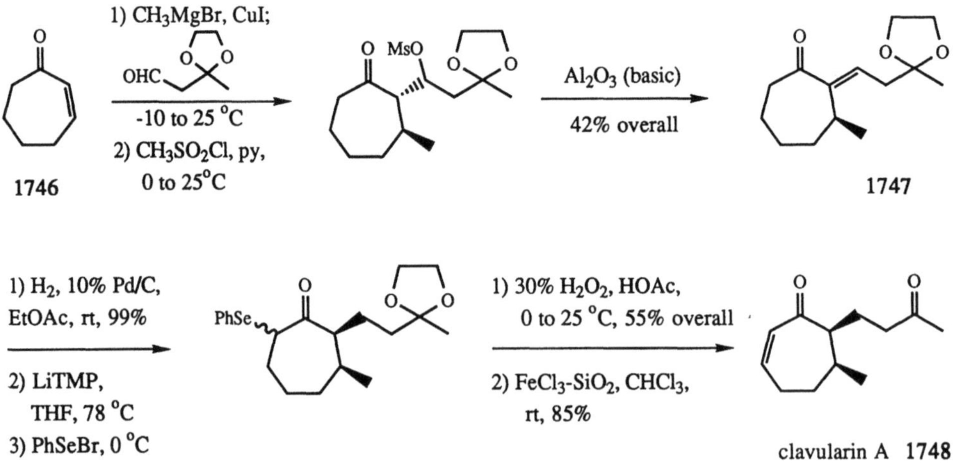

Scheme 307. Still Synthesis of clavularin A

4.4 Miscellaneous Metabolites

4.4.1 Acarnidines

A series of *N*-substituted polyamines known as acarnidines have been isolated [477] from the red-orange encrusting sponge *Acarnus erithacus*. All of these have the homospermidine skeleton, differing mainly in the alkyl chain lengths and positions of unsaturation. The acarnidines exhibit mild activity against *Herpes simplex* virus Type I, as well as broad spectrum antimicrobial activity. The only member of the class that has been subjected to total synthesis is **1754**. The only major problem to be handled in the synthesis of acarnidines is differentiation between the two amine groups in the 1,5-diamino functionality.

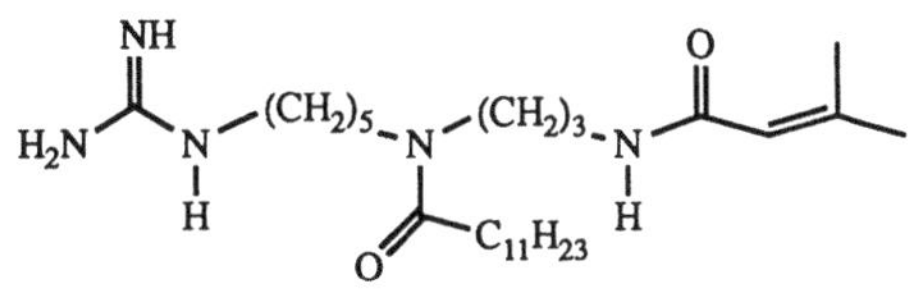

N-(5-Guanidinopentyl)-*N*-[3-(3-methylbut-2-eneamido)propyl]dodecanamide **1754**

The synthesis of **1754** by Golding is shown in Scheme 308 [478]. Monoderivatization of 1,5-diaminopentane occurs by reaction with **1750** to give **1751**, where protection of the guanidino functionality with an *N*-nitro group reduces its reactivity and basicity. Once the two amino groups have been differentiated, the synthesis proceeds in a straightforward fashion. Reductive amination of **1751** with **1752** gives the homospermidine skeleton **1753**. Acylation with *p*-nitrophenyl laurate followed by electrolytic removal of the guanidino nitro group affords **1754**, which was isolated as a dimethylpyrimidine derivative. The synthesis required approximately six steps and had a 14% overall yield.

A similar procedure was followed by Munro [479] as shown in Scheme 309. The problem of guanidine reactivity and basicity is circumvented by construction of the molecule in the reverse direction (as compared to Golding, Scheme 308), so that the guanidine moiety is attached last. Acylation of 3-aminopropanol and Swern oxidation gives aldehyde **1755**. Reductive amination of **1755** with the mono-BOC protected diamine **1756** affords the homospermidine skeleton **1757**. Acylation followed by BOC deprotection gives **1758**, which upon reaction with methylisothiouronium iodide yields the acarnidine **1754**. Synthesis of **1754** is completed in seven steps and 20% overall yield.

4.4.2 Pahutoxin

Pahutoxin is an ichthyotoxic and hemolytic substance isolated [480] from the epidermal mucus of the Hawaiian boxfish *Ostracion lentiginosus*. An enantiospecific synthesis by Tai [481] is shown in Scheme 310. Optical activity is

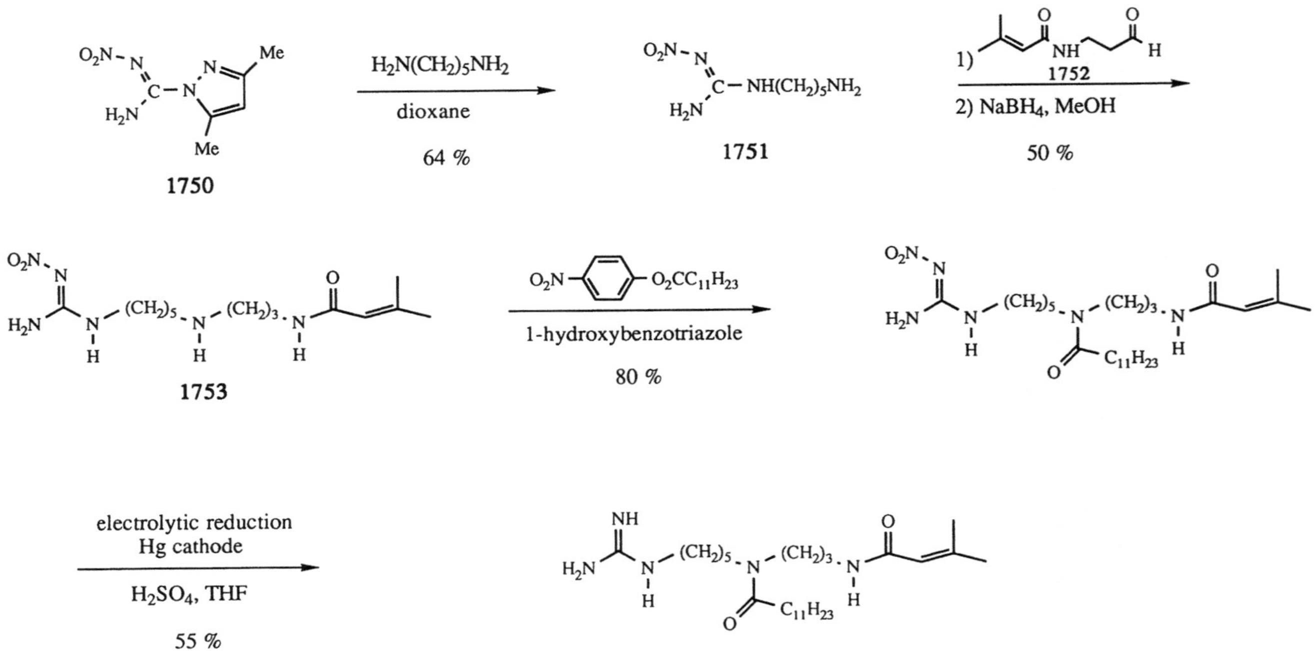

N-(5-guanidinopentyl)-N-[3-(3-methylbut-2-eneamido)propyl]dodecanamide **1754**

Scheme 308. Golding Synthesis of *N*-(5-guanidinopentyl)-*N*-[3-(3-methylbut-2-eneamido)propyl]dodecanamide

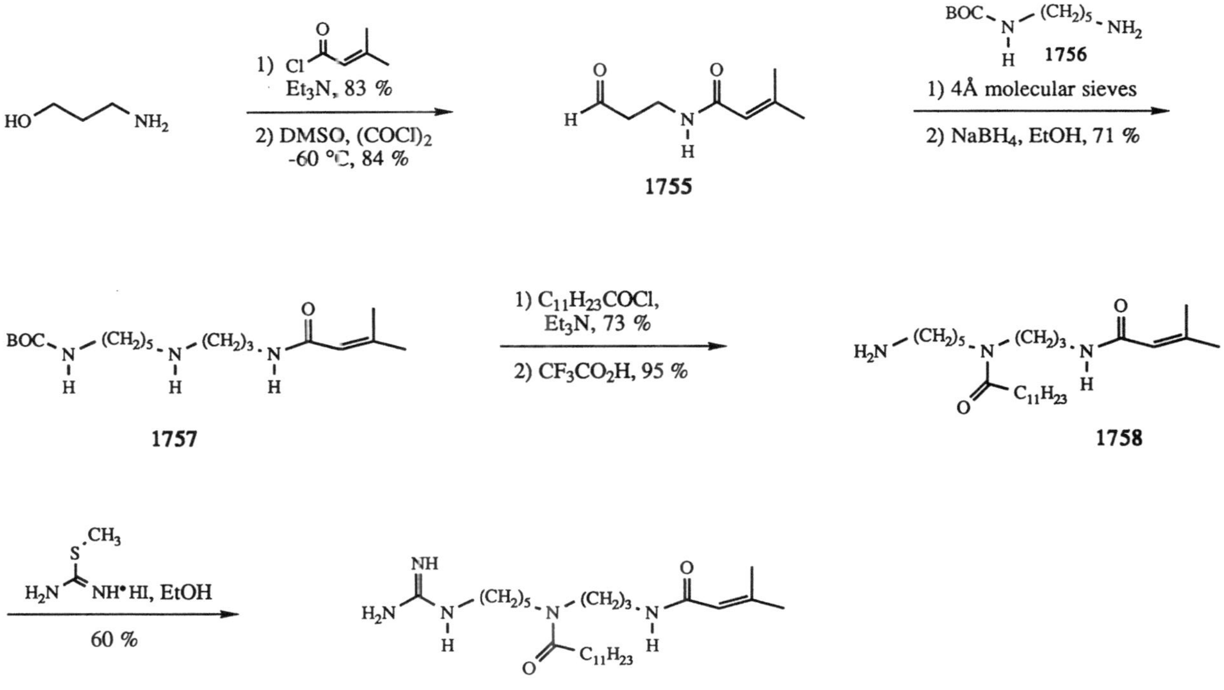

N-(5-guanidinopentyl)-N-[3-(3-methylbut-2-eneamido)propyl]dodecanamide **1754**

Scheme 309. Munro Synthesis of *N*-(5-guanidinopentyl)-*N*-[3-(3-methylbut-2-eneamido)propyl]dodecanamide

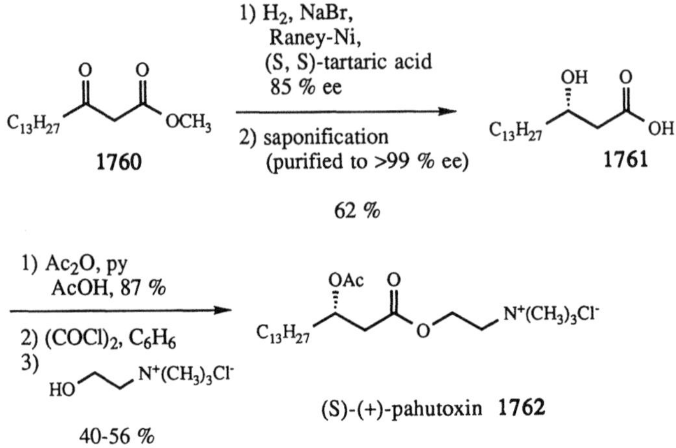

Scheme 310. Tai Synthesis of (S)-(+)-Pahutoxin

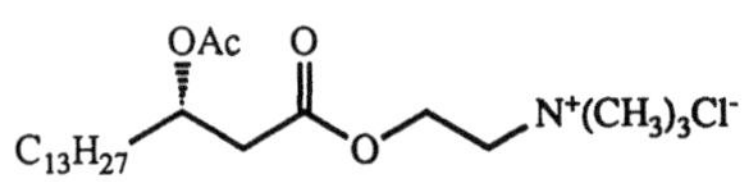

(S)-(+)-Pahutoxin **1762**

obtained in the first step by enantio-differentiating hydrogenation of β-ketoester **1760** over (*S*,*S*)-tartaric acid-sodium bromide modified Raney nickel. Saponification of the β-hydroxyester followed by multiple recrystallizations of the dicyclohexylammonium salt of **1761** affords optically pure **1762**. After acetylation of the alcohol, the acid is converted to the acid chloride and treated with choline chloride to provide (+)-pahutoxin in five steps and 34% yield overall.

4.4.3 D-*erythro*-1-Deoxydihydroceramide-1-Sulfonic Acid

The sulfolipid D-*erythro*-1-deoxydihydroceramide-1-sulfonic acid (**1769**) was isolated [482] from the alkali-stable lipids in the non-photosynthetic marine diatom *Nitzschia alba*. The only total synthesis of **1769** is that of Kamikawa [483], shown in Scheme 311. Selective acetalization of galactose (**1763**) followed

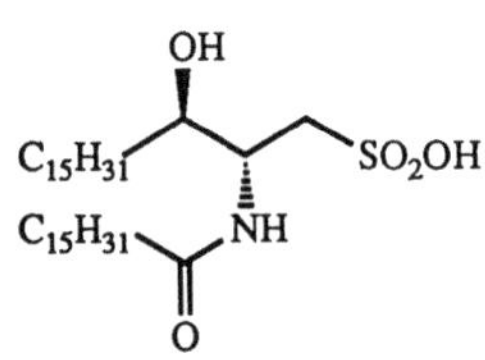

D-*erythro*-1-Deoxydihydroceramide-1-sulfonic acid 1769

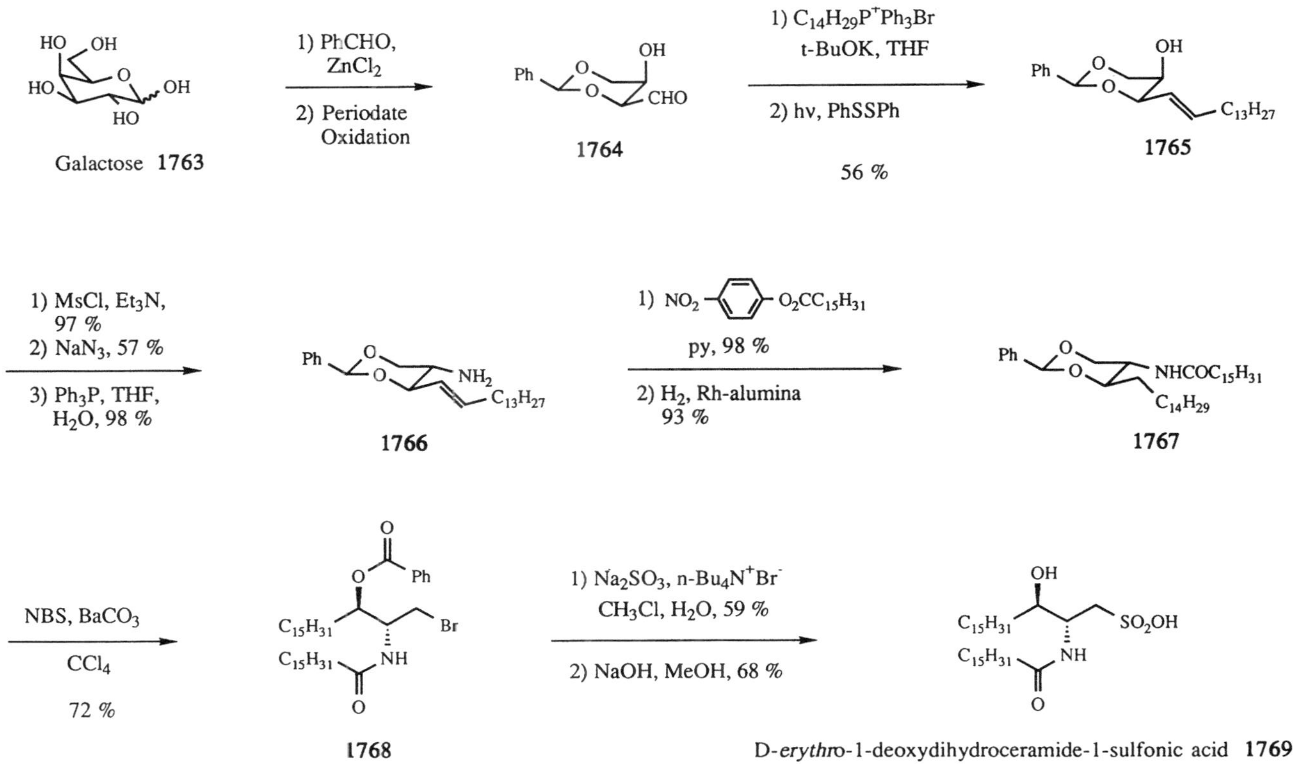

Scheme 311. Kamikawa Synthesis of D-*erythro*-1-Deoxydihydroceramide-1-sulfonic acid

by periodate oxidation gives aldehyde **1764**, which is subjected to a Wittig-photoisomerization sequence to afford the *trans* olefin **1765** in 56% yield. Mesylation of the axial alcohol, inversion with NaN_3 and reduction of the azide by treatment with triphenylphosphine yields the equatorial amine **1766**. *N*-Acylation of **1766** by treatment with *p*-nitrophenyl palmitate and catalytic hydrogenation of the alkene produces **1767**. Ring opening of the acetal with NBS provides the bromo benzoate **1768**. Treatment with sodium sulfite under phase transfer conditions and subsequent saponification of the benzoate affords **1769** in 8% overall yield via a 12 step synthesis.

4.4.4 Phosphonosphingoglycolipid from *Turbo cornutus*

A variety of phosphonosphingoglycolipids have been isolated from the tissues of marine Mollusca and Protostomia. Hayashi has isolated [484] several new phosphonosphingoglycolipids from the muscle tissues of the marine snail *Turbo cornutus*, the simplest of which is **1776**. Kamikawa [485] utilizes the protected ceramide **1771** from an earlier synthesis (Scheme 311) to develop an enantioselective synthesis of **1776** as shown in Scheme 312. The overall strategy involves coupling two optically pure fragments, both derived from galactose, to give **1771**. Hydrolysis of **1771** followed by silylation, benzoylation and desilylation affords the secondary benzoate **1772**. Glycosidation of **1772** with α-D-bromotetraacetylgalactose gives a mixture of components that can be converted to **1773** in 42% yield by treatment with trimethylsilyl triflate. The phosphonosphingoglycolipid **1776** is obtained by acetate saponification and ultrasound assisted coupling of **1774** and **1773** using carbodiimide **1775**. An overall yield of 3% is obtained in 15 steps from galactose.

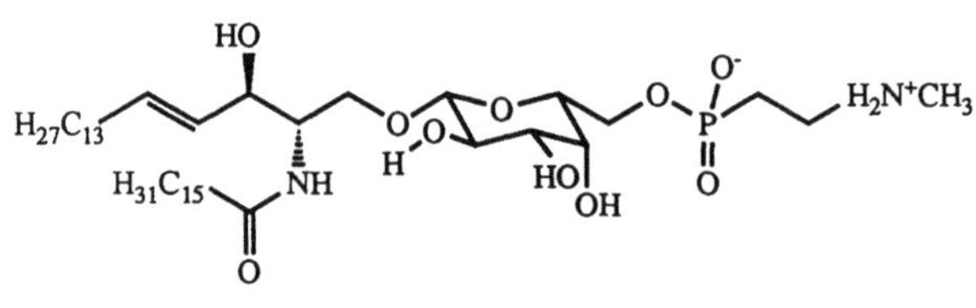

Phosphonosphingoglycolipid from *Turbo cornutus* **1776**

4.4.5 Metabolites of *Plexaura flava*

The butyrolactone containing metabolites **1779** and **1780** have been isolated [486] from the Gorgonian coral *Plexaura flava*. Font [487] has developed a diastereoselective enantiospecific synthesis of both **1779** and **1780** (Scheme 313). Acylation of commercially available ethyl (*S*)-lactate and intramolecular condensation yields lactone **1777**. Hydrogenation of **1777** provides compound **1778** in 76% yield (along with 13% of the *trans*-isomer). The *cis*-isomer can be

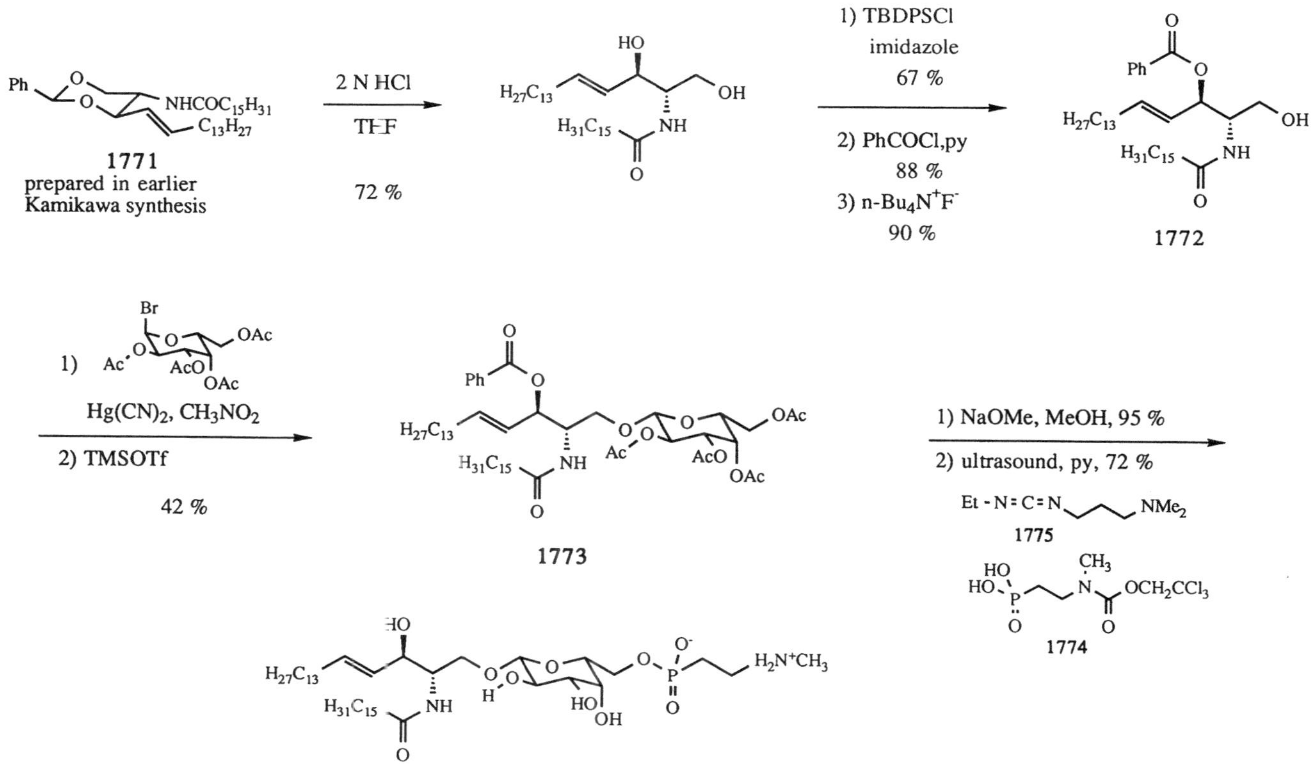

a phosphonosphingoglycolipid from *Turbo cornutus* **1776**

Scheme 312. Kamikawa Synthesis of Phosphonosphingoglycolipid from *Turbo cornutus*

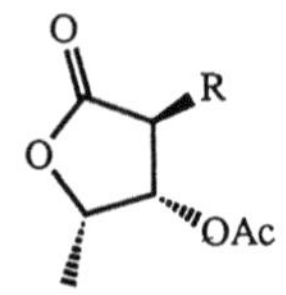

1779 R = $C_{16}H_{33}$

1780 R= $C_{14}H_{29}$

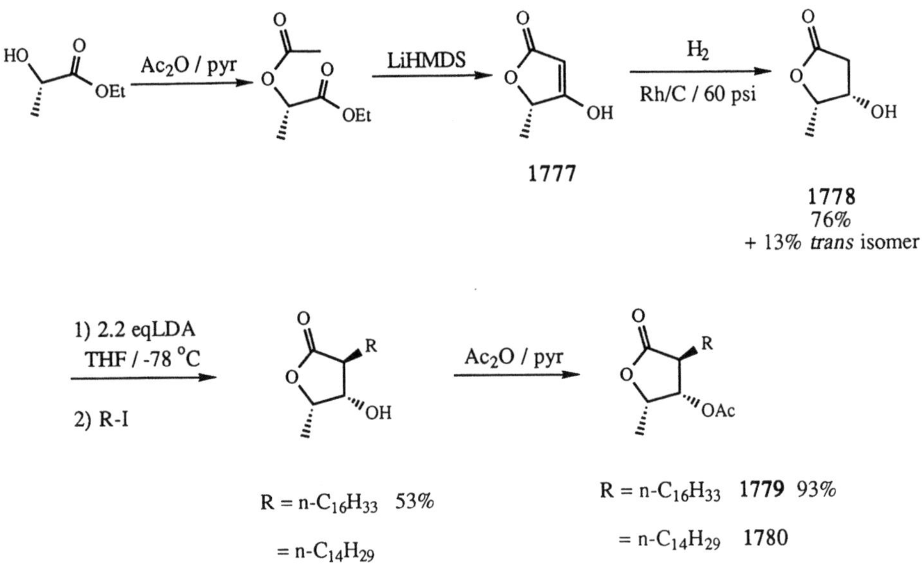

Scheme 313. Font Synthesis of *Plexaura flavus* Metabolites **1779** ans **1780**

converted in a straightforward manner to either **1779** or **1780** in a total of 6 steps.

4.4.6 *Notheia anomala* Metabolite

Williams et al. have developed methods for the stereocontrolled transformation of orthoester intermediates into substituted tetrahydrofurans. An interesting marine natural product suitable for demonstrating the utility of this methodology is the trisubstituted tetrahydrofuran **1787**. Metabolite **1787** has been found [488] as a constituent of the brown alga *Notheia anomala*. Its structure has been unambigously confirmed by single-crystal X-ray analysis. Williams's synthesis of **1787** [489] begins with the conversion of homopropargylic alcohol to the

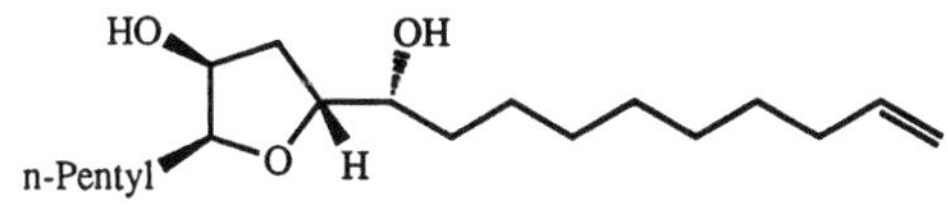

Notheia anomala Metabolite **1787**

protected *cis*-alkene diol **1782** (Scheme 314). Standard transformations provide a diastereomeric mixture of alcohols **1783**, which is separated to give alcohol **1784**. In the key transformation, oxidation of **1784** with NBS Stereospecifically provides substituted tetrahydrofuran **1786** via dioxenium cation intermediate **1785**. Desilylation, Swern oxidation and addition of 8-nonenylmagnesium bromide affords the natural product **1787** in 11 steps.

An enantiospecific synthesis of the *Notheia anomala* metabolite **1787** has been developed by Takano [490] and is presented in Scheme 315. Acetylide ring opening of diepoxide **1788**, available optically pure from (L)-diethyl tartrate, affords diyne **1789**. Conversion of **1789** to alcohol **1790** and treatment with phenylsulfenyl chloride gives substituted tetrahydrofurans **1791** and **1792** as a 3:1 mixture. Diimide reduction **1791** and **1792** and selenoxide elimination affords compound **1793**, which is oxidatively cleaved to give aldehyde **1794**. Separation of the diastereomers and Grignard reaction of the β-H isomer with 8-nonenylmagnesium bromide provides the natural product as a 3:1 mixture of diastereomers. The overall yield can be increased by inversion of the minor product (α-OH) in 80% yield. Metabolite **1787** is obtained in 11 steps.

4.4.7 Octacosadienoic Acids

Most marine sponges contain large amounts of long-chain C24-C30 fatty acids. A variety of structural types are found, including examples of straight or branched carbon skeletons with both terminal and internal methyl branching. The synthesis of one example of this class of lipids, 22*S*-methyl-5,9-octacosadienoic acid (**1801**) from the sponge *Aplysina fistularis* by Djerassi [491] is shown in Scheme 316. The synthesis of **1801** begins with 1,5-cyclooctadiene (**1796**). In four steps **1796** is converted to mono-THP protected diol **1797**. Tosylation of the free hydroxyl group and cuprate mediated coupling of the tosylate with 5-(trimethylsiloxy)pentylmagnesium bromide gives alcohol **1798**. A second tosylation and cuprate mediated coupling reaction with optically pure Grignard reagent **1799** (available from (+)-pulegone) affords THP ether **1800**. Elaboration of the other sidechain proceeds via formation of the aldehyde and

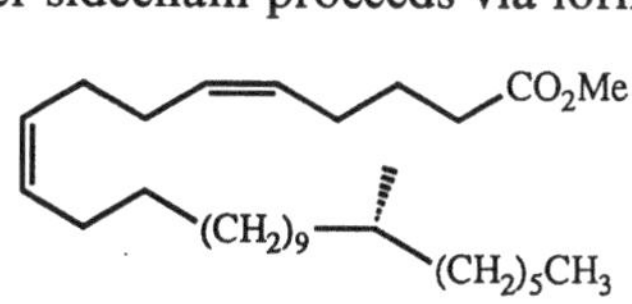

22*R*-Methyl-5,9-octacosadienoic acid 1801

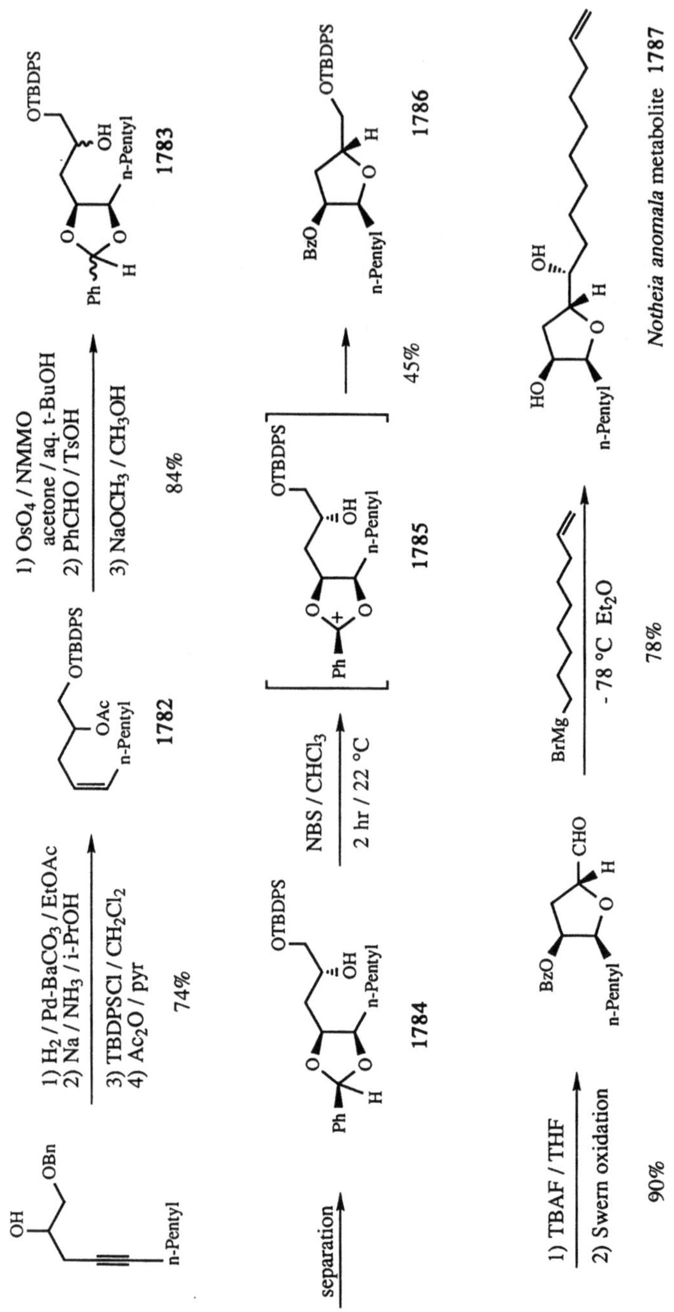

Scheme 314. Williams Synthesis of *Notheia anomala* Metabolite **1787**

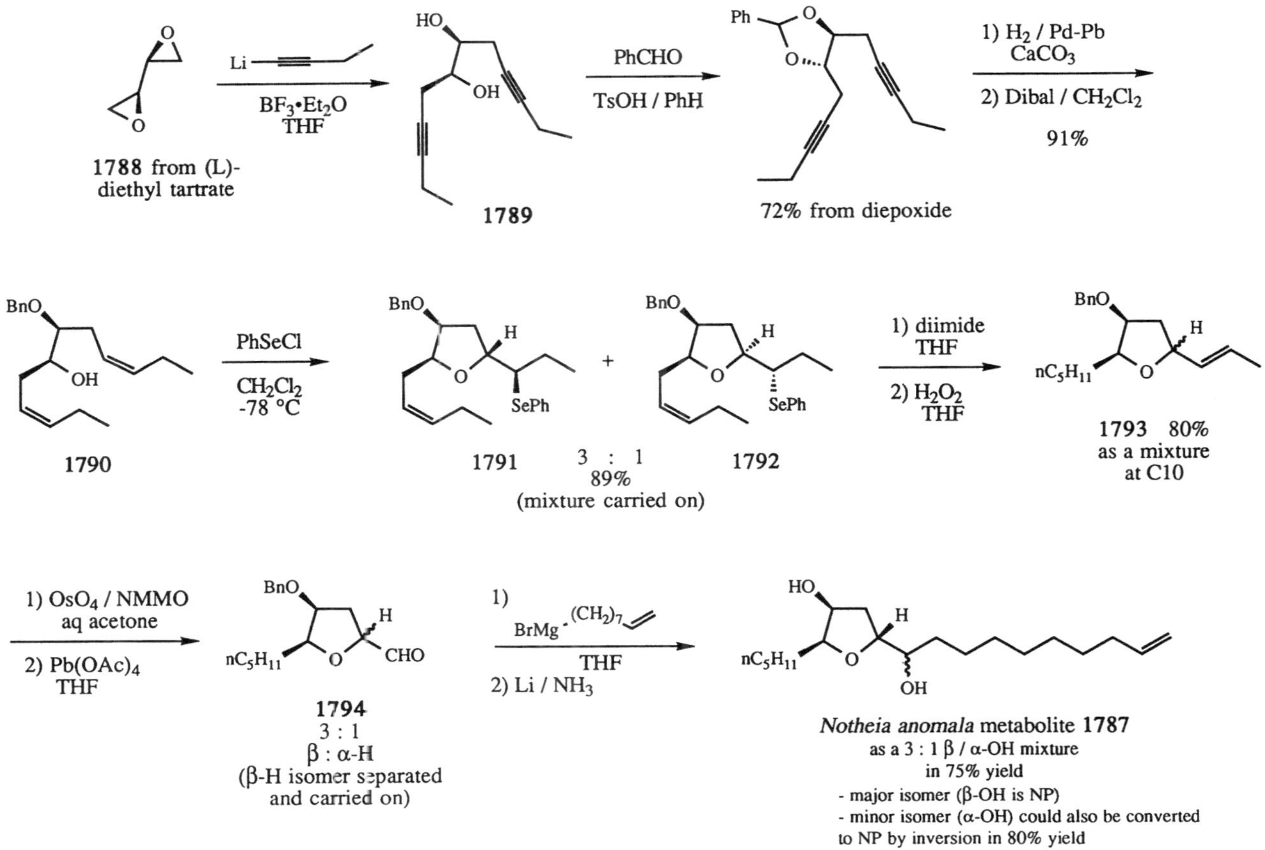

Scheme 315. Takano Synthesis of *Notheia anomala* Metabolite **1787**

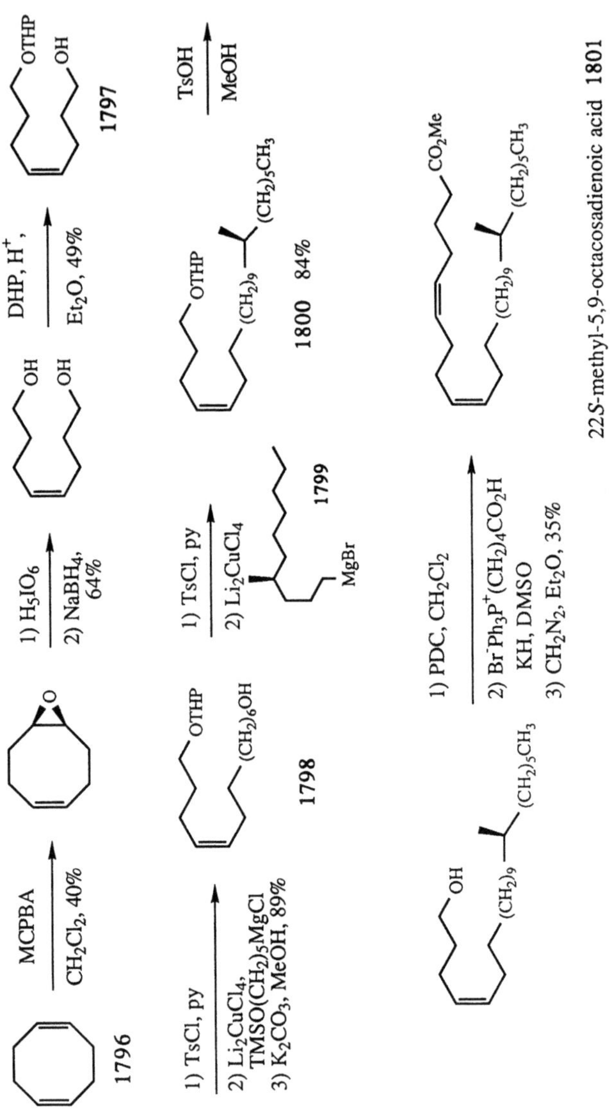
1796
MCPBA
CH2Cl2, 40%
1) H5IO6
2) NaBH4, 64%
DHP, H+,
Et2O, 49%
OTHP
OH
1797
1) TsCl, py
2) Li2CuCl4, TMSO(CH2)5MgCl
3) K2CO3, MeOH, 89%
(CH2)6OH
1798
1) TsCl, py
2) Li2CuCl4
MgBr
1799
(CH2)9
(CH2)5CH3
1800 84%
TsOH
MeOH
1) PDC, CH2Cl2
2) Br-Ph3P+(CH2)4CO2H
KH, DMSO
3) CH2N2, Et2O, 35%
CO2Me
22S-methyl-5,9-octacosadienoic acid 1801

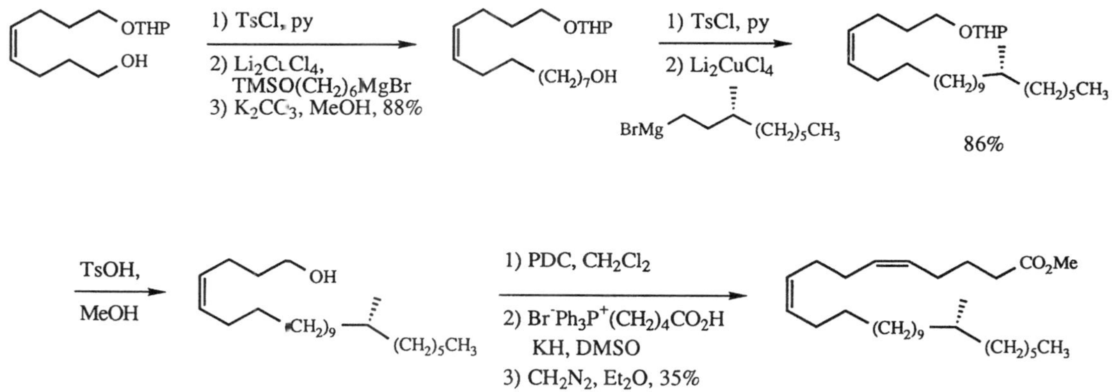

Scheme 316. Djerassi Synthesis of 22*R*- and 22*S*-Methyl-5,9-octacosadienoic Acid

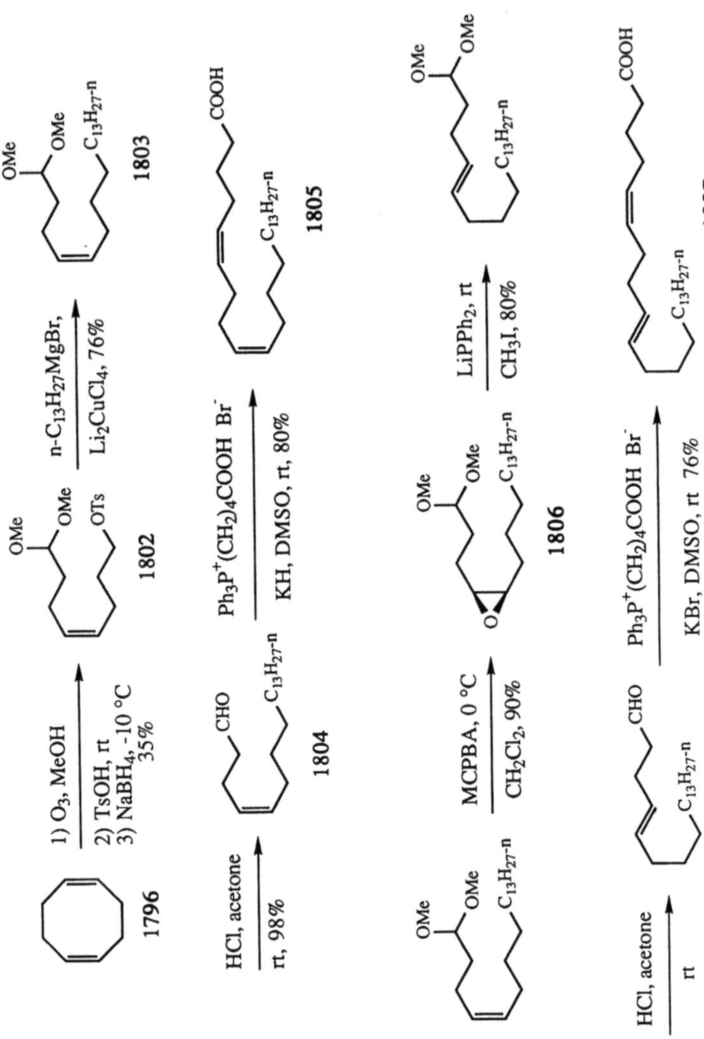
1796
1) O_3, MeOH
2) TsOH, rt
3) $NaBH_4$, -10 °C
35%
OMe
OMe
OTs
1802
n-$C_{13}H_{27}MgBr$,
Li_2CuCl_4, 76%
$C_{13}H_{27}$-n
1803
HCl, acetone
rt, 98%
CHO
1804
$Ph_3P^+(CH_2)_4COOH\ Br^-$
KH, DMSO, rt, 80%
COOH
1805
MCPBA, 0 °C
CH_2Cl_2, 90%
O
1806
$LiPPh_2$, rt
CH_3I, 80%
HCl, acetone
rt
$Ph_3P^+(CH_2)_4COOH\ Br^-$
KBr, DMSO, rt 76%
1807

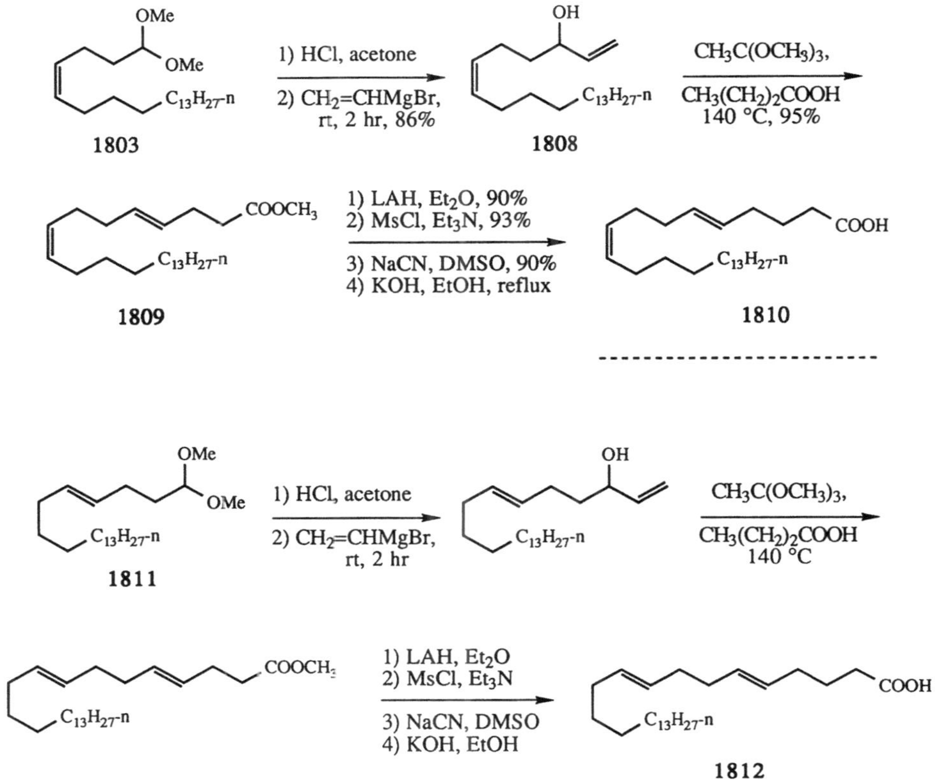

Scheme 317. Djerassi Synthesis of 5,9-Hexacosadienoic Acids

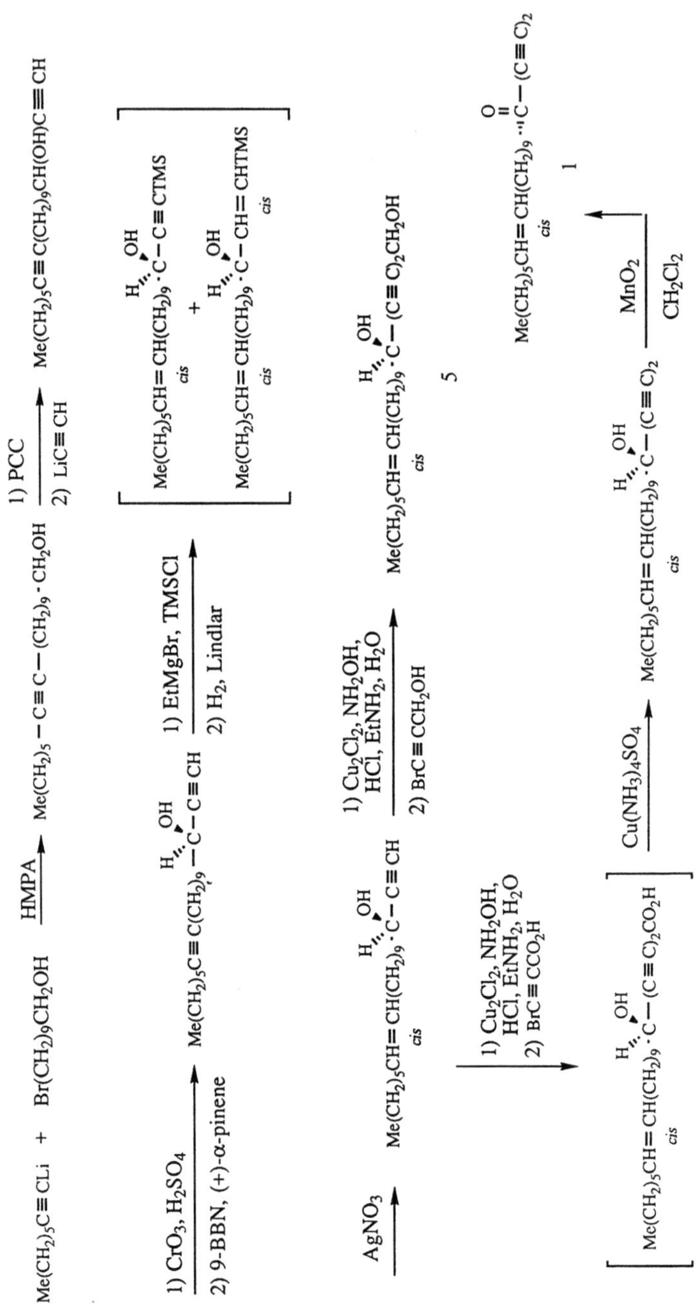

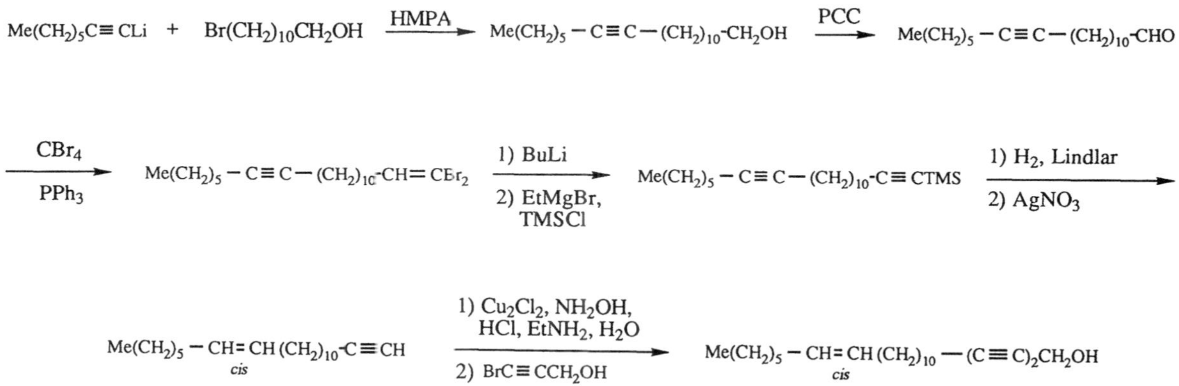

Scheme 318. Thaller Synthesis of C_{22} and C_{23} Diacetylenes from *Reniera fulva*

Wittig olefination to produce the methyl ester of 22*S*-methyl-5,9-octacosadienoic acid in 13 steps and 3% overall yield. The enantiomer of **1801** is prepared in a similar fashion.

4.4.8 Hexacosadienoic Acids

Djerassi has also prepared [492] the fatty acid (5*Z*,9*Z*)-5,9-hexacosadienoic acid (**1805**) as shown in Scheme 317. Selective ozonolysis of 1,5-cyclooctadiene (**1796**), tosylation, and reduction gives tosylate **1802** in which the two alkene substituents have been differentiated and the stereochemistry of the 9,10-double bond has been set. Cuprate catalyzed coupling of tridecylmagnesium bromide with tosylate **1802** yields the saturated sidechain. Wittig olefination of aldehyde **1804** provides the second double bond with correct geometry, thus giving **1805** in six steps and 21% yield. The *cis-trans* isomer was prepared in a similar manner, involving olefin isomerization via the epoxide **1806**. Treatment of **1806** with lithium diphenylphosphide using a standard sequence gave the *trans* isomer which was carried on to **1807** via a similar sequence. Scheme 317 also shows syntheses of the other two diene isomers. Intermediate **1803** was converted to the allylic alcohol **1808**. Claisen rearrangement provided the *trans* geometry about the new olefin in **1809**. Chain elongation led to the desired **1810**. Finally, the all *trans* isomer **1812** was prepared from the *trans* acetal **1811** by an analogous sequence.

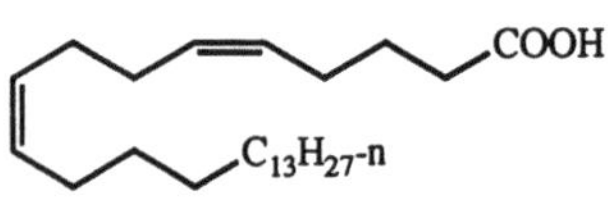

1805

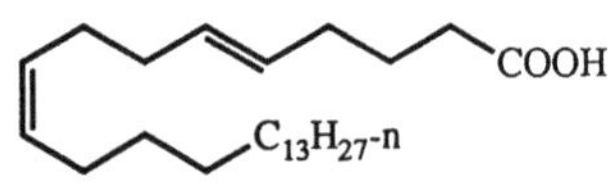

1810

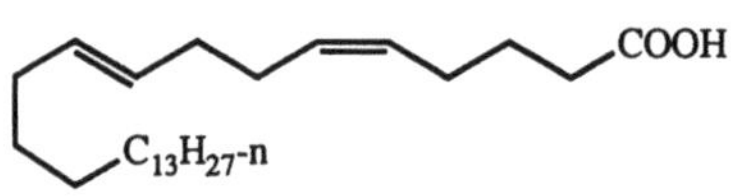

1807

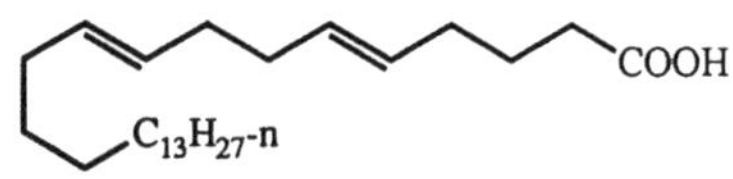

1812

4.4.9 Diacetylenes from *Reniera fulva*

Several long-chain diacetylenes containing a single propargylic stereocenter from the sponge *Reniera fulva* [493] have been synthesized by Thaller [494] in pure optically active form to confirm the assigned absolute configuration. The routes are shown in Scheme 318 and require very little comment. The

$$\mathrm{Me(CH_2)_5CH{=}CH(CH_2)_9{-}\overset{O}{\overset{\|}{C}}{-}(C{\equiv}C)_2}$$
cis

$$\mathrm{Me(CH_2)_5CH{=}CH(CH_2)_9{-}\overset{H\ \ OH}{C}{-}(C{\equiv}C)_2CH_2OH}$$
cis

$$\mathrm{Me(CH_2)_5{-}CH{=}CH\,(CH_2)_{10}{-}(C{\equiv}C)_2CH_2OH}$$
cis

asymmetry at the carbinol carbon (70% ee) was produced by reduction of an α,β-alkynyl ketone with 9-BBN-(+)·α-pinene complex. Based on the known *R*-selectivity of this reagent, the absolute configuration was confirmed.

5 Nitrogenous Metabolites

5.1 Indoles and Related Compounds

5.1.1 Flustramine B

Flustramine B (**1818**), along with several other brominated indole derivatives, has been isolated from the bryozoan *Flustra foliacea* [495]. Hino's synthesis [496] of **1818** is illustrated in Scheme 319. Acid-catalyzed cyclization of tryptamine derivative **1814** gives 5-nitropyrroloindole **1815**. Reduction to the amine, bromination and deamination provides the 6-bromo derivative. Ring opening of **1816**, prenylation, and ring closure gives the diprenylated *N*-carbomethoxy compound **1817**. Hydrolysis for 100 hours with sodium ethoxide in refluxing ethanol gives the free amine, which is converted to flustramine B by methylation. An overall yield of less than 0.5% is obtained over 9 steps.

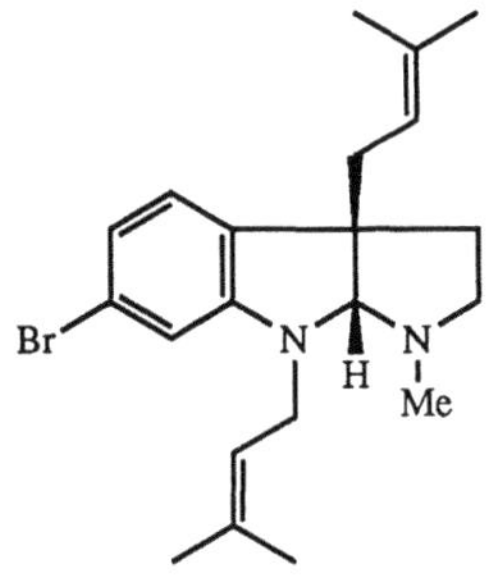

Flustramine B **1818**

5.1.2 Surugatoxins

The Japanese ivory shell *Babylonia japonica* elaborates a class of alkaloidal glycosides which exhibit potent toxicity upon ingestion. These spirooxindole- and pteridine ring-containing metabolites have been named surugatoxins [497]. Neosurugatoxin and prosurugatoxin have been characterized as the causative agents responsible for the toxic event. Inoue has described a basic approach to

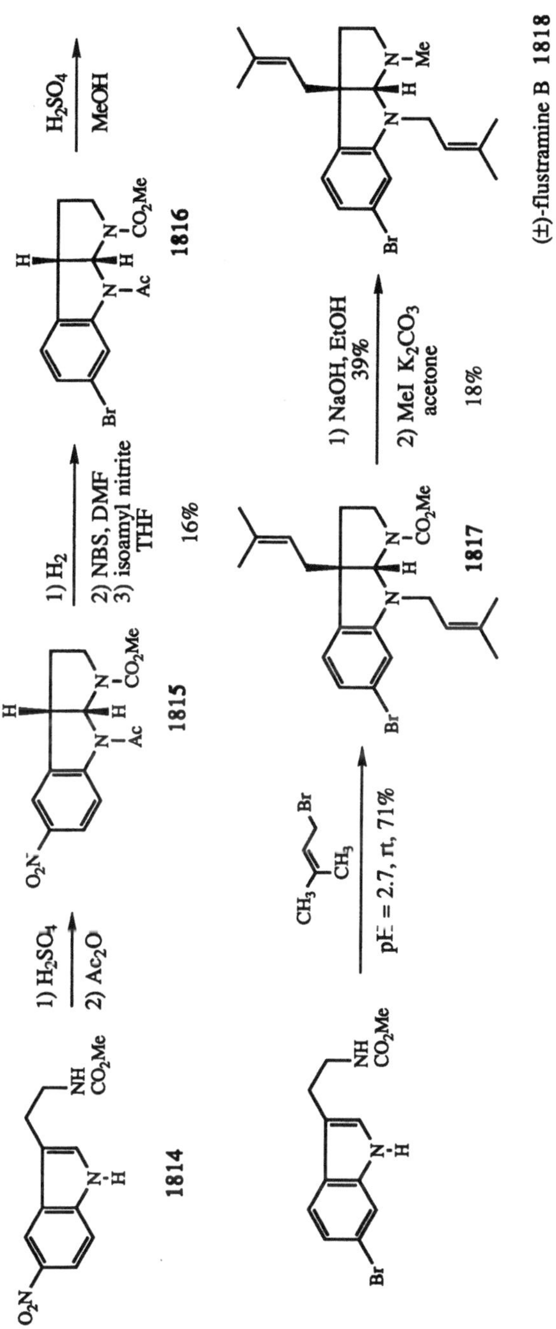

Scheme 319. Hino Synthesis of (±)-Flustamine B

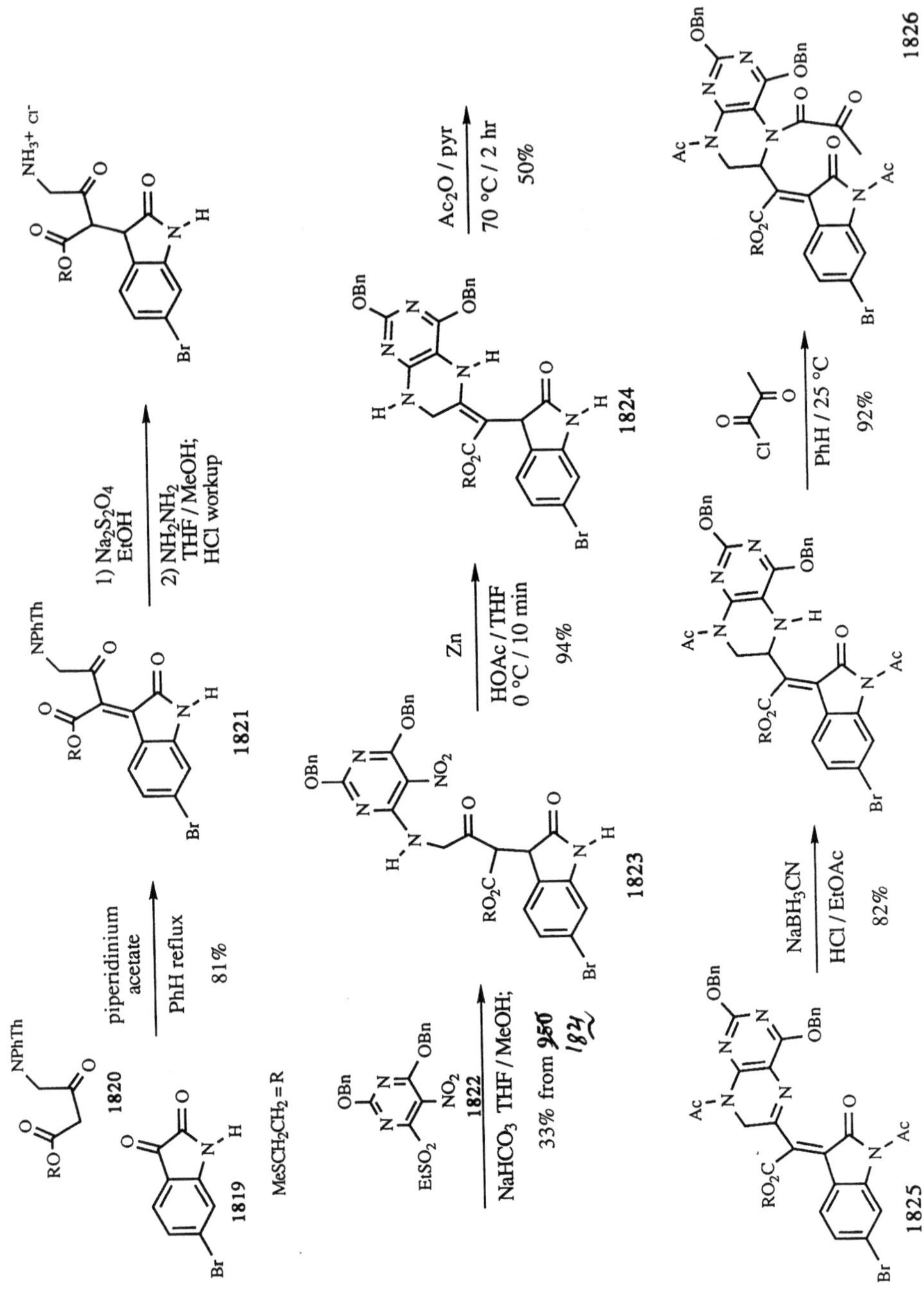

1819
1820
MeSCH2CH2 = R
piperidinium acetate
PhH reflux
81%
1821
1) Na2S2O4 EtOH
2) NH2NH2 THF / MeOH; HCl workup
1822
NaHCO3 THF / MeOH;
33% from 950
1823
Zn
HOAc / THF
0 °C / 10 min
94%
1824
Ac2O / pyr
70 °C / 2 hr
50%
1825
NaBH3CN
HCl / EtOAc
82%
PhH / 25 °C
92%
1826

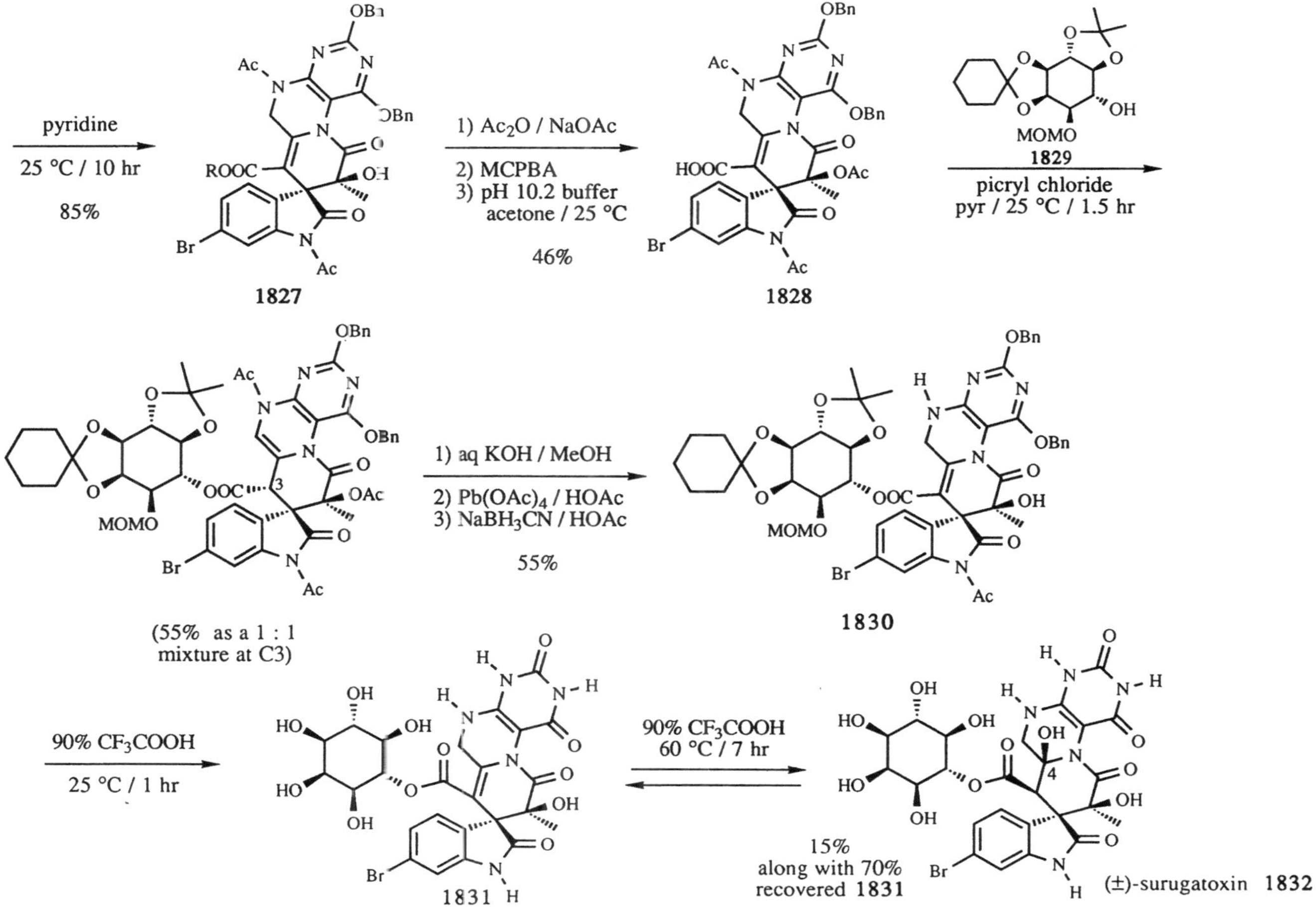

Scheme 320. Inoue Synthesis of (±)-Surugatoxin

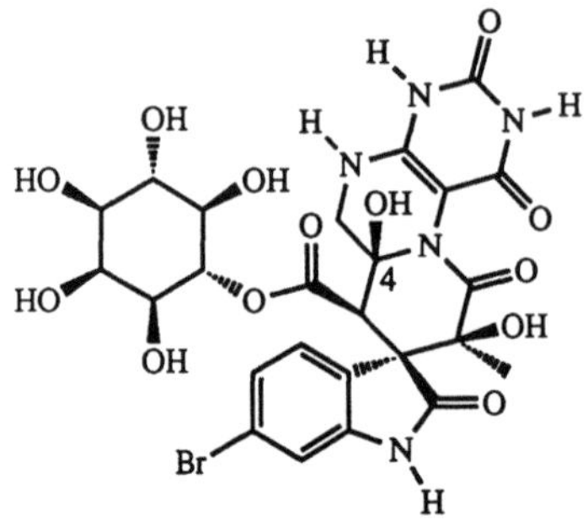

Surugatoxin **1832**

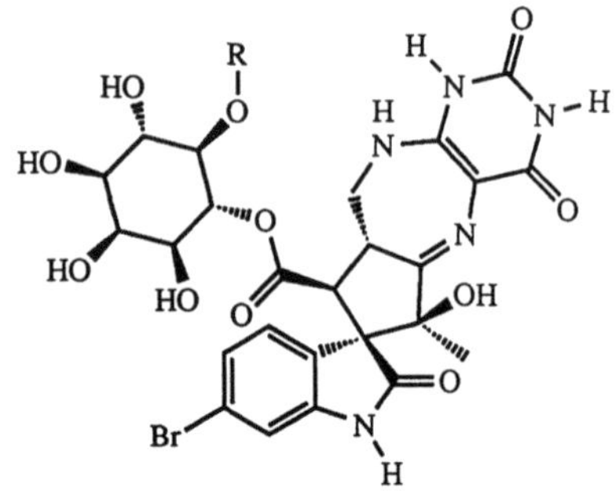

Prosurugatoxin **1853** R = H

Neosurugatoxin 1843 R =

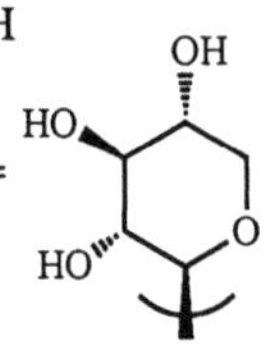

all three metabolites involving (as one might expect) production of separate aglycone and xylopyranosyl-*myo*-inositol components followed by coupling. Their approach is built around an acid-promoted equilibration late in the route. The synthesis of racemic surugatoxin [498] is shown in Scheme 320. Condensation of the isatin **1819** with **1820** yields **1821** as a single isomer. The amino group is freed and added to the highly functionalized pyrimidine **1822** resulting in substitution product **1823**. Reduction of the nitro group was accompanied by cyclization to **1824**. Oxidative dehydration led to **1825** in 50% yield. Construction of the final ring of the aglycone began with imine reduction and acylation to **1826**. Base-induced ring closure occurred to give **1827**, possessing the correct relative configuration at two of the eventual four centers of the aglycone. Acetylation and removal of the carboxyl protecting group produced **1828**, ready for the *myo*-inositol unit. Esterification with racemic **1829** was accompanied by olefin migration and the product mixture was characterized as a 55% yield of C3 epimers. Saponification of the acetyl groups and an oxidation-reduction sequence provided the desired isomer **1830**. Treatment with 90% CF_3COOH at 25 °C for 1 hour removed the various inositol and pteridine protecting groups providing the dehydro compound **1831**. Further treatment with TFA at 60 °C for 7 hours resulted in an equilibrium mixture from which a 70% yield of **1831** and 15% yield of racemic surugatoxin could be obtained.

Neosurugatoxin was produced using a similar strategy beginning from the previously synthesized adduct (Scheme 321) [499]. Construction of the remaining three rings began with Grignard addition to the ketone **1823** and refunctionalization to **1834**. Treatment with MCPBA to oxidize the sulfide in the carboxyl-protecting group to the sulfone and acetylation gave **1835**. Osmylation of the alkyne and reductive work-up in the presence of pyridine effected aldol cyclization of the presumed α-diketone intermediate giving **1836** as a mixture of

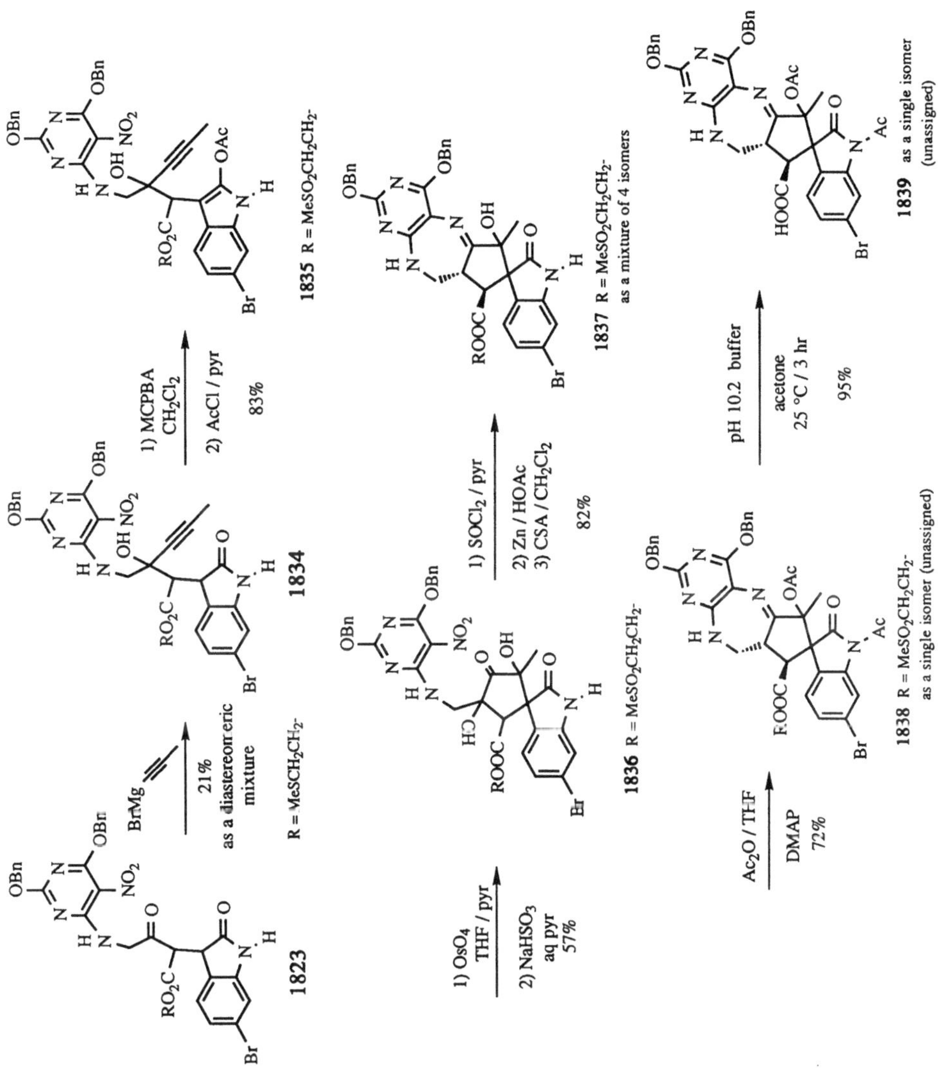

Scheme 321. Inoue Synthesis of Neosurugatoxin: Aglycone Fragment

four isomers. Dehydration of **1836** and reduction followed by brief treatment with CSA promoted cyclization and led to **1837** as a mixture of four isomers containing the basic neosurugatoxin skeleton. Acetylation and equilibration of the mixture at approximately pH 10 led in 95% yield to a single unassigned isomer **1839** with a free carboxyl group. At this point the optically active sugar-derived unit **1840** (Scheme 322, synthesis not divulged) was attached to the racemic aglycone fragment in low yield, providing a mixture of **1841** along with the all-*epi* aglycone isomer with two stereocenters still unassigned. Basic hydrolysis of the acetates and further equilibration led to an equilibrium mixture of four isomers, from which the desired isomer **1843** was separated (18%) for succeeding steps. Recycling of the three unwanted isomers improved

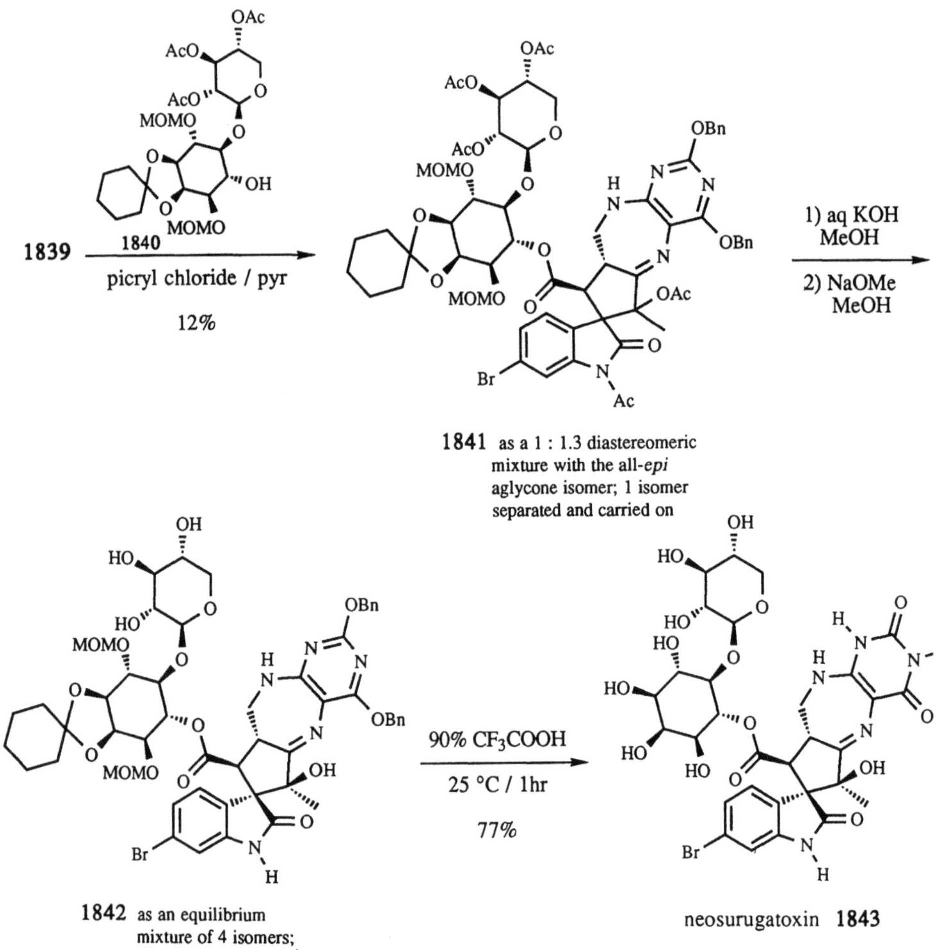

1841 as a 1 : 1.3 diastereomeric mixture with the all-*epi* aglycone isomer; 1 isomer separated and carried on

1842 as an equilibrium mixture of 4 isomers; this isomer separated (18%) and carried on

neosurugatoxin **1843**

Scheme 322. Inoue Synthesis of Neorsurugatoxin: Glycosylation

the overall yield somewhat. Final cleavage of the *myo*-inositol and pteridine protecting groups led to neosurugatoxin in 77% yield, completing a somewhat tortuous pathway. An alternative pathway to an intermediate in this process was recently described by Okada and Inoue [500] and is shown in Scheme 323. The Diels-Alder adduct **1845** was oxidatively cleaved and recyclized to the cyclopentene aldehyde **1846**. Conversion to the amine **1847** was straightforward and coupling to the pyrimidine **1822** as before yielded **1848**. OsO_4 oxidation of the olefin to the diol and selective protection with chloroacetic anhydride gave the mixture **1849**. Dehydration and removal of the chloroacetate gave **1850**. Oxidation of **1850** with phenylseleninic anhydride gave the diastereomeric mixture of α-hydroxyketones **1851**. Cyclization as before led to **1837** as a mixture of four isomers. Treatment with excess Ac_2O in THF with DMAP at 25 °C resulted in the neosurugatoxin intermediate **1838** in 72% yield. Presumably, equilibration of the tertiary carbinol center is occurring via a retro-aldol process under the basic conditions.

Finally, racemic prosurugatoxin could be obtained from the intermediate **1839** as shown in Scheme 324 [501]. Combination of **1839** (as one stereoisomer) with the racemic *myo*-inositol unit **1829** resulted in a low yield of a product mixture characterized as **1852** and its all-*epi* aglycone isomer. From this point the same technology that had been used in the neosurugatoxin synthesis was utilized with nearly the same results, eventually giving racemic prosurugatoxin. Interestingly, when prosurugatoxin was kept in 1% aqueous HOAc solution at room temperature for two days, surugatoxin was obtained. An O_2 oxidation was mechanistically implicated by running the reaction in $^{18}O_2$ such that a 40% yield of surugatoxin was obtained carrying the ^{18}O label at the C4 position.

5.1.3 Various Indoles

5.1.3.1 Trikentrins

From the sponge *Trikentrion flabelliforme* come a series of simple indoles containing a cyclopentane ring fused to the C6-C7 position. Some of these trikentrins [502] have been produced as the end products of three completely different approaches. A radical cyclization was used by MacLeod [503] (Scheme 325) in a synthesis of racemic *cis*-trikentrin A. 2-Bromoacetophenone was

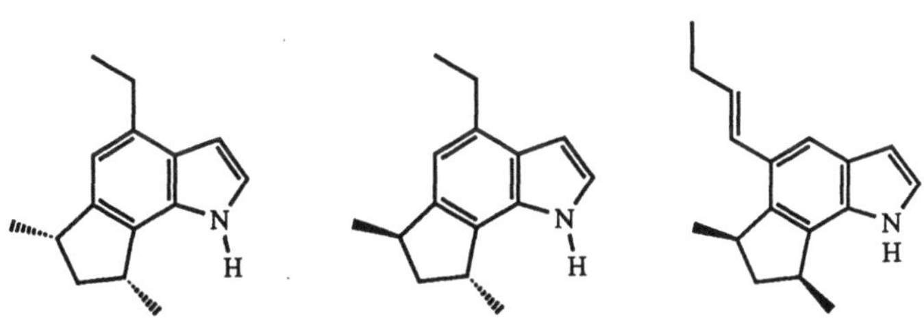

(-)-*cis*-Trikentrin A (-)-*trans*-Trikentrin A *cis*-Trikentrin B

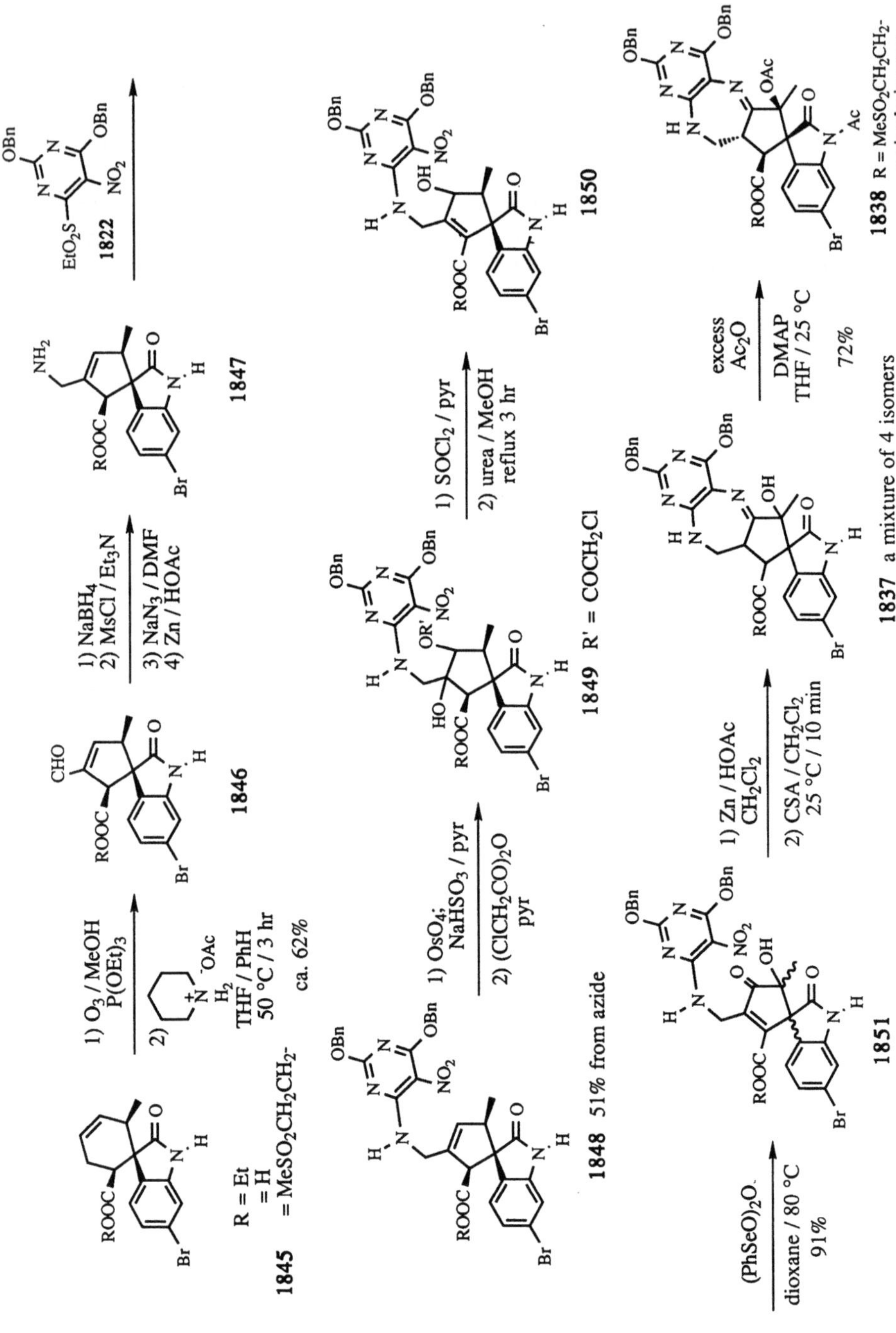

Scheme 323. Okada Synthesis of Neosurugatoxin Intermediate **1838**

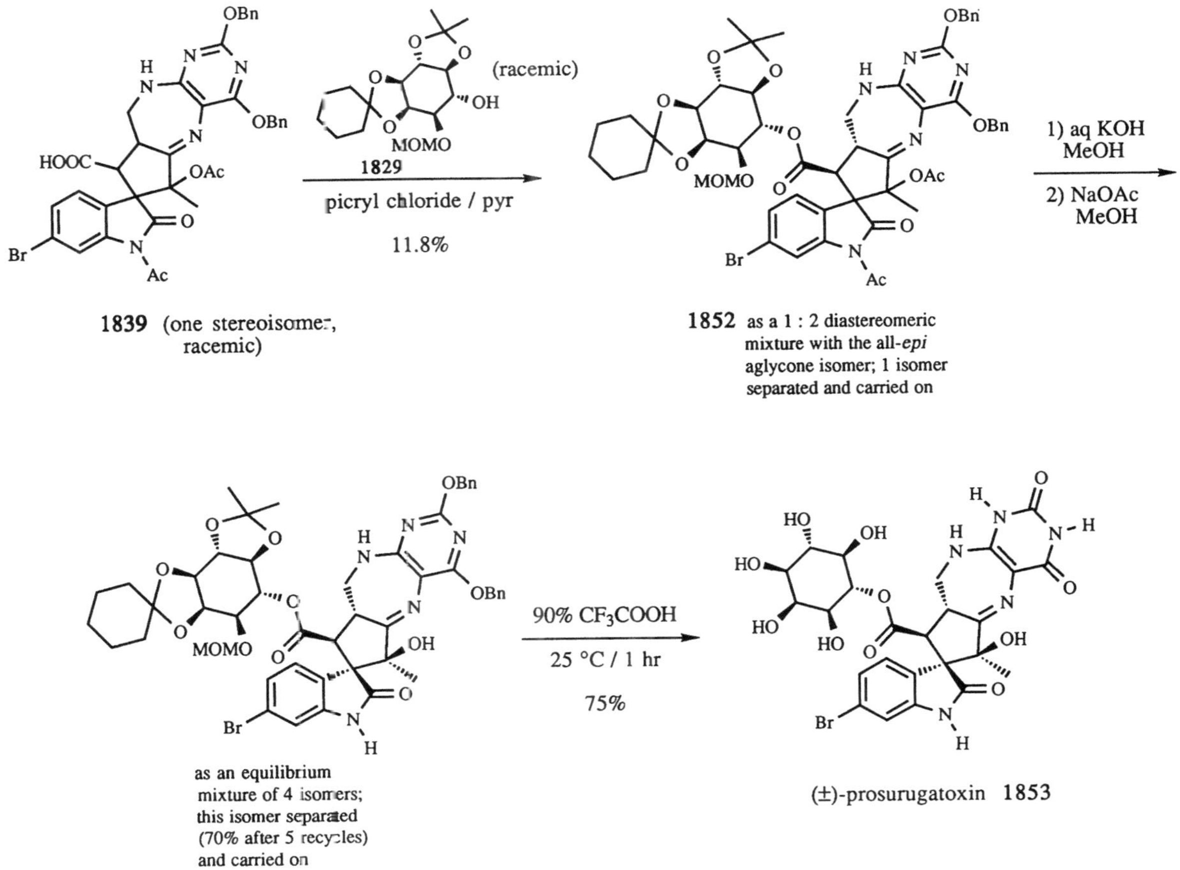

Scheme 324. Inoue Synthesis of (±)-Prosurugatoxin from **1839**

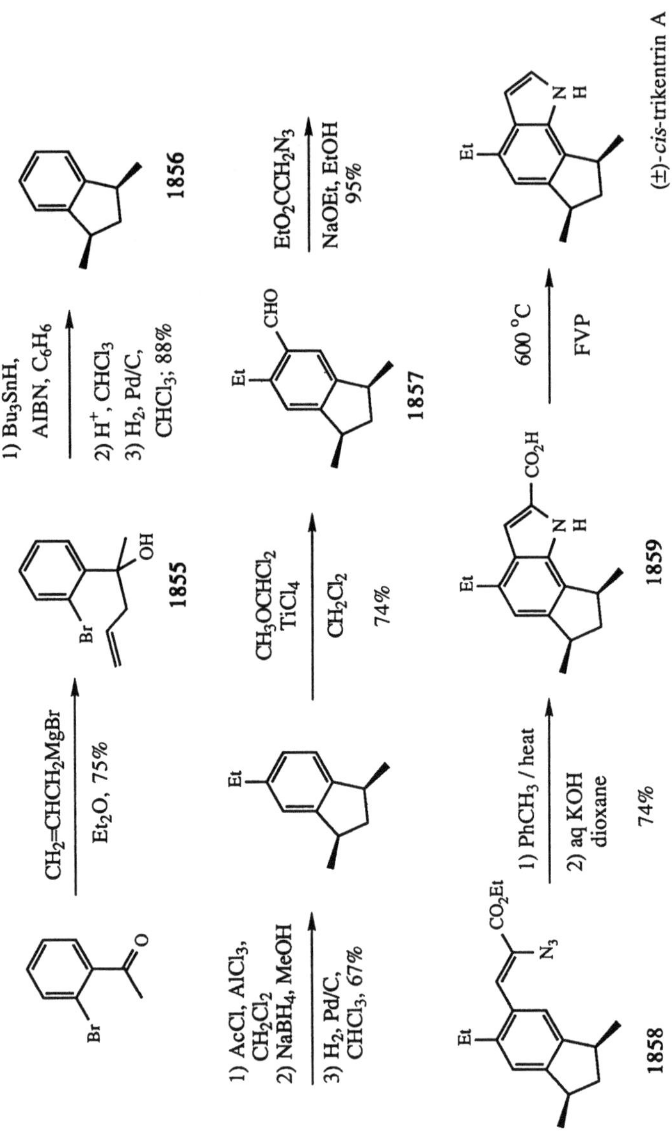

Scheme 325. MacLeod Synthesis of (±)-*cis*-Trikentrin A

allylated to **1855**. Generation of the benzene-type free radical resulted in cyclization to an intermediate aromatic which was dehydrated to an endocyclic olefin (not shown) and hydrogenated to produce the symmetrical *cis* isomer **1856**. Sequential functionalization of the aromatic ring established the ethyl group and the beginnings of a pyrrole. Condensation of the aldehyde **1857** with ethyl azidoacetate gave **1858**, which was thermally cyclized and saponified to the indole carboxylic acid **1859**. Decarboxylation via flash vacuum pyrolysis produced racemic *cis*-trikentrin A.

Syntheses of both *cis*- and *trans*-trikentrin A has been described by Natsume [504] beginning from (+)-pulegone (Scheme 326). Oxidation to the adipic acid **1860** and Dieckmann cyclization gave **1861** and **1862** as diastereomeric mixtures. Separation of **1861** and conversion to the mixture of *cis* and *trans* TMS enol ethers gave intermediate **1863**. Reaction of **1863** with the pyrrole endoperoxide **1864** promoted by $SnCl_2$ gave the ketone **1865**. Carbonyl addition of the *N,N*-dimethylhydrazone **1866** and acidic cyclization led to a 45% yield of the correctly substituted indole **1867**, still as a mixture. Saponification led to both *cis*- and *trans*-trikentrin A as the (−)-antipodes.

An interesting strategy has been reduced to practice by Kanematsu [505] in a synthesis of racemic *cis*-trikentrin B (Scheme 327). The racemic allenic ester **1869** was treated with cyclopentadiene to produce adduct **1870**. A series of routine transformations provided allenic dienamide **1871**. Intramolecular [4+2] cycloaddition and aromatization produced indole **1872**. Oxidation to the aldehyde and olefin osmylation led to **1873**. Wittig olefination gave a 2:1 (*E*/*Z*) mixture which was carried on to *cis*-trikentrin B prior to separation.

5.1.3.2 Indoles Related to Aplysinopsin

Several compounds related to aplysinopsin have been isolated from Dictyoceratid and Dendroceratid sponges. Pietra has described interesting photoisomerizations [506] involving metabolites in this class, which precipitated straightforward syntheses of the alkaloids **1875** and **1878** (Scheme 328). Condensation of the imidazolone **1874** with indole-3-carboxaldehydes gives the *Z*-isomers as the exclusive products. The *Z*-isomers undergo photochemical isomerization to mixtures enriched in the *E* isomers.

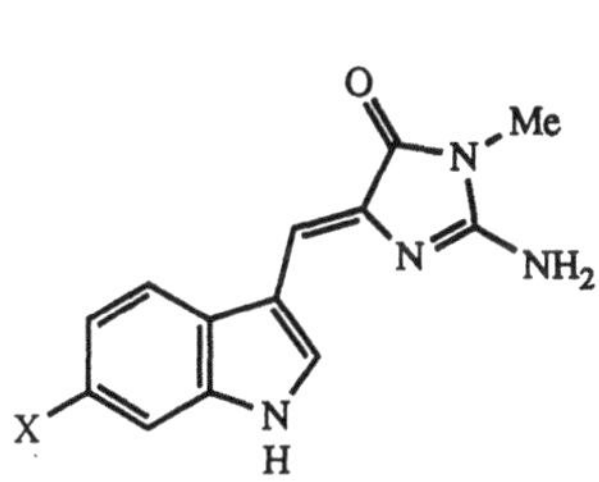

X = H **1825**
X = Br **1876**

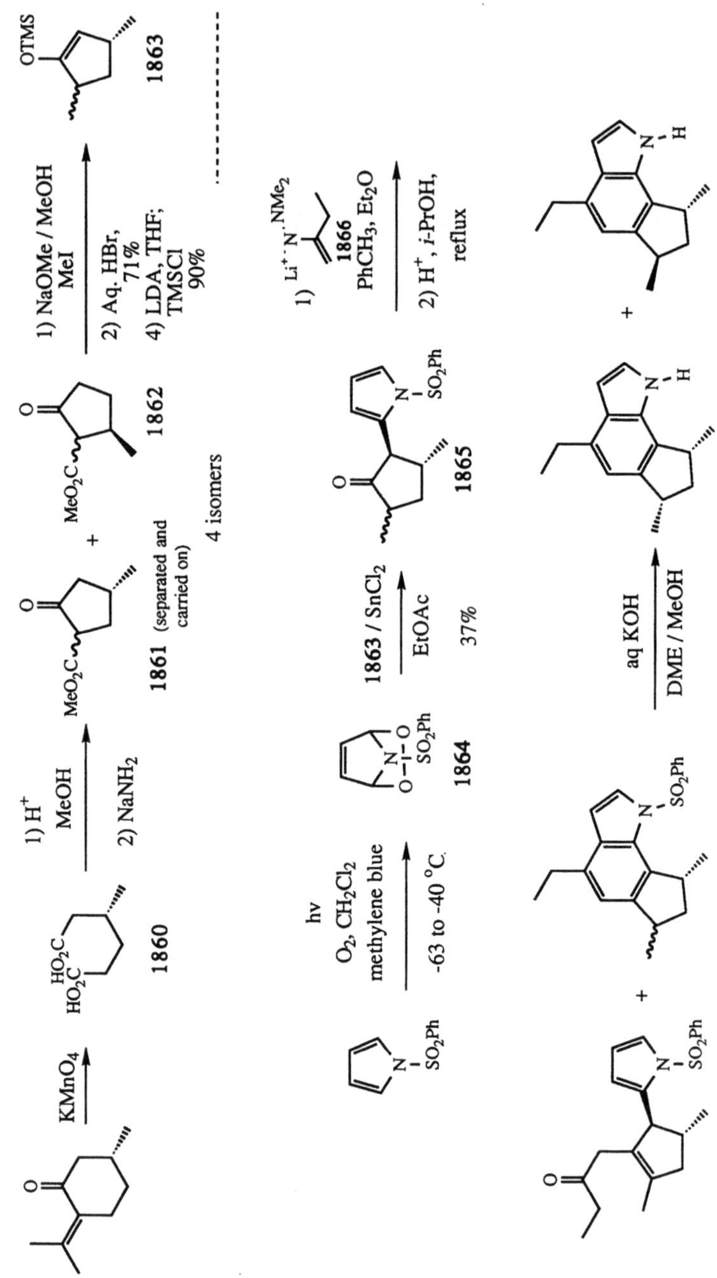

Scheme 326. Natsume Synthesis of (−)-*cis*- and (−)-*trans*-Trikentrin A

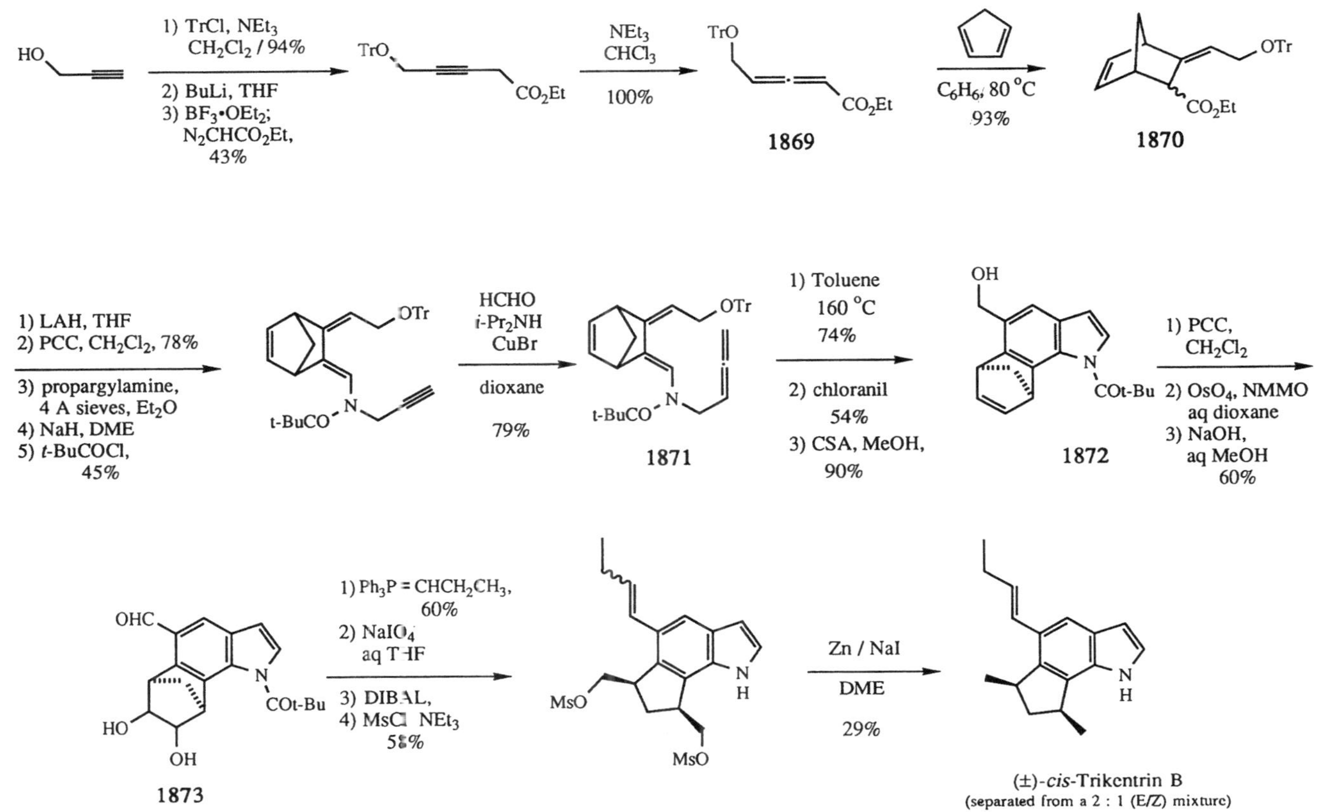

Scheme 327. Kanematsu Synthesis of (±)-*cis*-Trikentrin B

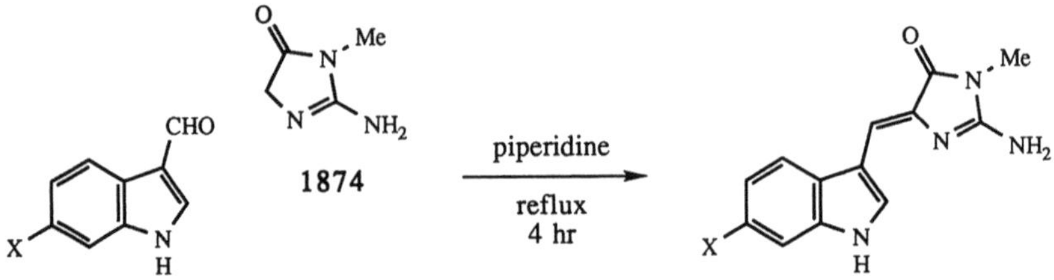

X = H **1875**
X = Br **1876**
nearly quantitative

Scheme 328. Pietra Synthesis of Aplysinopsin-like compounds from *Dendrophyllia sp.*

5.1.3.3 Dendrodoine

The marine tunicate *Dendroda grossular* (Styelides) found along the North Brittany coast provides the cytotoxic metabolite dendrodoine (**1879**). Although incorporation of the indole unit of tryptamine is common among metabolites, the further elaboration of the sidechain into a 1,2,4-thiadiazole moiety is rare [507]. Sainsbury's synthesis of **1879** is short and to the point (Scheme 329) [508]. Warming of indole with oxalyl chloride and copper(I) cyanide yields the acyl cyanide **1878** via corresponding acid chloride. The 1,3-dipolar cycloaddition of

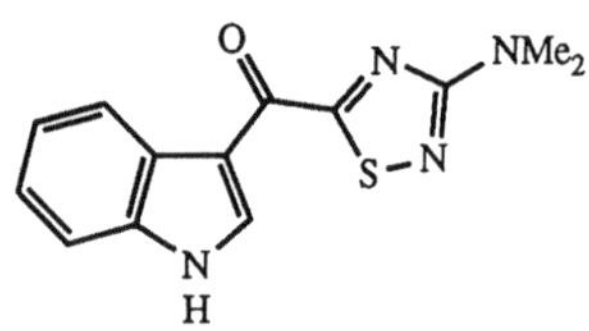

Dendrodoine **1879**

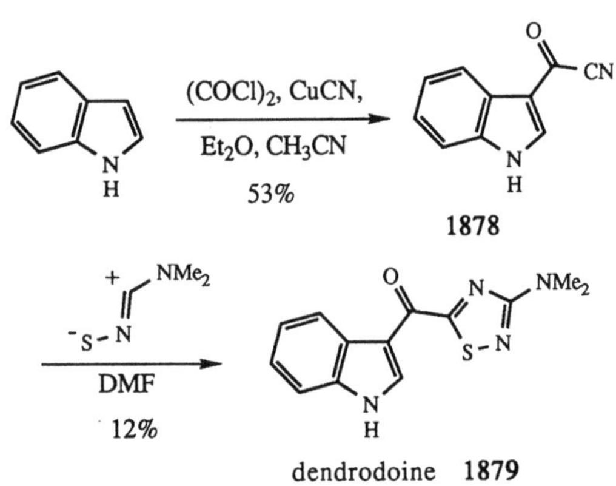

Scheme 329. Sainsbury Synthesis of Dendrodoine

1878 with *N,N*-dimethylaminonitrile sulfide (generated *in situ* through the thermolysis of 5-(*N,N*-dimethylamino)-1,3,4-oxathiazol-2-one) provides dendrodoine in two steps and 6.4% overall yield.

5.1.3.4 *Tetrahalogenoindoles*

Crude extracts of the New Zealand marine alga *Rhodophyllis membranacea* (Harvey) exhibit strong antifungal activity. This activity is due to the presence of a variety of polyhalogenated indoles [509]. Several of the 2,3,4,7-tetrahalogenated members of this class have been prepared by Somei [510] (Scheme 330). Oxidation of 4,7-dihalogenated indole **1880** with sodium chlorite in the presence of *tert*-butanol gives acid **1881**. Decarboxylation of **1881** affords key intermediate indole **1882**. Bromination with NBS yields the 3,4-dibromo compounds **1883a** and **1883b**, whereas chlorination with sulfuryl chloride provides the 3,4-dichloro metabolites. Sequential chlorination and bromination of 4,7-dibromoindole **1882a** yields 3-chloro-2,4,7-tribromoindole (**1884**).

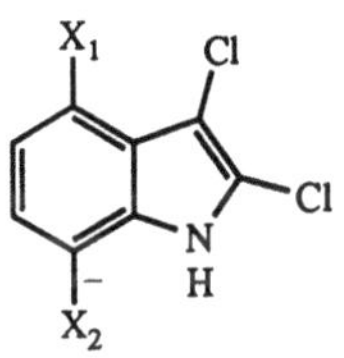

$X_1 = X_2 =$ Cl
$X_1 =$ Cl, $X_2 =$ Br
$X_1 =$ Br, $X_2 =$ Cl
$X_1 = X_2 =$ Br

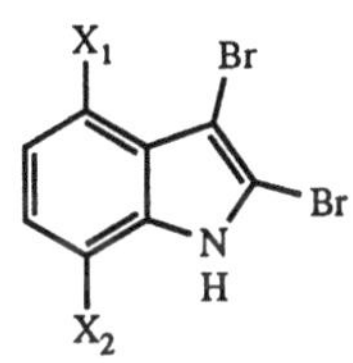

$X_1 = X_2 =$ Br **1883a**
$X_1 = X_2 =$ Cl **1883b**

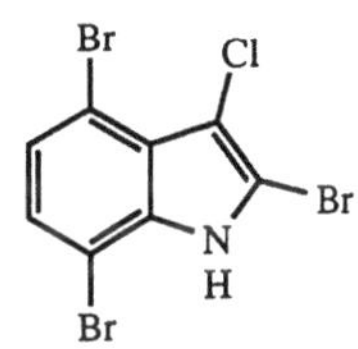

1884

5.1.3.5 *(E)-3-(6-Bromoindole-3-yl) Prop-2-enoate*

Bromoindole **1888** has been isolated from a sponge of the genus *Iotrochota*, collected off Freemantle, Western Australia. Sargent isolated, characterized and confirmed the structure of **1888** through synthesis (Scheme 331) [511]. Condensation of 4-bromo-2-nitrotoluene with *N,N*-dimethylformanide dimethyl acetal followed by catalytic reduction gives 6-bromoindole (**1886**). Formylation of **1886** and reaction of the resulting aldehyde **1887** with monomethyl malonate

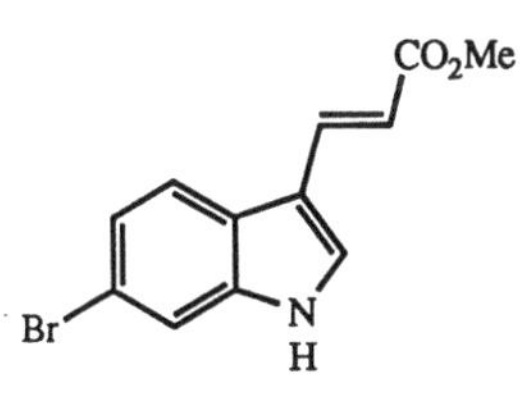

1888

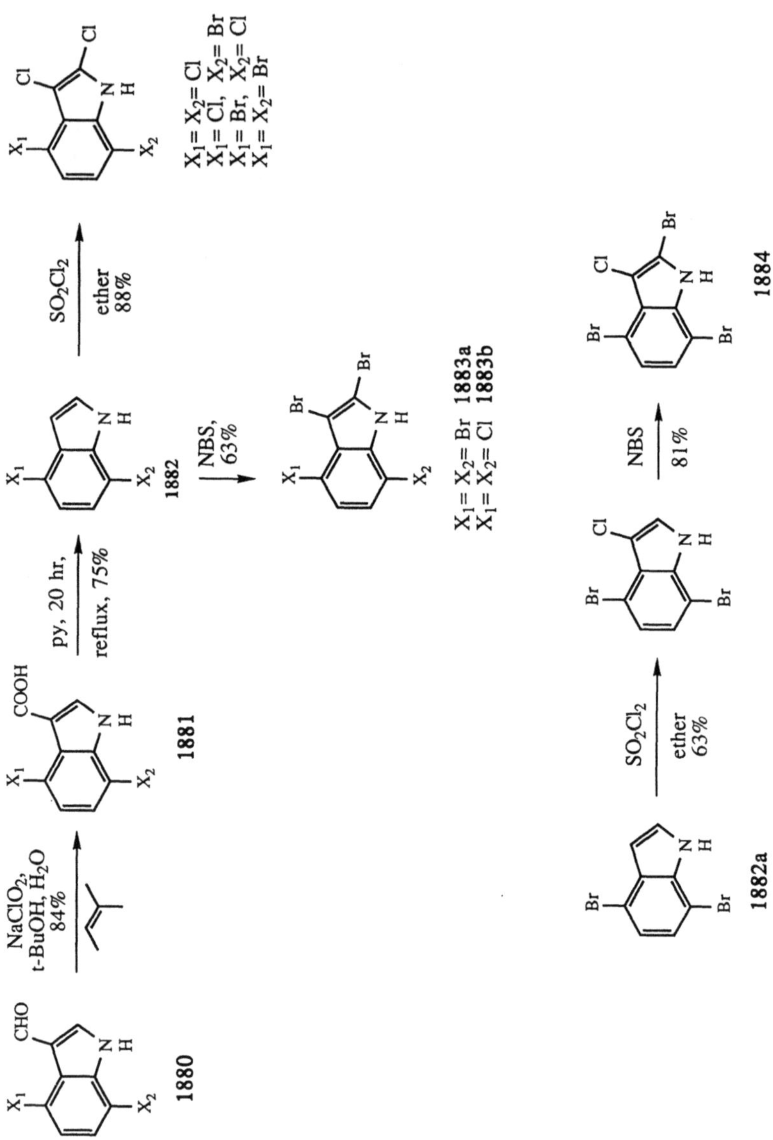

Scheme 330. Somei Synthesis of Tetrahalogenoindoles

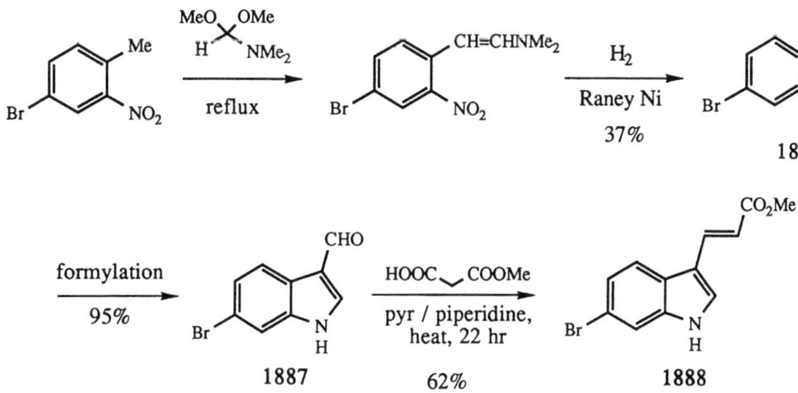

Scheme 331. Sargent Synthesis of an Indole from Asutralian *Iotrochota*

under the conditions of the Doebner reaction gives (*E*)-3-(6-bromoindole-3-yl)prop-2-enoate **(1888)** in four steps and 22% overall yield.

5.2 Carbazoles

5.2.1 Hyellazole and 6-Chlorohyellazole

The unusual carbazole alkaloid hyellazole (**1894**) has been isolated from the Hawaiian blue-green alga *Hyella caespitosa* [512]. All four syntheses of hyellazole that have appeared involve construction of the substituted C ring onto a pre-existing indole ring system. The first of these syntheses, by Kano [513] (Scheme 332) begins with the condensation of lithio-*N*-(benzenesulfonyl)indole **1889** with propiophenone to yield alcohol **1890**. Elimination of the alcohol and formylation of indole **1891** gives aldehyde **1892**. Wittig olefination provides conjugated triene **1893** which undergoes cyclization and aromatization, upon heating in the presence of palladium on carbon. Hyellazole is obtained in five steps and 14% overall yield. The overall yield is lowered somewhat by the poor yield of the cyclization step.

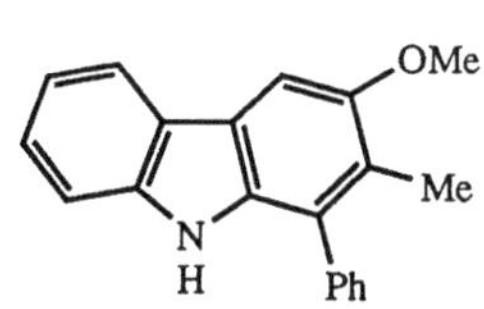

Hyellazole **1894**

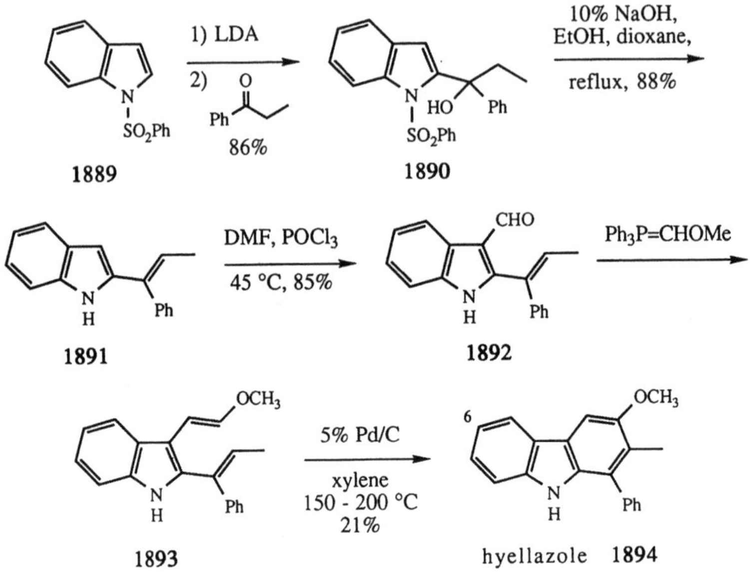

6-Chlorohyellazole was also synthesized in an analogous manner

Scheme 332. Kano Synthesis of Hyellazole

Takano's [514] synthesis of **1894** is initiated by the formation of enamine **1897** from 2,3-disubstituted indole **1896** (Scheme 333). Cyclization of **1897** by treatment with acetic anhydride and acetic acid affords acid **1898**, after saponification. Conversion of the carboxyl group to a methoxy group requires three steps. Formation of the isocyanate by reaction with diphenylphosphoryl azide (DPPA) and addition of water yields urea **1899**. Hydrolysis of the urea gives amine **1900**, which is converted to the natural product by diazotization in the presence of methanol. Hyellazole is obtained in 5.5% overall yield in a total of 6 steps from indole **1897**. The low overall yield can be attributed entirely to the poor yield obtained in the final step. 6-Chlorohyellazole is prepared in a similar fashion.

Sakamoto [515] begins the synthesis of hyellazole with the Wittig reaction of **1902** with 3-methoxyindolin-3-one **1901** to give indole **1903** (Scheme 334). Conversion of **1903** to the 3-buta-1,3-dienylindole sets the stage for electrocyclic ring closure, giving a 4:1 mixture of **1904** and **1905**. Treatment of this mixture with tetrabutylammonium fluoride provides **1905** in 56% yield. Methylation of **1905** gives hyellazole in a total of five steps and 24% yield.

Moody's synthesis [516] of hyellazole is initiated by Diels-Alder reaction of ethyl 3-trimethylsilylpropynoate with indole **1907** (derived from commercially available indol-3-ylacetic acid) occurring with concomitant loss of carbon dioxide to give **1908** (Scheme 335). The resulting carbazole is reduced to 2-methylcarbazole **1909** and subjected to mercurio-desilylation yielding

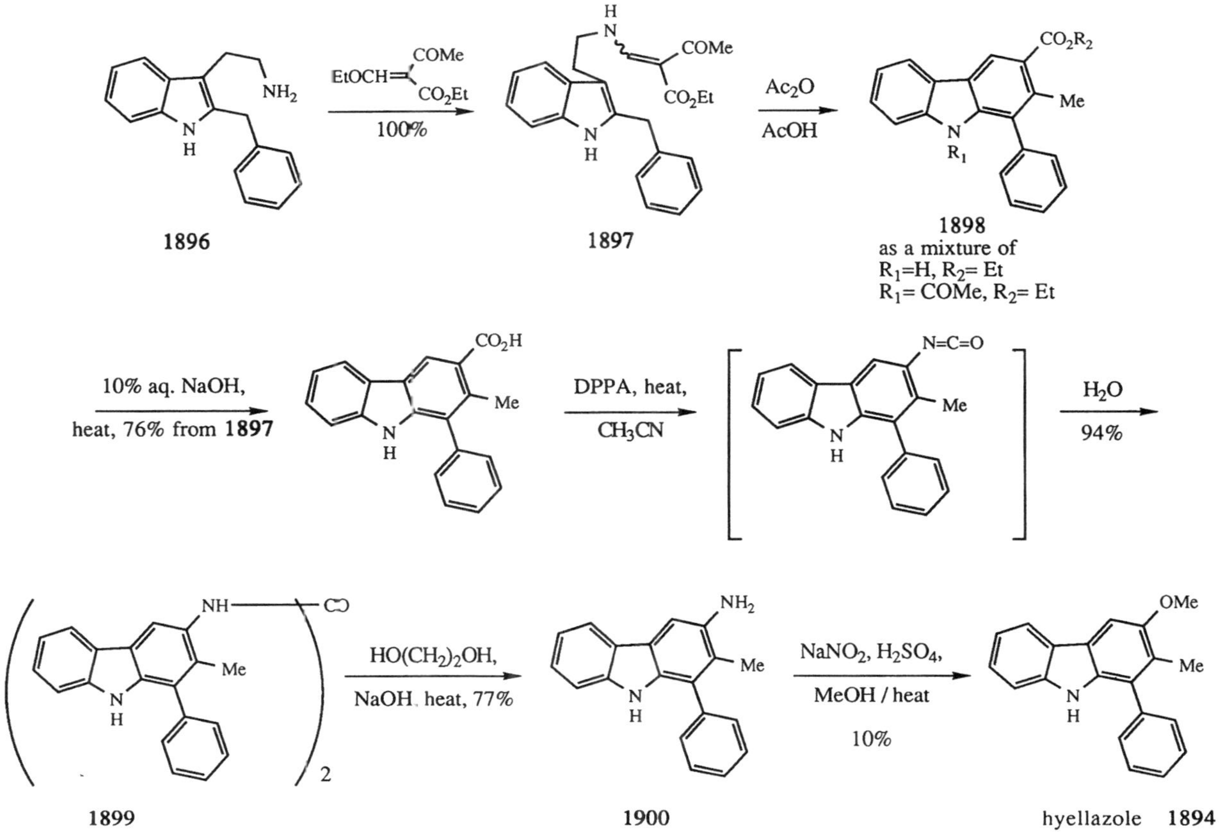

Scheme 333. Takano Synthesis of Hyellazole

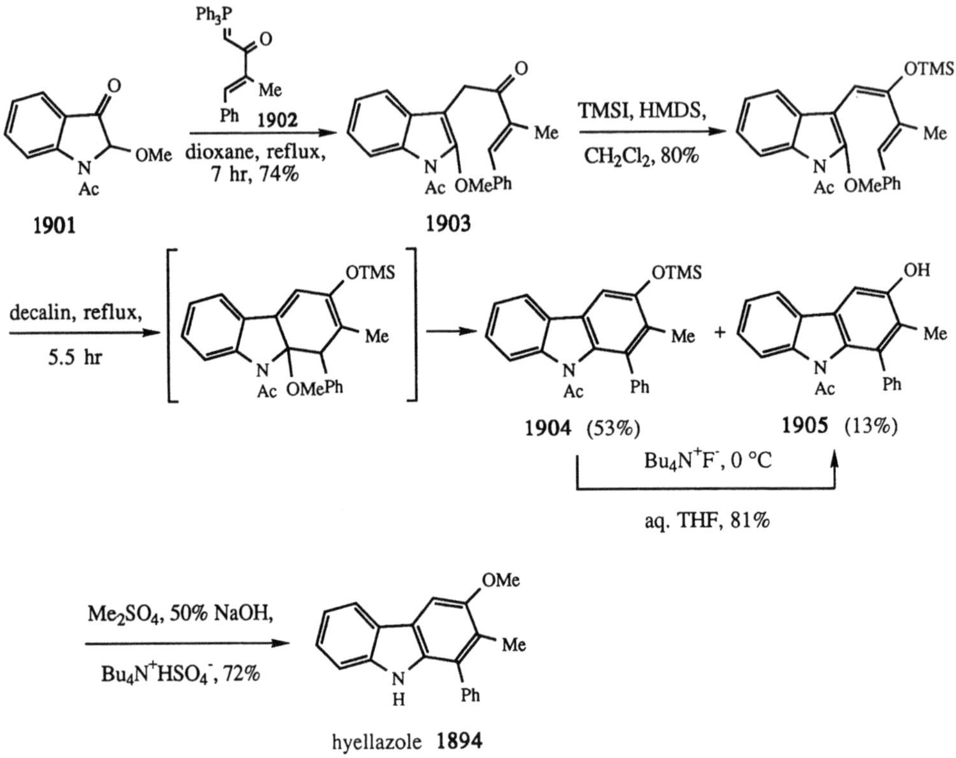

Scheme 334. Sakamoto Synthesis of Hyellazole

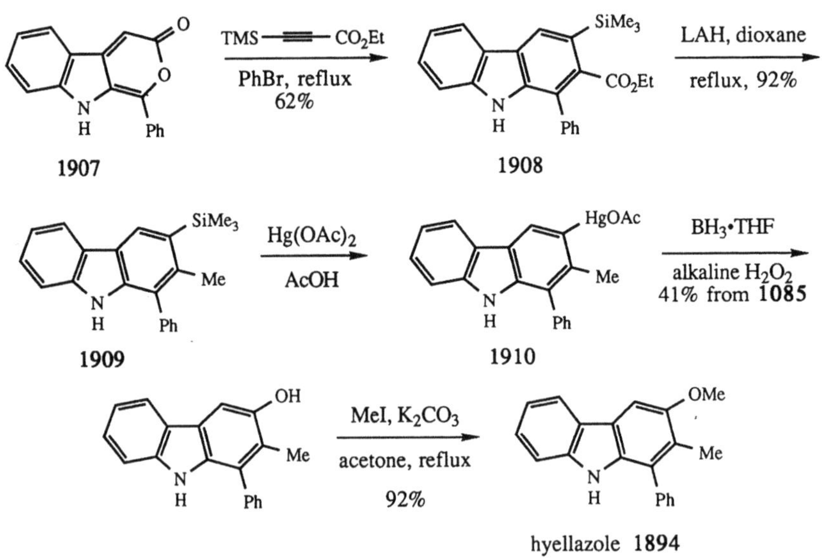

Scheme 335. Moody Synthesis of Hyellazole

arylmercury compound **1910**. Hydroboration-oxidation of **1910** and methylation provides hyellazole in five steps and 22% overall yield.

5.2.2 Eudistomins

A series of interesting carbazoles exhibiting diverse antiviral and antimicrobial activity (including potent activity against *Herpes simplex* virus type I) have been found in the Caribbean tunicate *Eudistoma olivacea* [517]. Many members in this family possess the simple carbazole structure shown below in which R may be pyrrole, phenylacetyl or some other structurally simple group. The benzene moiety may contain no further substitution or may hold hydroxy and/or bromide groups. The more interesting metabolites in this series resemble eudistomin L, possessing an unusual heterocyclic ring system.

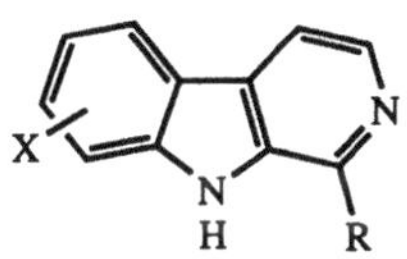

1911

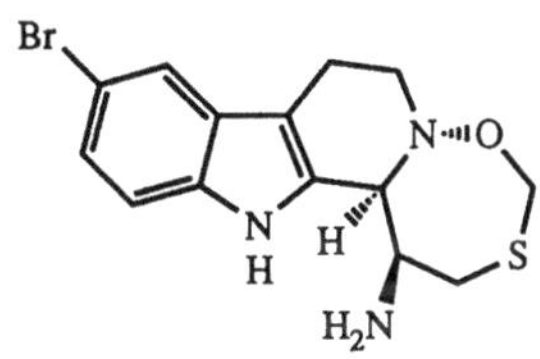

Eudistomin L

5.2.2.1 Eudistomin A

Murakami [518] utilizes Fischer indolization as a key reaction in the preparation of eudistomidin-A (Scheme 336). The regioselectivity of the cyclization step is improved by tosylation of the free hydroxyl. Thus, treatment of **1912** with polyphosphoric acid yields indole **1913** in 41% yield. Decarboxylation, formylation, and condensation with nitromethane provides 3-(nitrovinyl) indole **1914**. Conversion of **1914** to the amide **1915** and Bischler-Napieralski reaction followed by dehydrogenation in low yield gives the tosyl-β-carboline **1916**. Hydrolysis of **1916** affords eudistomin A in 10 steps and 2.0% overall yield.

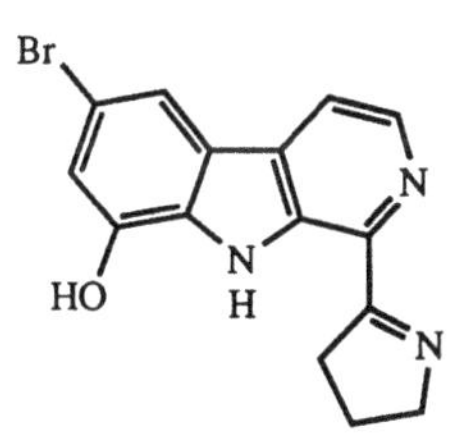

Eudistomin A

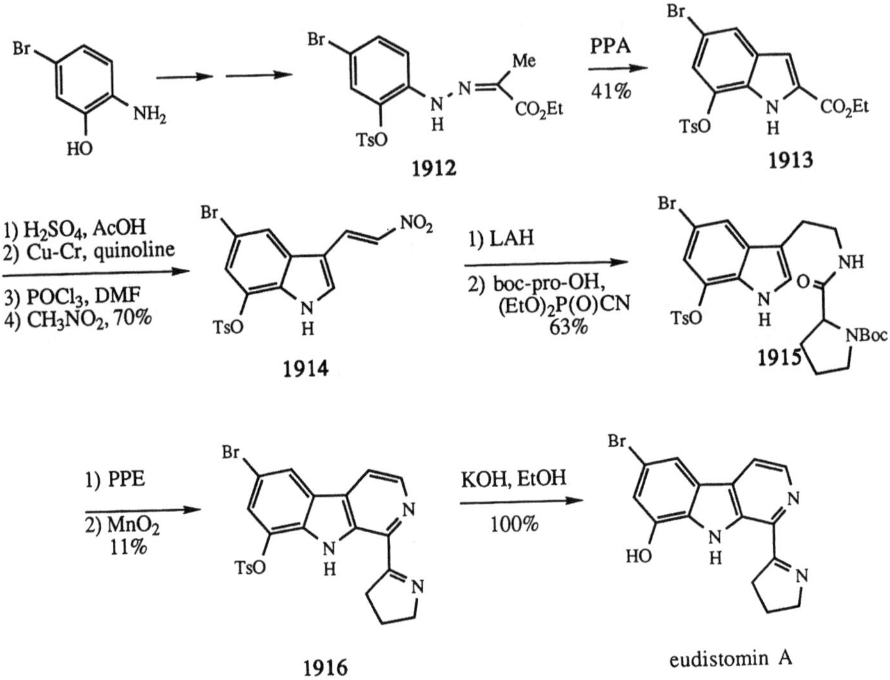

Scheme 336. Murakami Synthesis of Eudistomin A

5.2.2.2 *Eudistomins H, I, P*

Hino [519] has prepared eudistomins H and I by utilizing the Bischler-Napieralski reaction of *N*-(*N*-*tert*-butoxycarbonylpropyl)tryptamine **1918** to obtain dihydro-β-carboline **1919** (Scheme 337). Dehydrogenation of **1919** with

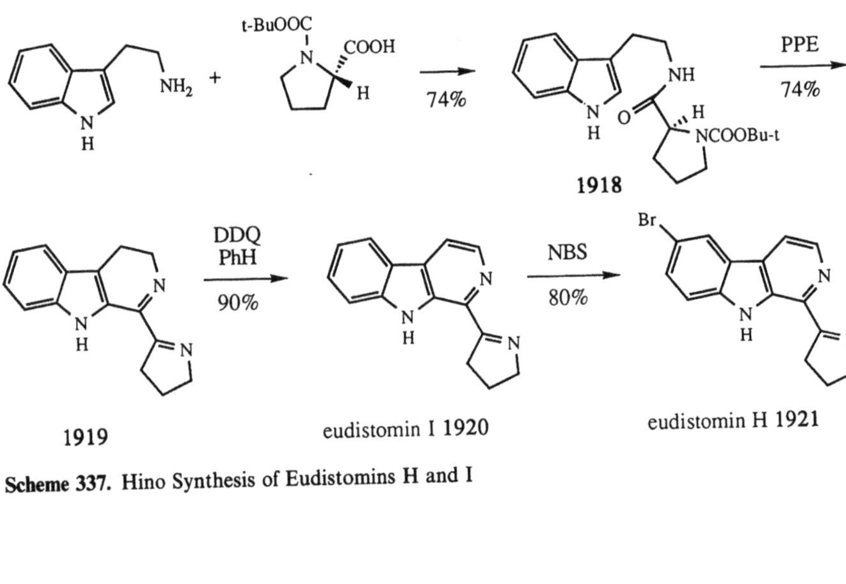

Scheme 337. Hino Synthesis of Eudistomins H and I

DDQ provides eudistomin I (**1920**) in three steps and 63% overall yield. Bromination of **1920** with NBS yields eudistomin H (**1921**) with an overall yield of 50%. Preparation of eudistomin P requires 6-bromo-5-methoxy-tryptamine (**1922**) as the Bischler-Napieralski precursor (Scheme 338). This is prepared in five steps utilizing standard methodology. Reaction of **1923** with *N*-benzyloxy-carbonyl-L-prolyl chloride followed by treatment with phosphorus oxychloride gives the eudistomin ring system **1924**. Elimination of the *N*-benzyloxycarbonyl protecting group, dehydrogenation and demethylation provides eudistomin P in 10 steps and 7.6% overall yield.

5.2.2.3 Other Eudistomins

Rinehart [520] has synthesized six of the seventeen known eudistomins. Eudistomins H and I are prepared in an identical fashion starting with either 1-cyano-β-carboline **1926a** or 6-bromo-1-cyano-β-carboline **1926b** respectively (Scheme 339). Grignard addition to **1926** and sodium borohydride reduction of the intermediate imine provides amino acetal **1927**. Deprotection of the acetal and cyclization to the imine provides isomeric eudistomin **1928**. Removal and reinsertion of the double bond in its proper position gives eudistomin H 32% over five steps. Eudistomin I is obtained in a similar fashion in 20% overall yield. The synthesis of eudistomin M begins with 1-cyano-6-methoxy-β-carboline (**1929**) (Scheme 340). Grignard addition to the nitrile and mild hydrolysis affords ketone **1930**. Formation of the pyrrole, by reaction of **1930** with ammonium acetate and acetic acid, followed by demethylation yields eudistomin M in three steps and 36% overall yield. The preparations of eudistomins N, D and O are outlined in Scheme 341. Eudistomin N can be prepared in one step from β-carboline by treatment with bromine in THF. In similar fashion, bromination of 6-methoxy-β-carboline and demethylation with BBr_3 gives eudistomin D in 54% yield. The synthesis of eudistomin O requires a more circuitous route. In four steps, 4-amino-2-nitrotoluene is converted to 6-bromoindole **1932**. Alkylation with aziridinium tetrafluoroborate gives amine **1933** which is converted to eudistomin O by glyoxylation and dehydrogenation. The natural product is obtained in eight steps and 7.2% overall yield.

Eudistomin I along with eudistomin T have been prepared by Cardellina [521] as shown in Scheme 342. Acylation of *N*-protected indole isonitrile **1934** with phenylacetyl chloride followed by treatment with silver tetrafluoroborate affords dihydro-β-carboline **1935** in 92% yield. Treatment of **1935** with elemental sulfur at 200 °C for four minutes results in deprotection and dehydration to give eudistomin T. Eudistomin I is prepared by modification of this procedure in 12% yield.

Wasserman [522] utilizes a flexible synthetic strategy that provides eudistomins T and I, along with the methyl ether of eudistomin M (Scheme 343). The key transformation in these syntheses is the condensation of the 1,2,3-tricarbonyl component **1937** with tryptamine **1938** in the presence of trifluoro-

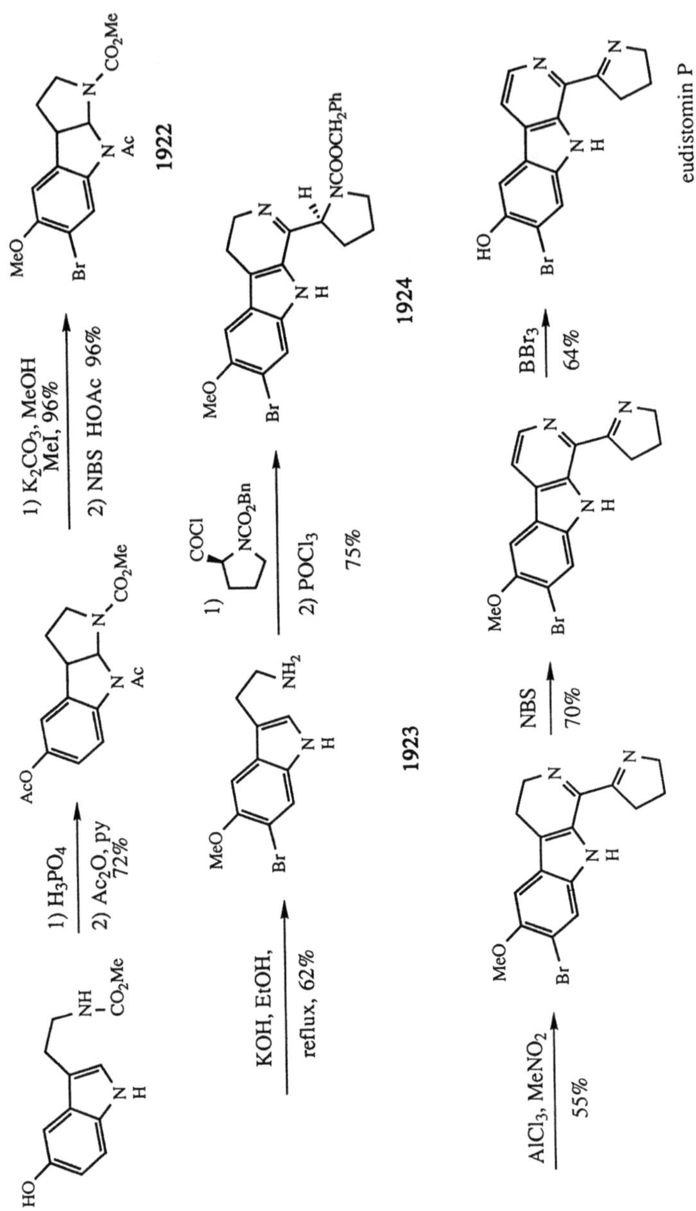

Scheme 338. Hino Synthesis of Eudistomin P

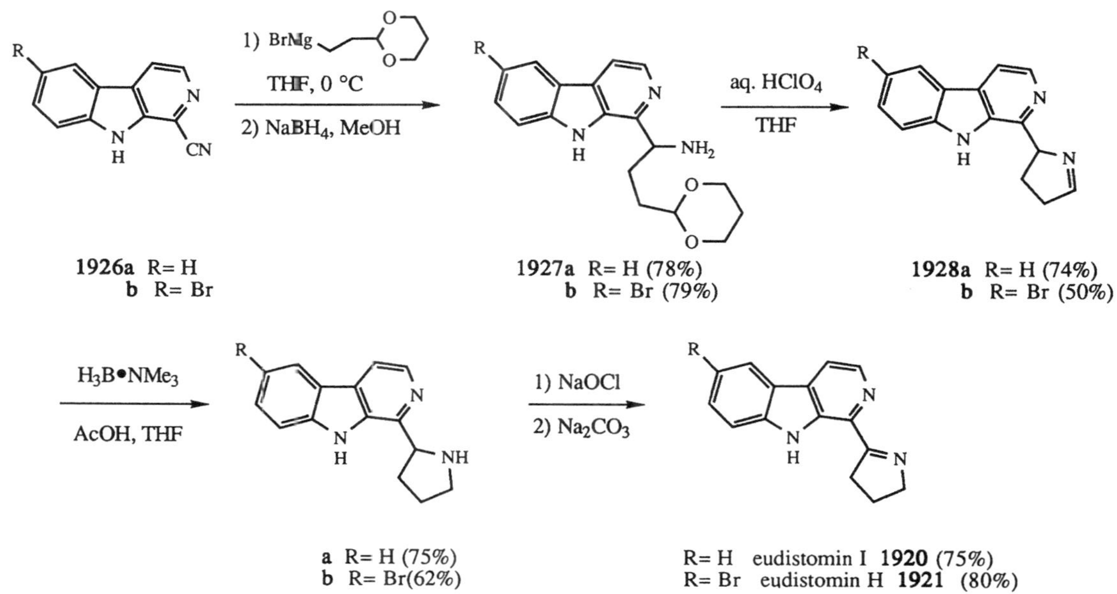

Scheme 339. Rinehart Synthesis of Eudistomins H and I

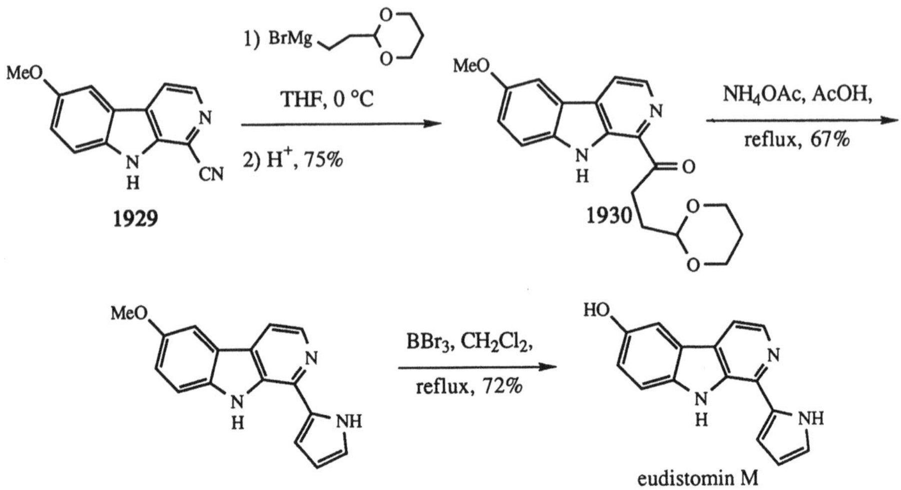

Scheme 340. Rinehart Synthesis of Eudistomin M

acetic acid. Condensation at the central carbonyl to form the imine, Pictet-Spengler cyclization, and subsequent decarboxylation affords dihydro-β-carboline **1939**. Dehydrogenation of **1939a** affords eudistomin T (**1940a**) (four steps and 40% overall yield), while dehydrogenation of **1939b** and reaction with hydrazine gives eudistomin I (five steps and 44% overall yield). Conversion of **1940c** to eudistomin M methyl ether requires reduction, Swern oxidation to form the γ-ketoaldehyde, and treatment with ammonia to form the pyrrole **1941**. Demethylation to give the natural product has already been reported. Methyl ether **1941** is obtained in 7 steps and 32% overall yield.

Eudistomins T and S have been prepared by Still [523] from tryptamine **1943a** and 5-bromotryptamine **1943b**, respectively (Scheme 344). Pictet-Spengler cyclization of **1943** with glyoxylic acid and esterification gives **1944**. Aromatization of **1944** by heating with elemental sulfur and addition of benzylmagnesium chloride in the presence of lithium chloride provides the eudistomins in 4 steps. Eudistomins T and S are obtained in 39% and 25% yields, respectively.

Nakagawa's [524] strategy for the preparation of (−)-eudistomin L and (−)-debromoeudistomin L is outlined in Scheme 345. By careful control of reaction conditions optically active nitrone **1946** can be cyclized to either the normal Pictet-Spengler product **1947** or to the tetracyclic adduct **1948**. Thus, cyclization of optically active nitrone **1946** with trifluoroacetic acid at low temperatures provides Pictet-Spengler product tetrahydro-β-carboline **1947** with high diastereoselectivity. Oxidation of **1947** with NCS yields the oxathiazepine **1948** in poor yield. Deprotection of **1948** gives (−)-debromoeudistomin L in four steps and 6.1% overall yield. Cyclization of **1949** at room temperature for five minutes gives tetracyclic hydroxylamine **1950** in 70% yield, along with 21%

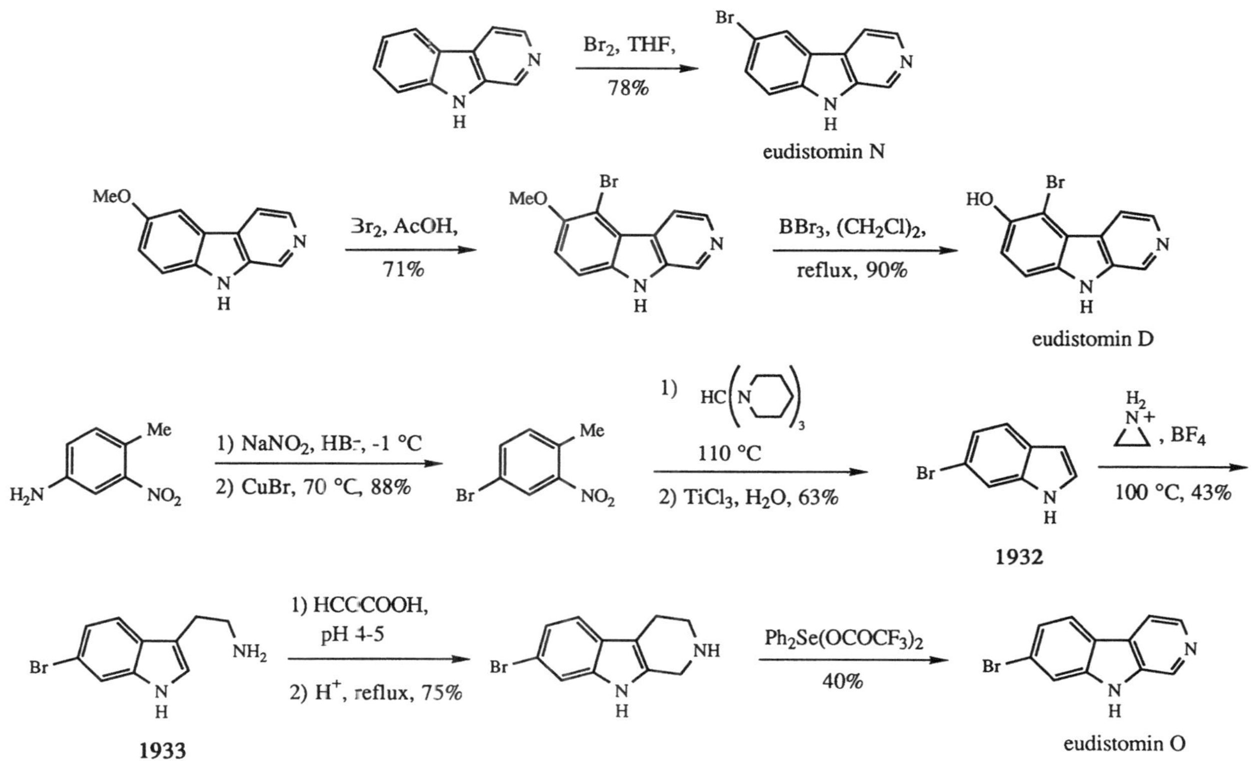

Scheme 341. Rinehart Synthesis of Eudistomins N, D and O

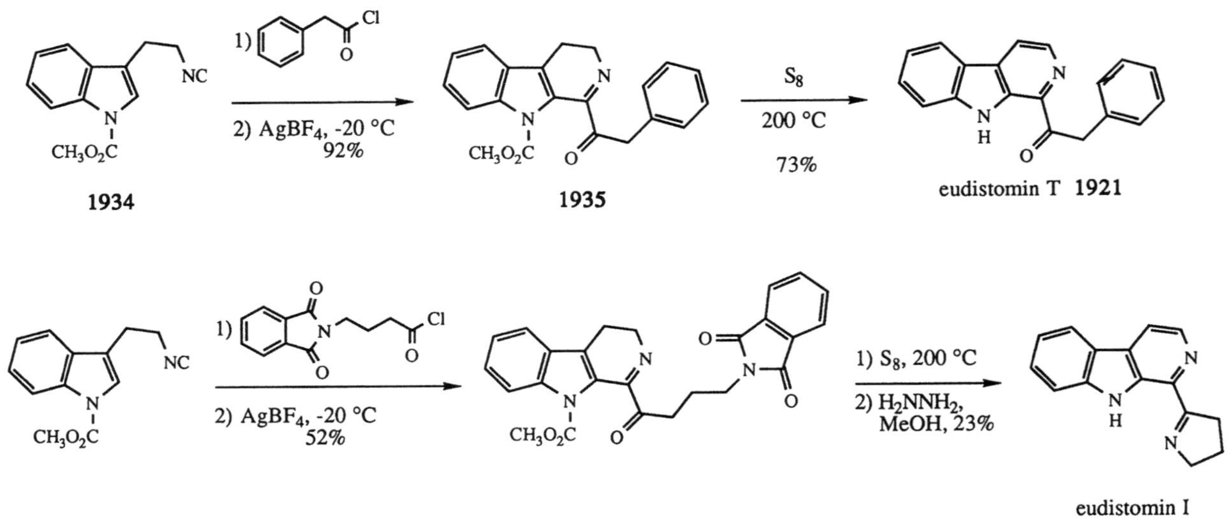

Scheme 342. Cardellina Synthesis of Eudistomins I and T

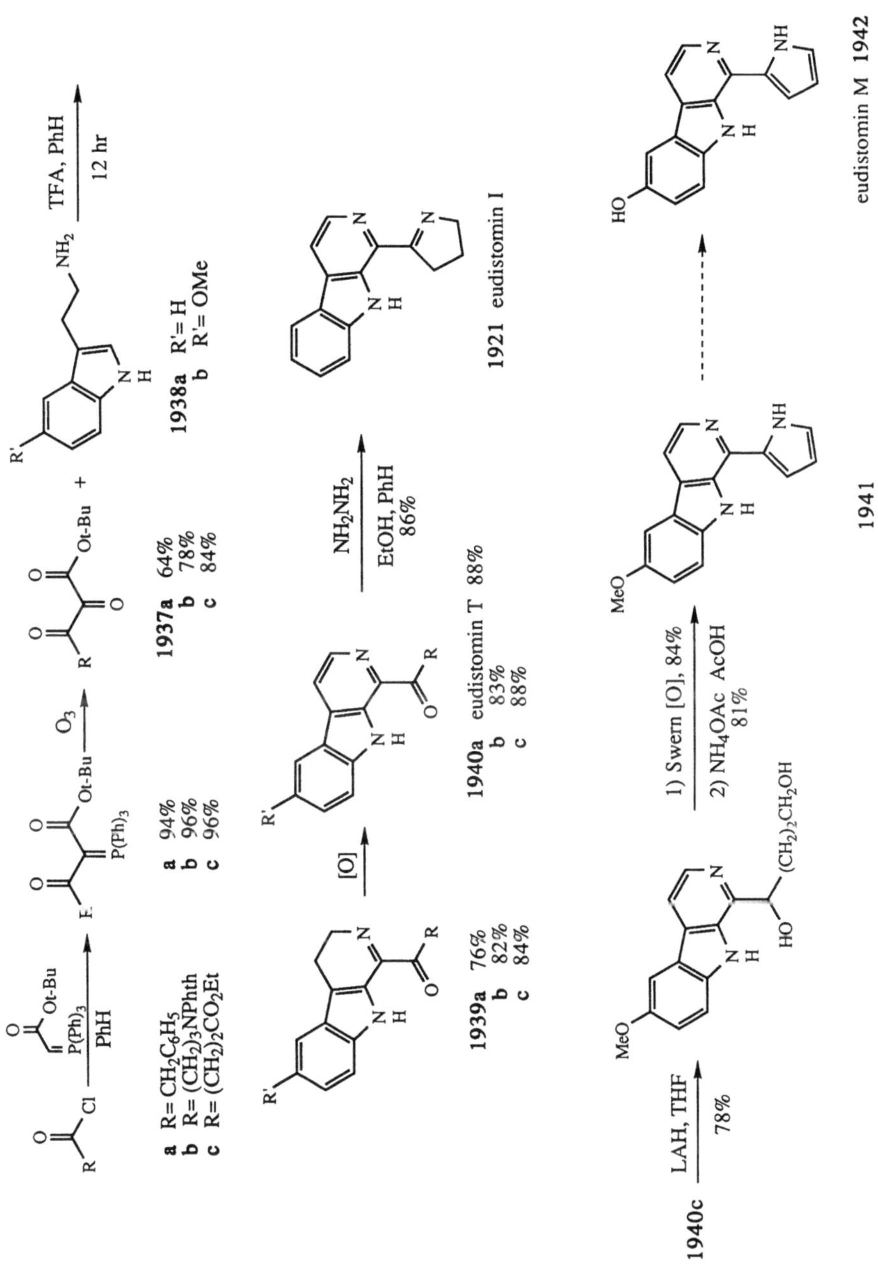

Scheme 343. Wasserman Synthesis of Eudistomin T, I and M

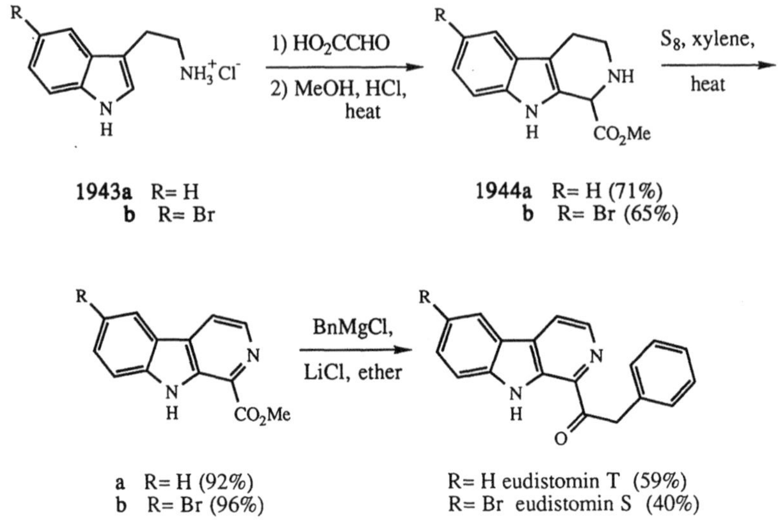

Scheme 344. Still Synthesis of Eudistomins S and T

of the normal Pictet-Spengler product. Acylation, bromination and deacylation of **1950** provides the brominated heterocycle **1951**. Rearrangement of **1951** with trifluoroacetic acid gives the Pictet-Spengler product **1952**, which is converted to (−)-eudistomin L in poor yield. The natural product is obtained in 0.69% yield over 6 steps.

Still has prepared the *N*(10)-acetyl derivative of eudistomin L, but unfortunately was unable to remove the acetyl protecting group to obtain the natural product (Scheme 346). Bromoindole **1953** is converted to hydroxylamine **1955** via nitroolefin **1954** in four steps. Coupling of **1955** with aldehyde **1956** (prepared as shown in the upper half of the scheme) yields nitrone **1957**. Treatment of **1957** with trifluoroacetic acid gives the Pictet-Spengler product **1958**. Cyclization of **1958** occurs via a sila-Pummerer reaction of the diastereomeric mixture of sulfoxides obtained upon treatment of **1958** with MCPBA. *N*(10)-acetyl-eudistomin L is obtained in eight steps and 1.9% overall yield.

5.2.3 Manzamine C

Manzamine C is a novel β-carboline alkaloid bearing an azacycloundecene ring. It is the simplest member of the manzamine family and has been isolated from an Okinawan sponge [525]. Manzamine C exhibits potent antitumor activity. Hino's synthesis [526] of **1963** is shown in Scheme 347. The azacycloundecenyl portion (**1961**) is prepared via Lindlar reduction of alkyne **1960** and cyclization of the ditosylate to give **1961** in 28% overall yield. Acylation of tryptamine followed by Bischler-Napieralski cyclization leads to substituted β-carboline

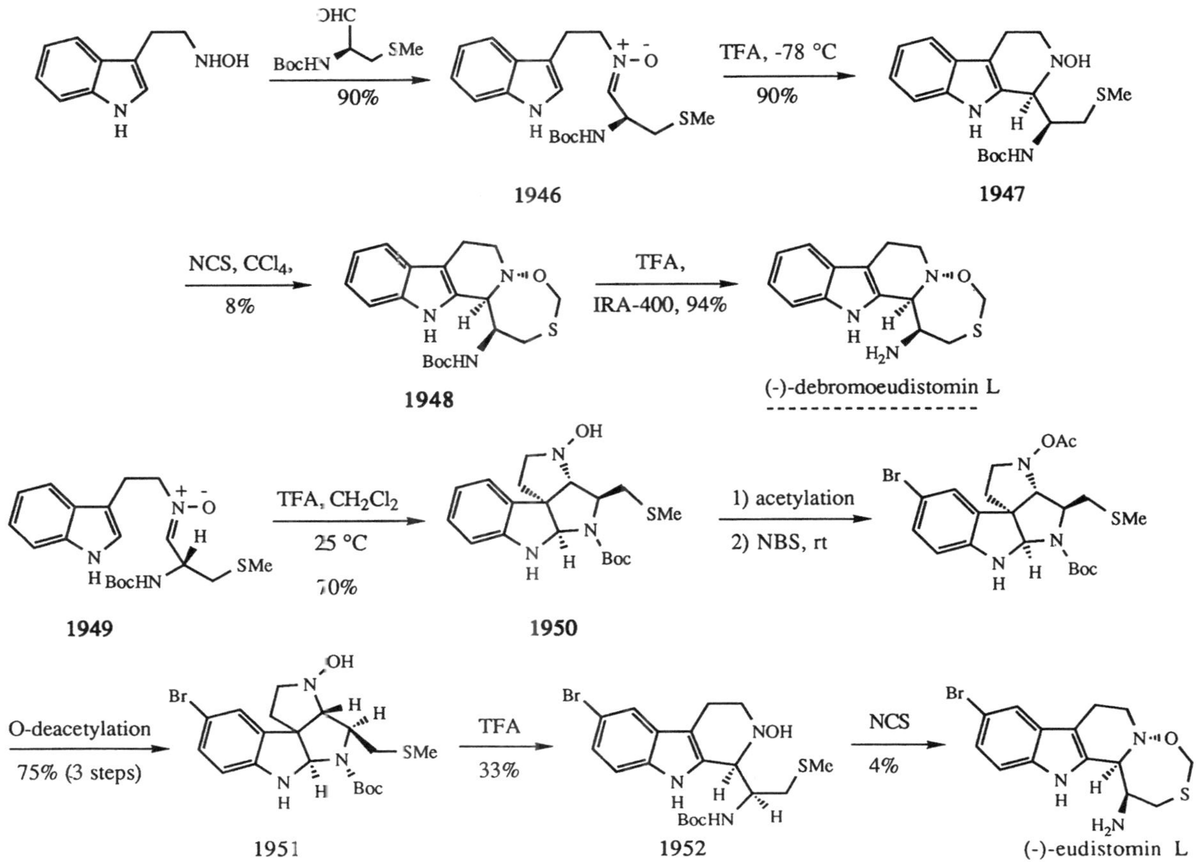

Scheme 345. Nakagawa Synthesis of (−)-Eudistomin L and (−)-Debromoeudistomin L

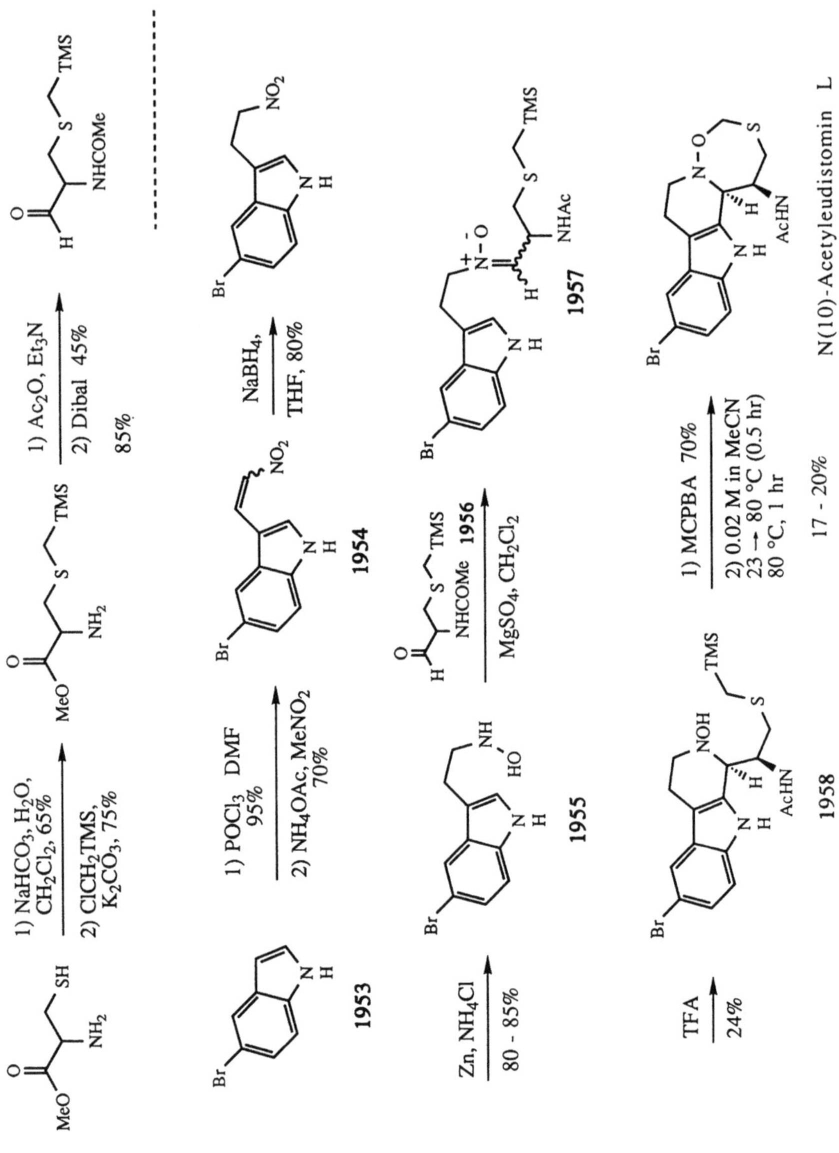

Scheme 346. Still Synthesis of *N*(10)-Acetyleudistomin L

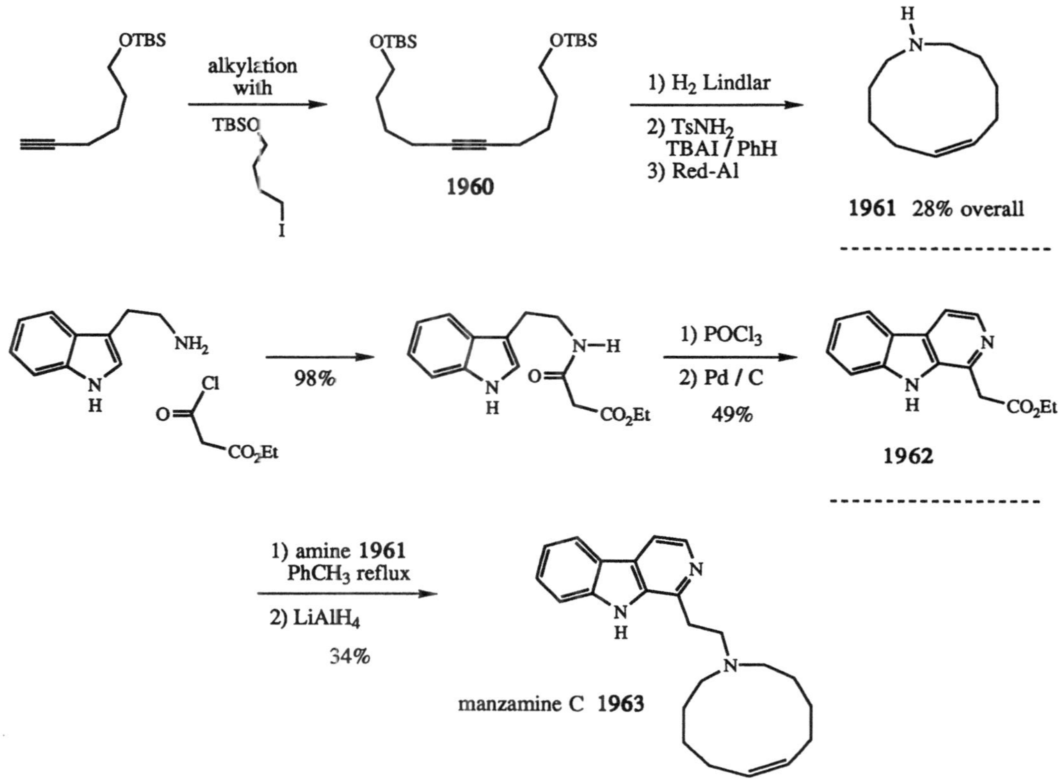

Scheme 347. Hino Synthesis of Manzamine C

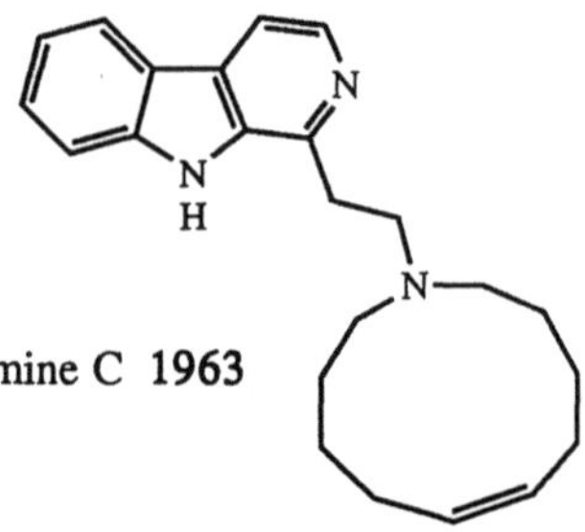

Manzamine C **1963**

1962. Coupling of **1962** with amine **1961** provides manzamine C in six steps and 16% overall yield.

5.3 Pyridines

5.3.1 Navenone A

Huang's synthesis [527] of navenone A is shown in Scheme 348. Due to a variety of complications involved in the use of normal phosphonium salts in the preparation of conjugated *E*-polyenes, triphenylarsonium salts were used in the preparation of navenone A. Reaction of triphenylarsonium salt **1965** with 3-pyridinecarboxaldehyde in the presence of potassium carbonate and a trace of

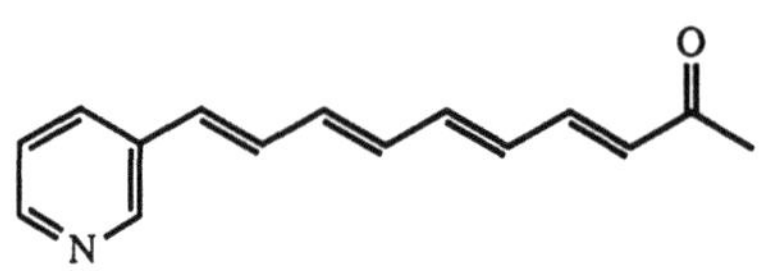

Navenone A **1967**

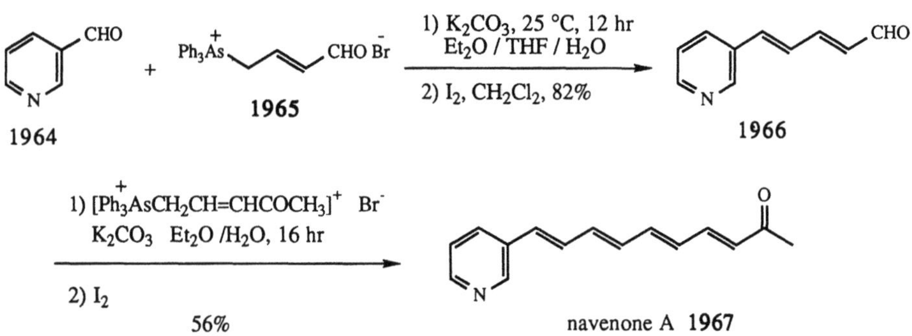

Scheme 348. Huang Synthesis of Navenone A

water gives a mixture of isomers, treatment of which with iodine in sunlight gives the (*E*,*E*)-isomer **1966** in 82% yield. Repeating this sequence with the triphenylarsonium bromide prepared from 5-bromo-3-penten-2-one gives navenone (**1967**) in four steps and 46% overall yield.

5.3.2 Pulo'upone

Pulo'upone (**1976**) is a minor metabolite of the Hawaiian opisthobranch mollusk *Philinopsis speciosa.* It is an uncommon pyridine derivative substituted at C2 by a bicyclic C_{16}-polyketide. The natural product most closely related to **1976** is navenone A (**1967**) (Scheme 348). As only small quantities of Pulo'upone could be isolated, biological testing was not possible [528]. Pulo'upone has been synthesized by three different groups. In all three cases, the general strategy utilized involves formation of the hydrindene nucleus followed by addition of the pyridine moiety.

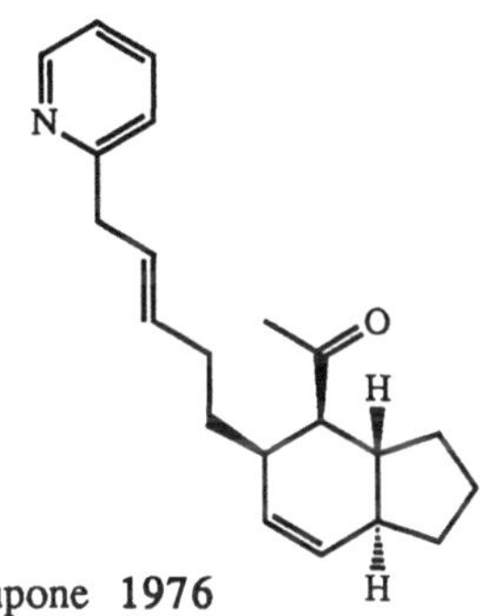

Pulo'upone 1976

Burke has prepared Pulo'upone beginning with aldehyde **1969** (Scheme 349) [529]. Homologation of **1970** gives diene **1971** in four steps and 56% yield. Reduction of diene **1971** gives **1972a** and **1972b** in 69% and 29% yields respectively. Separation and conversion of each to lactone **1973** proceeds with a combined yield of 88%. In the key transformation, formation of the trimethylsilyl ketene acetal of **1973** followed by heating gives hydrindene **1974** via a retro hetero Diels-Alder—intramolecular Diels-Alder pathway. Hydrindene **1974** is obtained as a 4.2:1 mixture of diastereomers, reflecting the *endo* and *exo* cycloaddition modes. The synthesis is completed by transformation of the carboxyl to secondary alcohol **1975** and homologation of the vinyl iodide to give the pendant pyridyl moiety. Pulo'upone (**1976**) is obtained in 14 steps and 14% overall yield.

Oppolzer's approach [530] to (−)-Pulo'upone involves a bornane-10,2-sultam—directed, Lewis acid-accelerated, intramolecular Diels-Alder reaction (**1981** → **1982**) as the key transformation (Scheme 350). Homologation of aldehyde **1977** provides acetate **1978** with the needed (*E*,*E*) geometry. Further

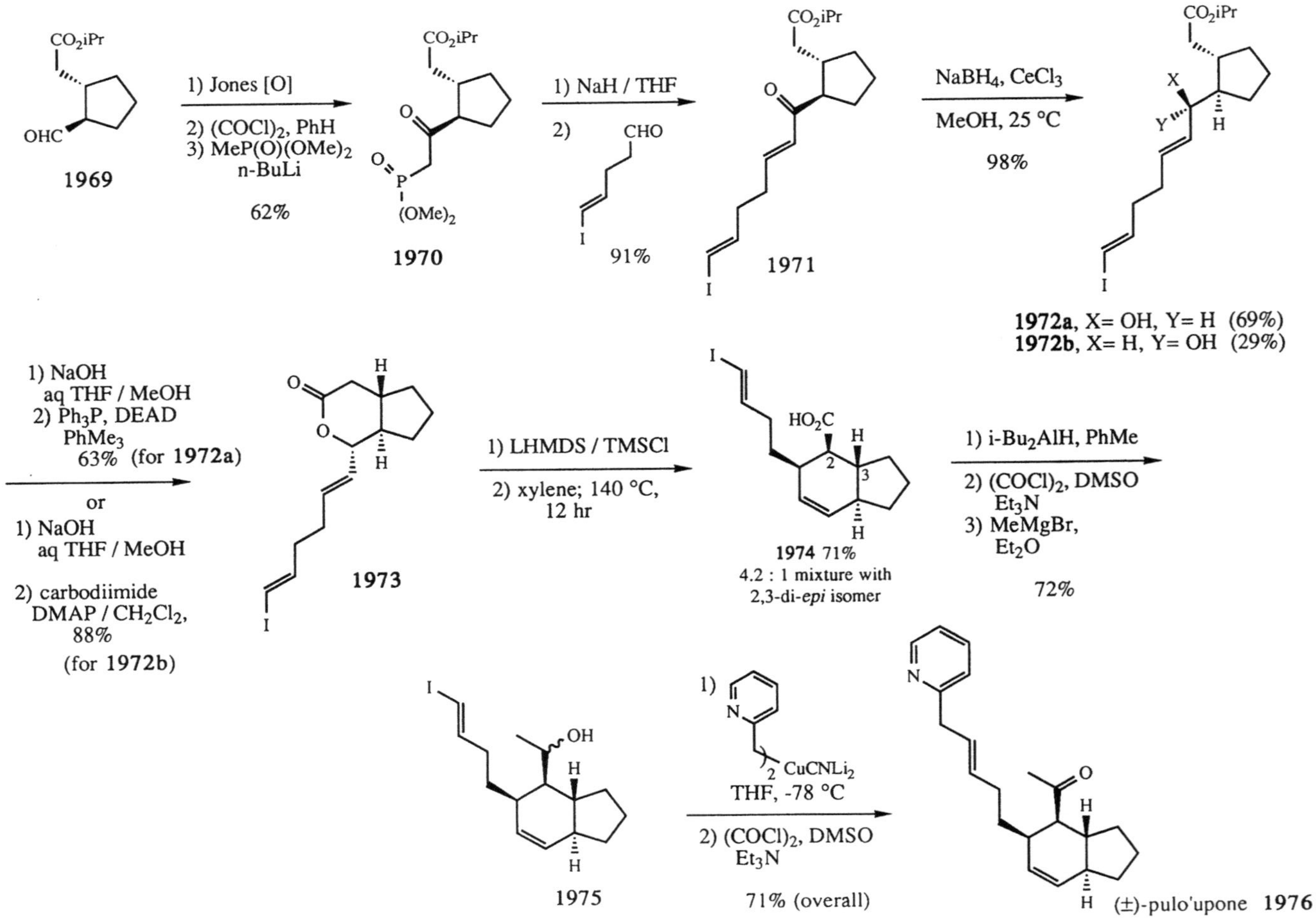

Scheme 349. Burke Synthesis of Racemic Pulo'upone (1976)

homologation, by cuprate-assisted displacement of the acetate with Grignard reagent **1979** and addition of chiral acylphosphonate **1980** provides sultam **1981** and gives the required (*E*)-geometry about the newly formed double bond. Addition of the mild Lewis acid Me_2AlCl affords the *endo*-cycloaddition product in 71% yield and 93% d.e. The diastereomeric excess is raised to 100% by recrystallization. Desilylation and intramolecular displacement of the sultam provides lactone **1983** which is ring-opened and converted to nitrile **1984**. Reduction of the nitrile and Wittig olefination leads to acetate **1985** which in turn provides the natural product **1976** upon displacement with lithium di-(2-pyridyl)cuprate. Pulo'upone is obtained in 2.6% yield over 20 steps.

Takano [531] utilizes an approach to (−)-pulo'upone (Scheme 351) that is very close to that of Oppolzer (Scheme 350). The key step is the preparation of the *trans*-hydrindene nucleus from chiral trienimide **1991** by Evans's asymmetric Diels-Alder reaction [532]. Treatment of the aldehyde derived from **1987** with lithio triethyl 4-phosphonocrotonate gives diene ester **1988** with (*E,E*)-geometry. Homologation of **1988** by cuprate displacement and addition of methyl diisopropylphosphonoacetate to the deprotected aldehyde from **1989** gives triene acid **1990**. Addition of 3-lithio-(*S*)-(+)-oxazolidinone, available from (*S*)-phenylalanine, to the acid chloride of **1990** gives chiral triene **1991**. Treatment of **1991** with 1.3 equivalents of Me_2AlCl gives the *trans*-hydrindene **1992** in 57% yield. Removal of the chiral auxiliary by the normal methods was unsuccessful due to steric crowding. However, oxidative removal of the *p*-anisyl group with ceric ammonium nitrate and intramolecular displacement of the chiral auxiliary affords lactone **1993**, which is transformed into aldehyde **1994**. The synthesis is completed with a lack of diastereoselectivity in the final Wittig reaction to give (−)pulo'upone in 11% yield along with 34% of the undesired *cis*-isomer. Pulo'upone (**1976**) is obtained in 0.16% yield over 18 steps.

5.3.3 Ascididemin

The pentacyclic alkaloid ascididemin (**1999**) is isolated in low yield from a *Didemnum* sp. of an Okinawan tunicate. It is of interest due to its antileukemic properties [533]. The only synthesis of **1999** to-date is that of Bracher [534], outlined in Scheme 352. Dichromate oxidation of quinoline **1995** gives quinone **1996** which is converted to tetracyclic quinone **1997** via regioselective oxidative amination with *o*-aminoacetophenone. Reaction of **1997** with dimethylform-

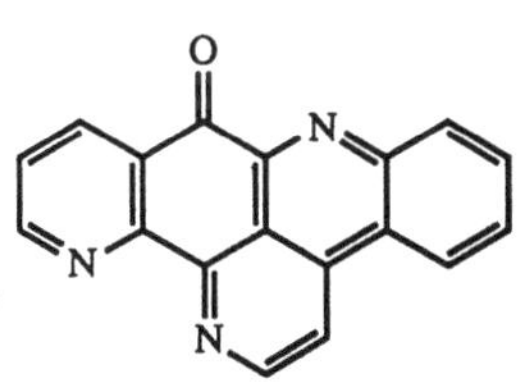

Ascididemin **1999**

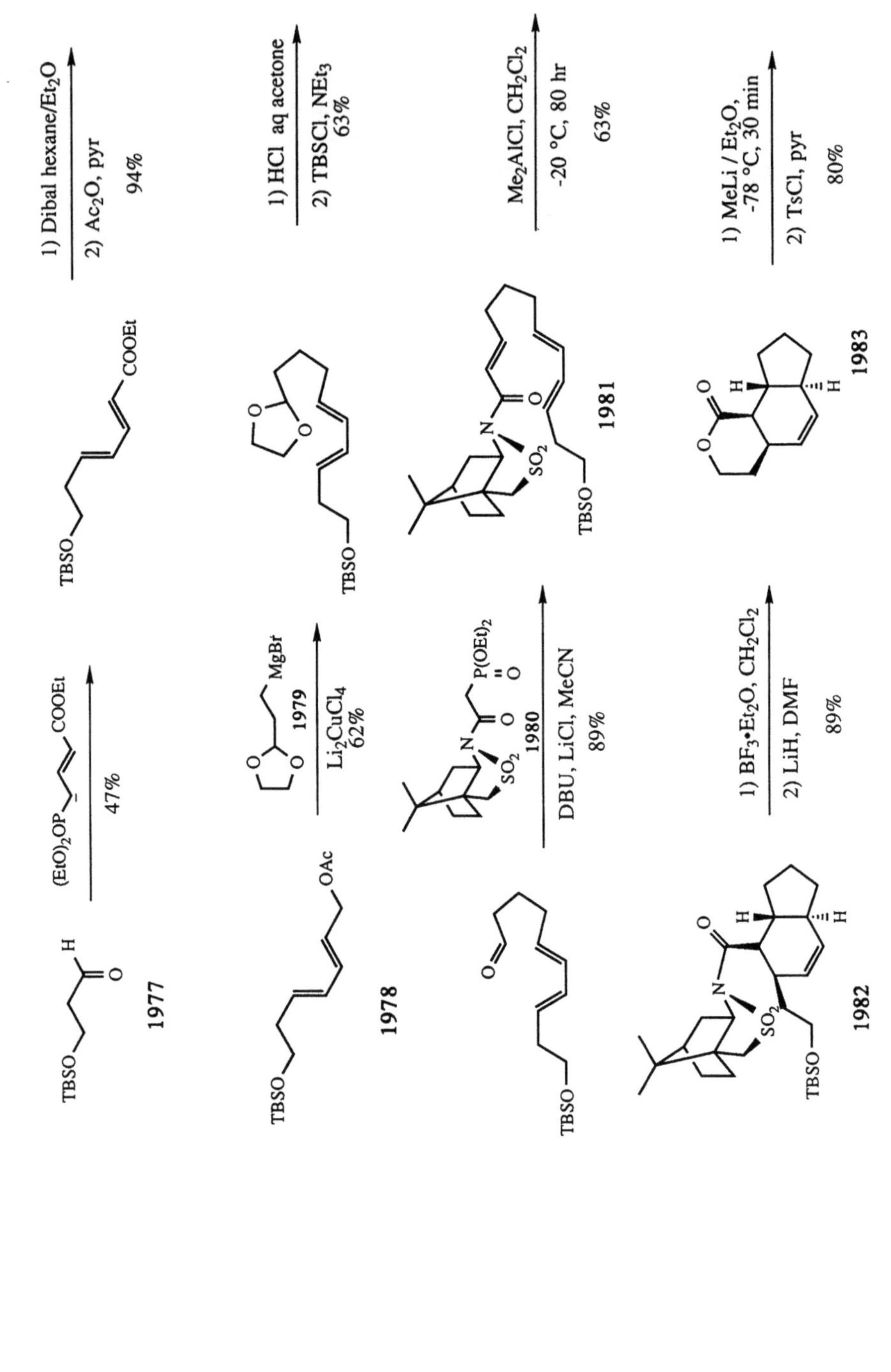

TBSO
1977
(EtO)2OP
COOEt
47%
1) Dibal hexane/Et2O
2) Ac2O, pyr
94%
TBSO
OAc
1978
MgBr
1979
Li2CuCl4
62%
1) HCl aq acetone
2) TBSCl, NEt3
63%
P(OEt)2
SO2
1980
DBU, LiCl, MeCN
89%
Me2AlCl, CH2Cl2
-20 °C, 80 hr
63%
1981
1) BF3•Et2O, CH2Cl2
2) LiH, DMF
89%
1982
1) MeLi / Et2O, -78 °C, 30 min
2) TsCl, pyr
80%
1983

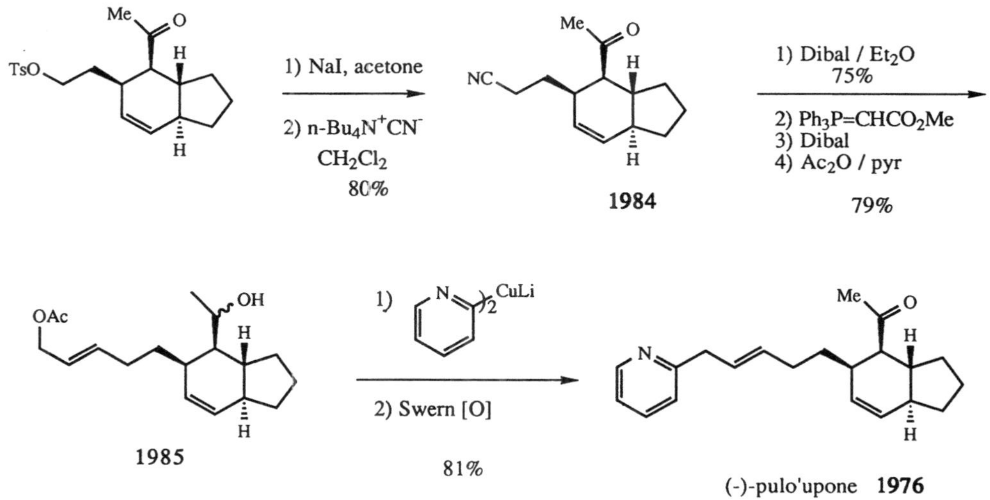

Scheme 350. Oppolzer Synthesis of (−)-Pulo'upone

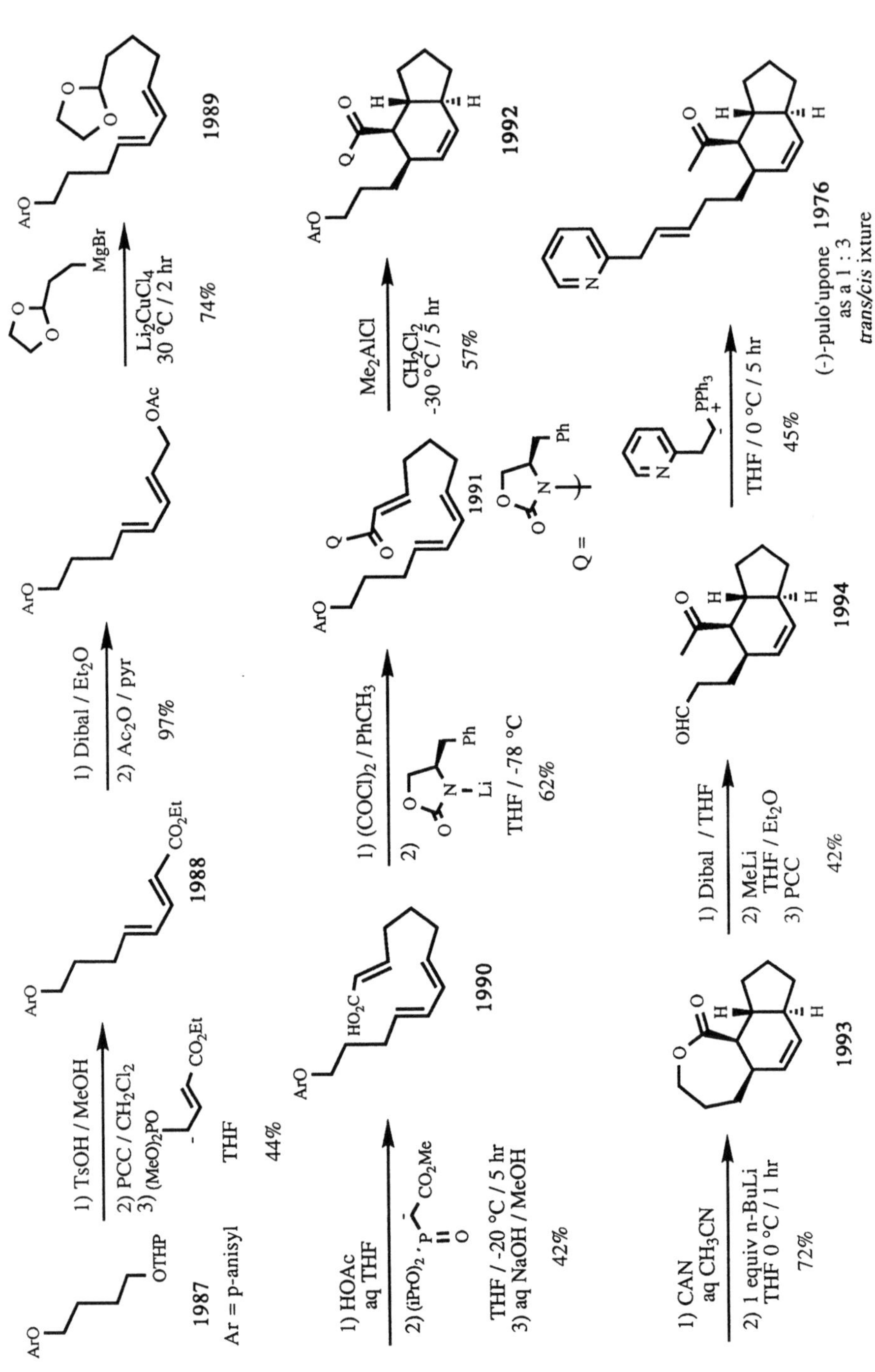

Scheme 351. Takano Synthesis of (−)-Pulo'upone (1976)

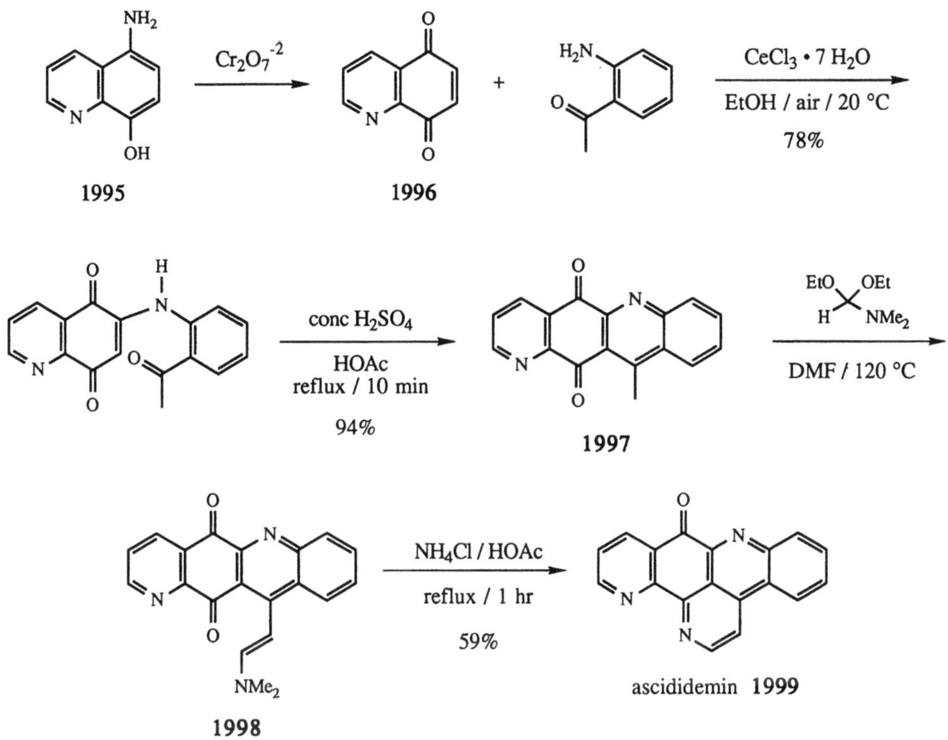

Scheme 352. Bracher Synthesis of Ascididemin

amide diethyl acetal yields enamine **1998** which is cyclized directly to ascididemin (**1999**) by treatment with ammonium chloride in refluxing acetic acid. Ascididemin is obtained in four steps, from quinone **1996** in 43% overall yield.

5.3.4 Aaptamine and Demethyloxyaaptamine

The aqueous ethanol extracts of the Okinawan sponge *Aaptos aaptos* yield an unusual heterocycle given the name aaptamine (**2006**). This bright yellow compound is the first example of the 1*H*-benzo[*de*]-1,6-naphthyridine ring system. Also isolated from the ethanolic extracts of *A. aaptos*, is the related

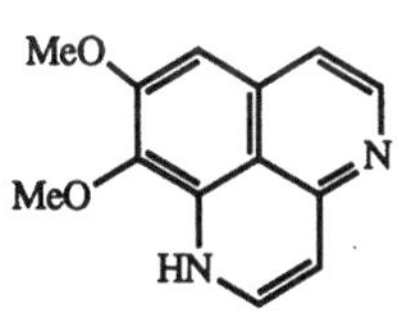

Aaptamine **2006**

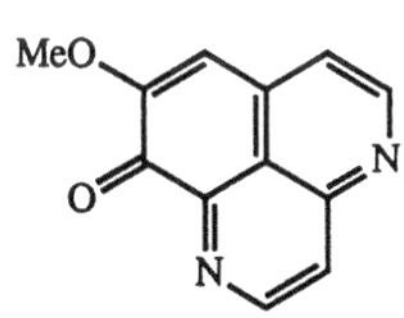

Demethyloxyaaptamine **2009**

demethyloxyaaptamine (**2009**). Aaptamine exhibits α-adrenoceptor blocking activity, while demethyloxyaaptamine possesses antitumor and antimicrobial activity [535].

The first reported synthesis of aaptamine (**2006**) is that of Cava [536] in 1985 (Scheme 353). Selective nitration of dihydroisoquinoline **2001** provides compound **2002**. Heating **2002** with the monoethyl ester of malonic acid followed by re-methylation of the free hydroxyl gives ester **2003**, which can be converted to lactam **2004** by hydrogenation under acidic conditions. Removal of the amide carbonyl and dehydrogenation gives aaptamine in 38% yield along with 45% of imine **2005**. Isolation of aaptamine as its hydrochloride gives an overall yield of 12% in eight steps. The same synthetic strategy is also used for the preparation of demethyloxyaaptamine (**2009**) (Scheme 354). Instead of methylating the free hydroxyl group before cyclization, nitro compound **2007** is reduced and cyclized under acidic conditions and then doubly protected to give lactam **2008**. Removal of the amide carbonyl and dehydrogenation with concomitant debenzylation gives demethyloxyaaptamine (**2009**) in a total of nine steps and 5.2% overall yield.

Kelly's synthesis [537] of aaptamine (Scheme 355) begins with the ortholithiation of veratrole and treatment with trimethylsilylmethyl azide to afford amine **2011**. Conjugate addition of **2011** to methyl propiolate followed by thermal cyclization gives quinolone **2012**. Chlorination of **2012** and reaction with aminoacetaldehyde dimethylacetal affords compound **2013**. Although the original synthetic strategy was based on inducing **2013** to undergo an intramolecular Pomeranz-Fritsch type reaction, this proved not to be viable. Treatment of **2013** with a mixture of chlorosulfonic acid and antimony pentafluoride produces a 1:1 mixture of aaptamine and pyrrole derivative **2014**. Aaptamine is obtained in 8.5% yield over five steps.

The synthesis of aaptamine by Yamanaka [538] is iterative in nature. It begins with the conversion of aldehyde **2015** to nitrile **2016** (Scheme 356). Palladium-catalyzed coupling of trimethylsilylacetylene with **2016** gives acetylene **2017** which is converted to chloroisoquinoline **2018** in four steps. A second palladium-catalyzed coupling with trimethylsilylacetylene followed by cyclization gives aaptamine as its hydrochloride in four steps. Aaptamine is obtained in 6.6% yield over 12 steps.

Tollari's synthesis [539] of aaptamine involves intramolecular cyclization of a 1-vinylnitrene isoquinoline to obtain **2006** very quickly. Condensation of nitromethane with aldehyde **2020** and elimination gives vinyl nitro compound **2021** in 85% yield (Scheme 357). Treatment of **2021** with refluxing triethylphosphite gives aaptamine (**2006**) in 49% yield over 2 steps. The intermediate nitrene is presumably obtained as a mixture of *E*/*Z* isomers that undergo thermal isomerization before cyclization. Conjugation of the nitrene reduces the amount of rearrangement products.

Most recently, Raphael [540] has prepared aaptamine (**2006**) as shown in Scheme 358. Conversion of nitro aldehyde **2022** to silyl protected cyanohydrin **2023** followed by reduction gives amine **2024**. Condensation of **2024** with

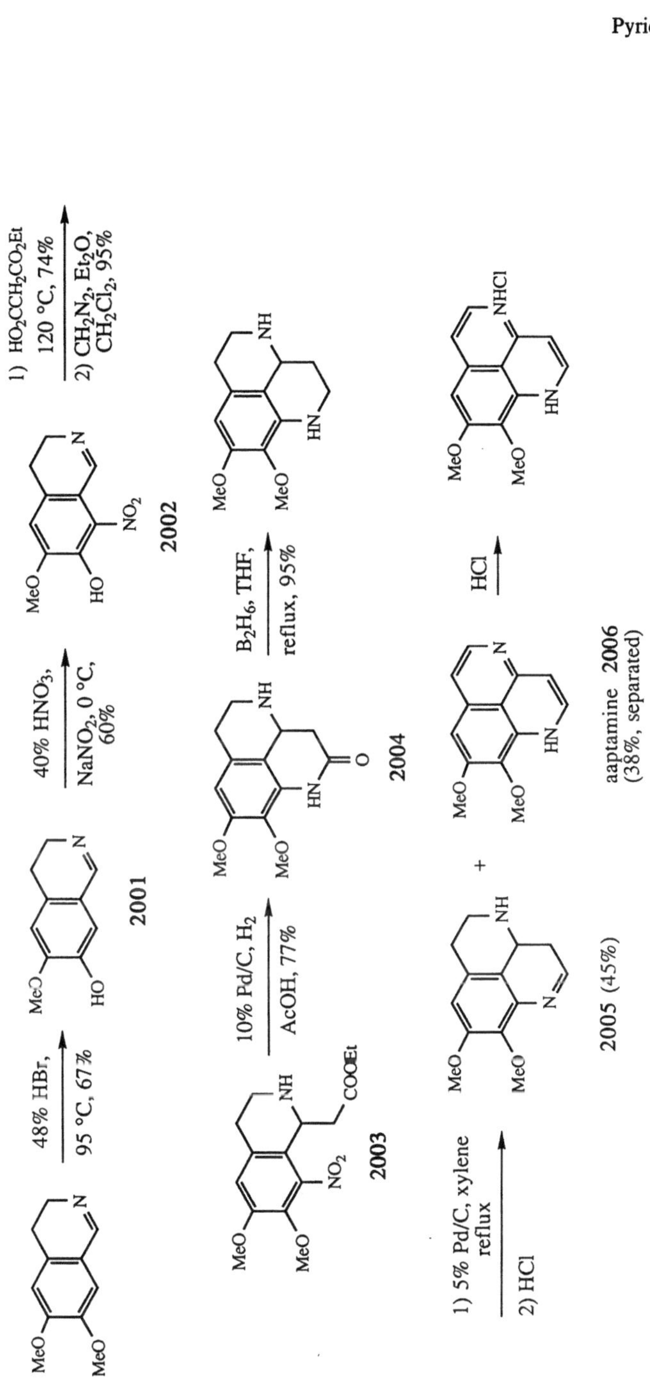

Scheme 353. Cava Synthesis of Aaptamine

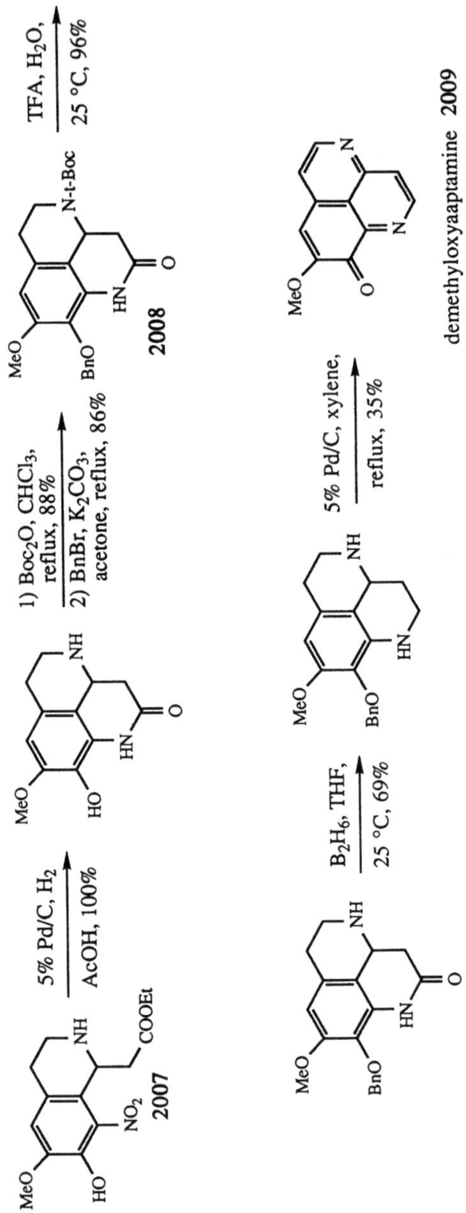

Scheme 354. Cava Synthesis of Demethyloxyaaptamine

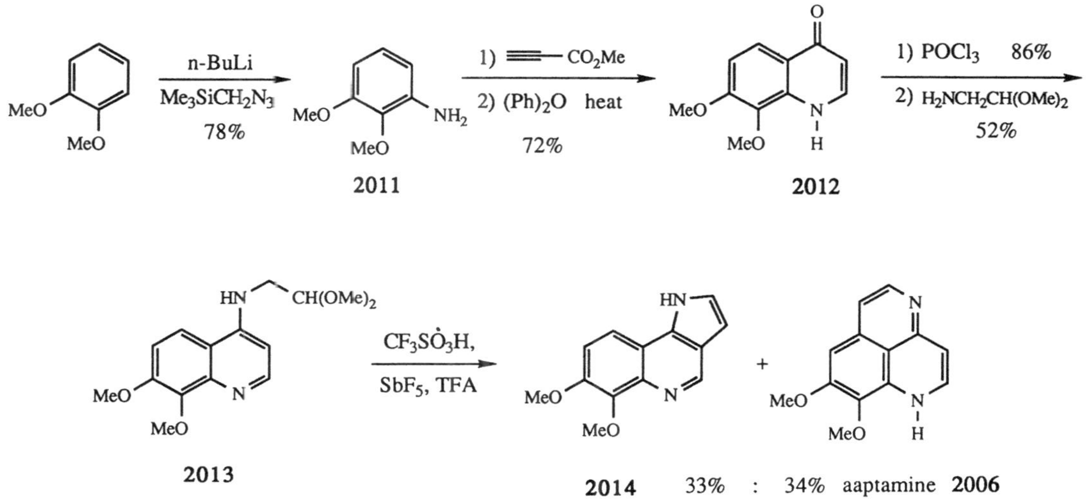

Scheme 355. Kelly Synthesis of Aaptamine

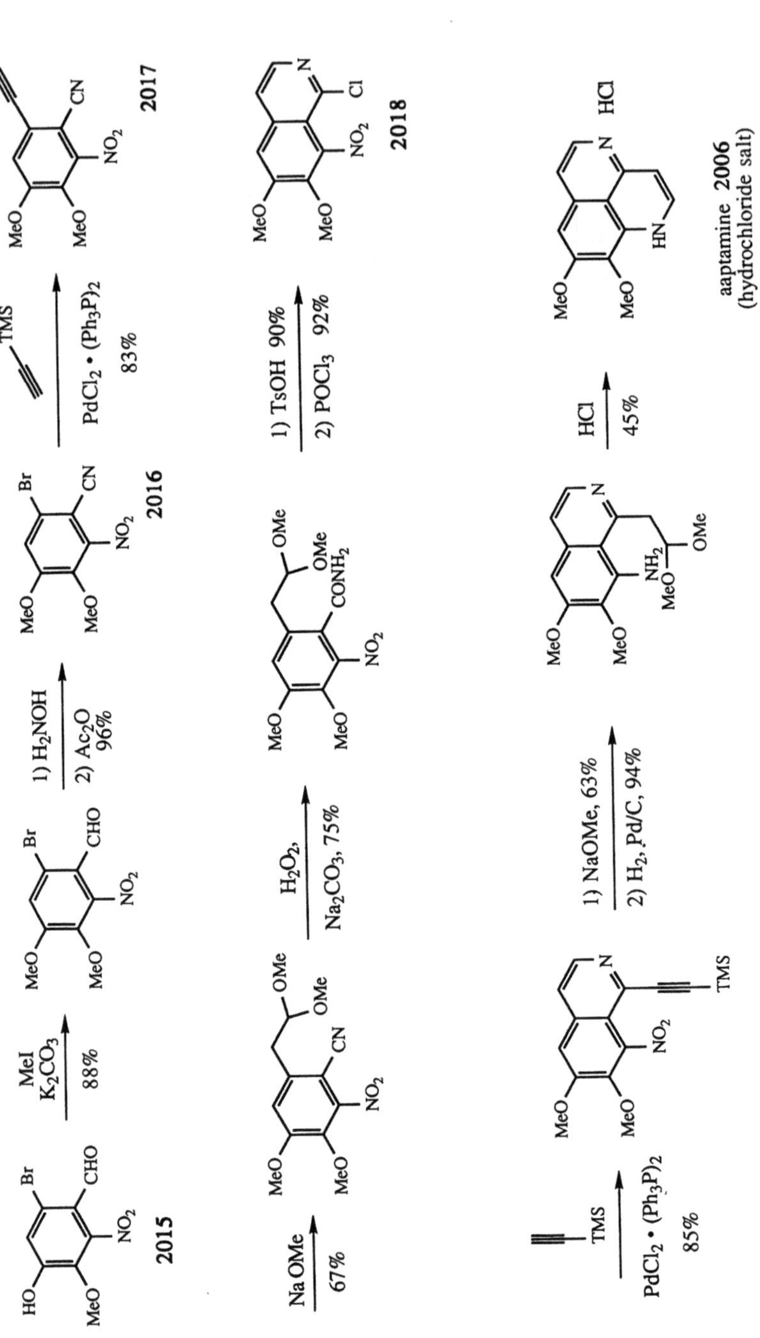

Scheme 356. Yamanaka Synthesis of Aaptamine

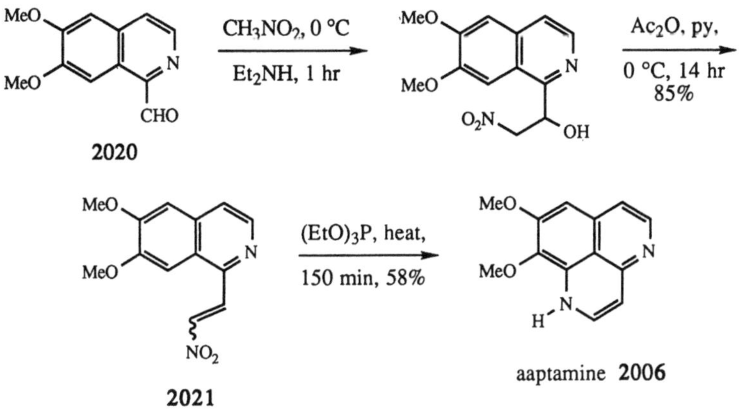

Scheme 357. Tollari Synthesis of Aaptamine

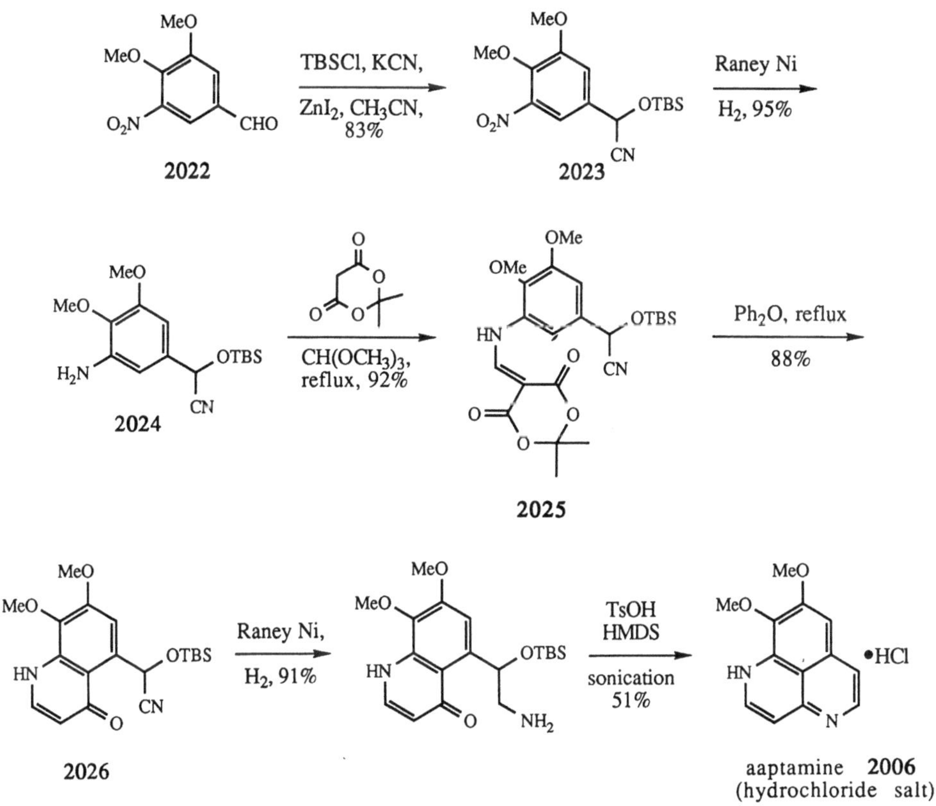

Scheme 358. Raphael Synthesis of Aaptamine

trimethyl *ortho* formate and Meldrum's acid affords enamide **2025** which cyclizes to the quinolone **2026** upon heating. Reduction of the nitrile **2026** and intramolecular condensation gives aaptamine **2006**. Isolation of aaptamine as its hydrochloride provides the natural product in six steps and 30% overall yield.

5.3.5 Amphimedine

In 1983 Schmitz reported the isolation of the pentacyclic alkaloid amphimedine (**2033**) from an *Amphimedon* sp. sponge found near Guam island [541]. Amphimedine exhibits general cytotoxicity and is distantly related to the mimosamycin-type family of antibiotics [542].

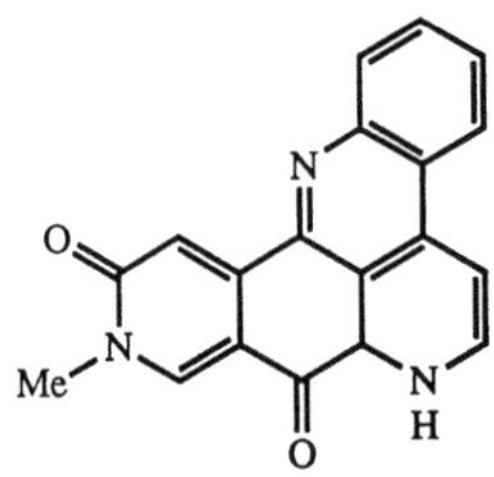

Amphimedine **2033**

Kubo [543] utilizes a non-regioselective approach to prepare amphimedine, as shown in Scheme 359. Condensation of 2,5-dimethoxyaniline (**2028**) with β-ketoester **2029** affords amide **2030**. Cyclization of **2030** followed by chlorination of the resulting 2-quinolone gives 2-chloroquinoline **2031**. Oxidative demethylation of **2031** with ceric ammonium nitrate and a non-selective Diels-Alder reaction with 2-aza-1,3-bis(*tert*-butyldimethylsilyloxy)-1,3-butadiene leads to 7% of the desired adduct **2032**, along with 8% of the other regioisomer. Hydrogenation of **2032** occurs with cyclization and loss of chloride to give amphimedine (**2033**) in seven steps and 0.25% overall yield.

A somewhat similar approach is followed by Stille [544] in the preparation of **2033** (Scheme 360). Conversion of 4-quinolone **2034** to triflate **2035** followed by palladium-catalyzed cross-coupling with organostannane **2036** provides quinoline **2037**. Exchange of protecting groups to give **2038** and monodemethylation gives **2039**. Bromination of **2039** and oxidative demethylation with ceric ammonium nitrate yields bromoquinone **2040**. The presence of the bromine solves the problem of regioselectivity. Diels-Alder cycloaddition of **2040** with azadiene **2041** gives a 48% yield of quinone **2042** as the only regioisomer, after treatment with pyridinium hydrofluoride. Interestingly, the use of dry acid-free chloroform as solvent gives spiro compound **2043**, a heretofore unprecedented mode of reactivity in these types of cycloadditions. Acid-catalyzed cyclization of

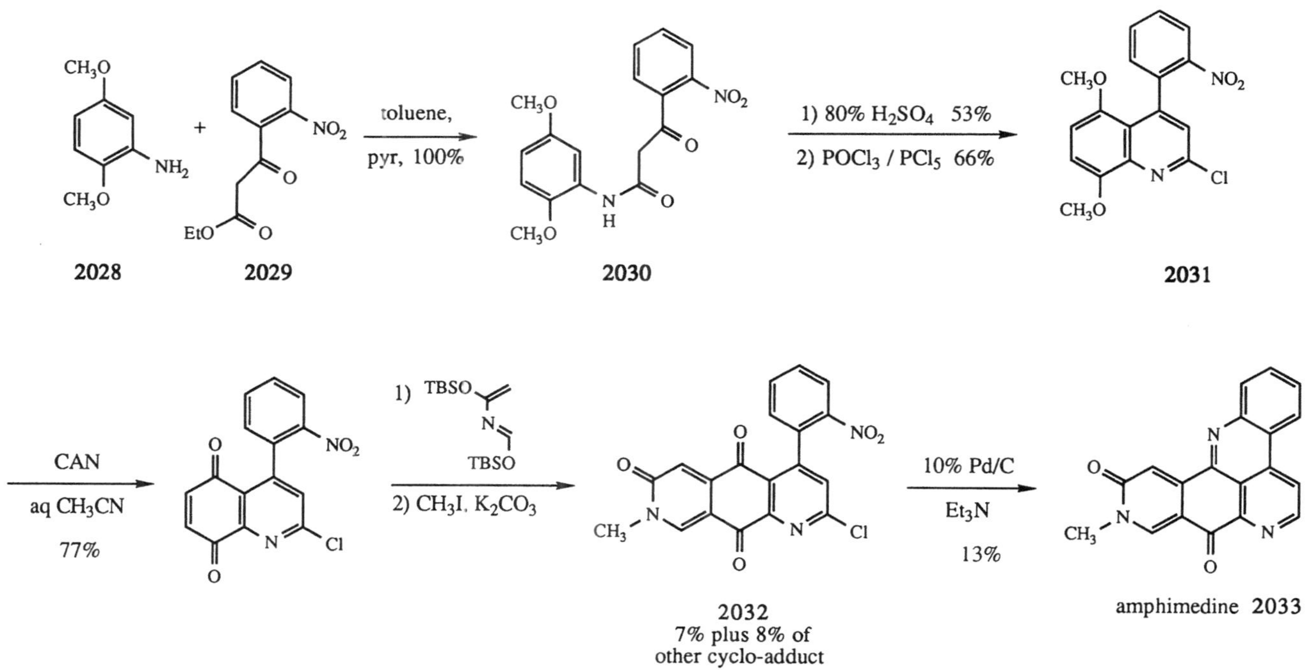

Scheme 359. Kubo Synthesis of Amphimedine

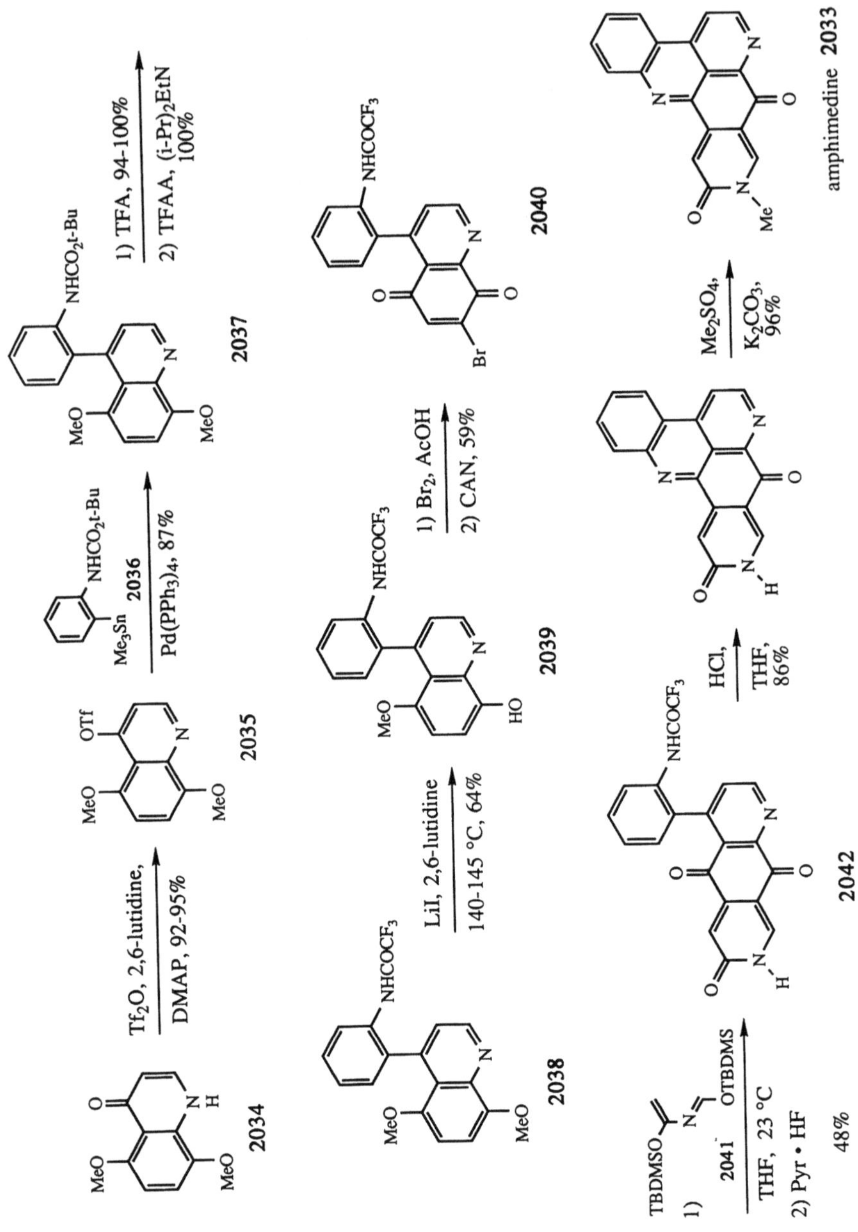

Scheme 360. Stille Synthesis of Amphimedine

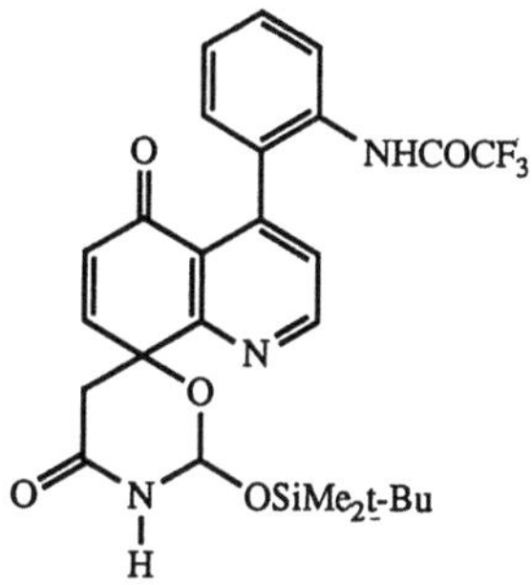

2043

2042 and *N*-methylation completes the synthesis to afford amphimedine (**2033**) in 11 steps and 12% overall yield.

A different approach to amphimedine (**2033**) is followed by Prager [545], as illustrated in Scheme 361. Addition of 4-pyridyllithium to silyl-protected fluorenone **2045** gives fluorenol **2046**. Treatment of **2046** with hydrazoic acid gives substituted quinoline **2047** via migration of the more electron-rich aromatic ring. Chlorination of **2047** yields **2048** which is converted to pyridone **2049** by *N*-methylation and oxidation of the pyridine ring. Nucleophilic displacement of

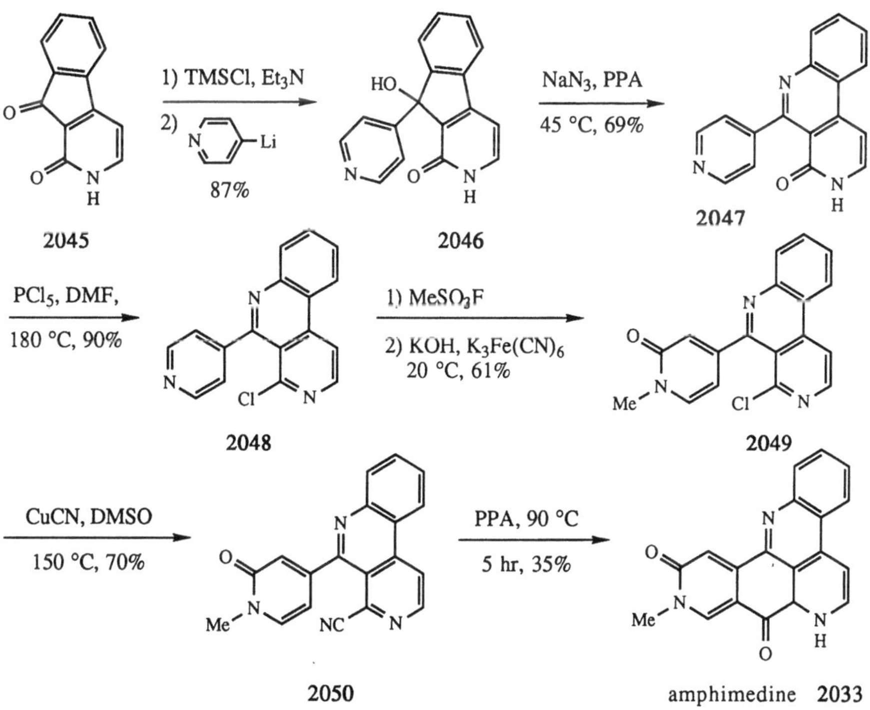

Scheme 361. Prager Synthesis of Amphimedine

chloride to give nitrile **2050** and cylization and hydrolysis with polyphosphoric acid gives amphimedine (**2033**) in eight steps and 8.1% overall yield.

5.4 Guanidine-Containing and Related Metabolites

5.4.1 Tetrodotoxin

The history of tetrodotoxin (**2064**) is as interesting as it is convoluted [546]. A major food toxin, **2064** is found in several species of puffer fish (genus *Spheroides*) as well as a number of diverse organisms. This suggests a microorganismic source of this toxin, which has been identified as *Pseudomonas* [547]. The substance is an extremely powerful neurotoxin and is a useful neuropharmacological tool. On the dark side, several fatalities are recorded each year in countries where the puffer is considered a delicacy.

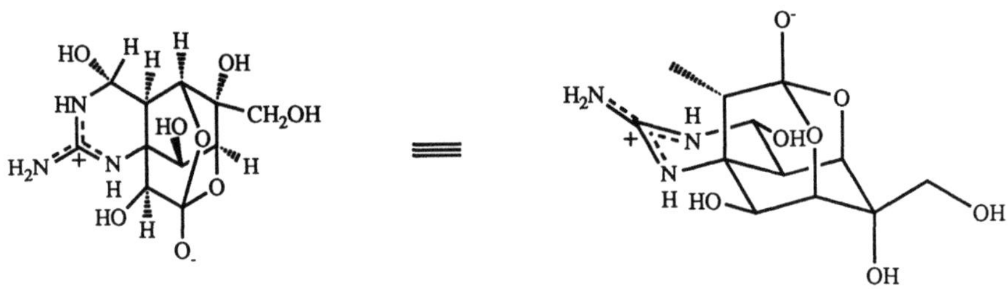

Tetrodotoxin **2064**

The general strategy followed by Kishi in the synthesis of tetrodotoxin is outlined in Scheme 362. Lewis acid-catalyzed Diels-Alder reaction of quinone **2051** and butadiene leads to bicyclic oxime **2052** [548]. Beckmann rearrangement of the oxime, regio- and stereospecific reduction of the less hindered carbonyl, and epoxidation leads to tricyclic decalone **2053**. Further transformations provide triacetate **2057**, containing all six stereocenters of tetrodotoxin with the correct relative stereochemistry. Triacetate **2057** is then converted in four steps to the acetylated tetrodoamine equivalent **2060** [549, 550]. The synthesis is completed by addition of the guanidine and hemiortho ester functionality [551] to give tetrodotoxin in a total of approximately 38 steps from commercially available starting materials.

The detailed stepwise conversion of tricyclic decalone **2053** to the acetylated tetrodamine equivalent **2060** is illustrated in Scheme 363. Standard transformations stereospecifically provide acetate **2054** in four steps from **2053**. Allylic oxidation and epoxidation are key steps that lead to the formation of ketone **2055**. Epoxidation of enol ether **2056**, obtained through elimination of the diethyl ketal, provides triacetate **2057**. Baeyer-Villiger oxidation of **2057**

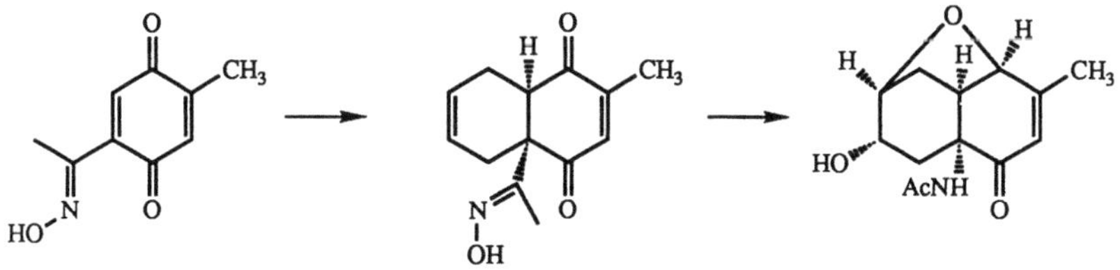

2051 2052 2053

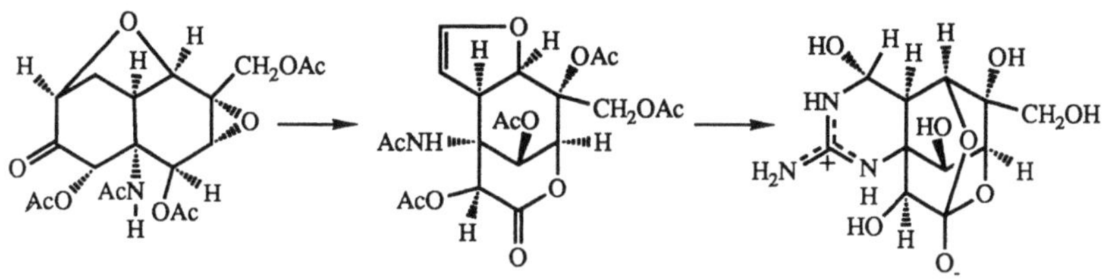

2057 2060 Tetrodotoxin 2064

Scheme 362. Kishi's General Synthetic Strategy for the Preparation of Tetrodotoxin

regiospecifically provides the seven-membered ring lactone **2058** which rearranges via saponification and intramolecular epoxide opening to give six-membered ring lactone **2059**. Acetylation and thermal elimination provides acetylated tetrodamine equivalent **2060** in approximately 18 steps.

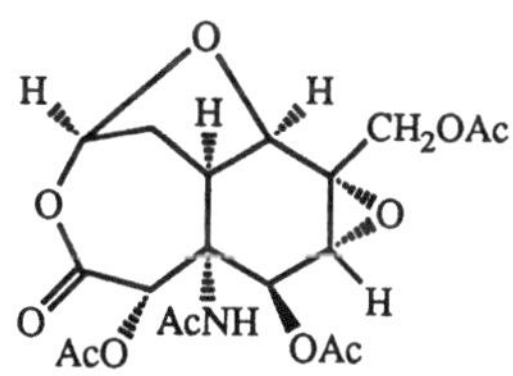

2058

Conversion of **2060** to tetrodotoxin occurs as shown in Scheme 364. Deacylation of the amide functionality in **2060** provides amine **2061** which is converted to diacetylguanidine **2062** in two steps. Treatment of **2062** with ammonia followed by osmylation yields the monoacetylguanidine diol **2063**. Quenching of the excess oxidant with ethylene glycol followed by hydrolysis with aqueous ammonia gives the guanidine functionality and hemiortho ester, thus providing tetrodotoxin in 8 steps from **2060**.

5.4.2 Saxitoxin

Saxitoxin (**2075**), one of the most toxic nonprotein substances known, has been isolated from a variety of sources. Among these are the Alaska butter clam

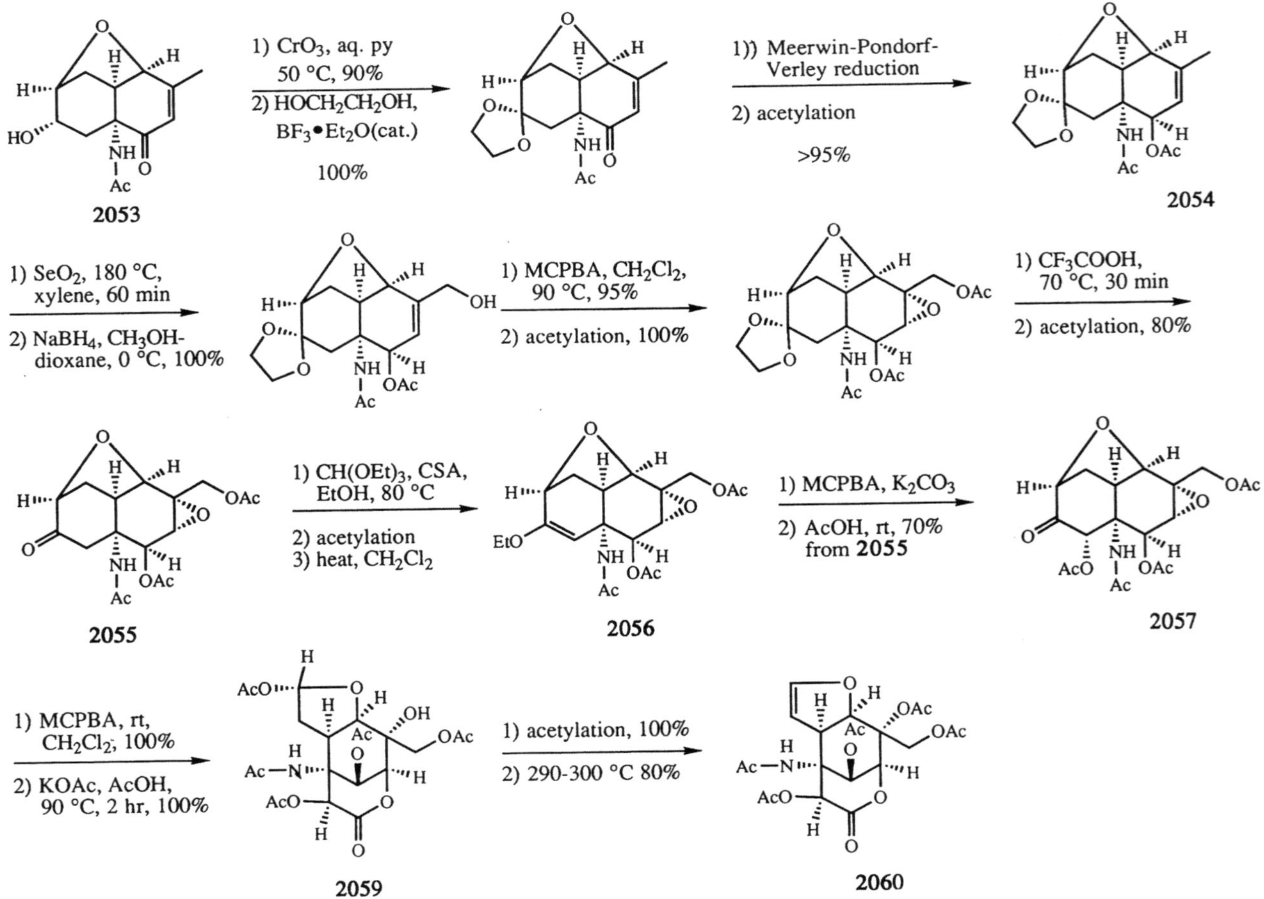

Scheme 363. Kishi Synthesis of an Equivalent of Acetylated Tetrodamine

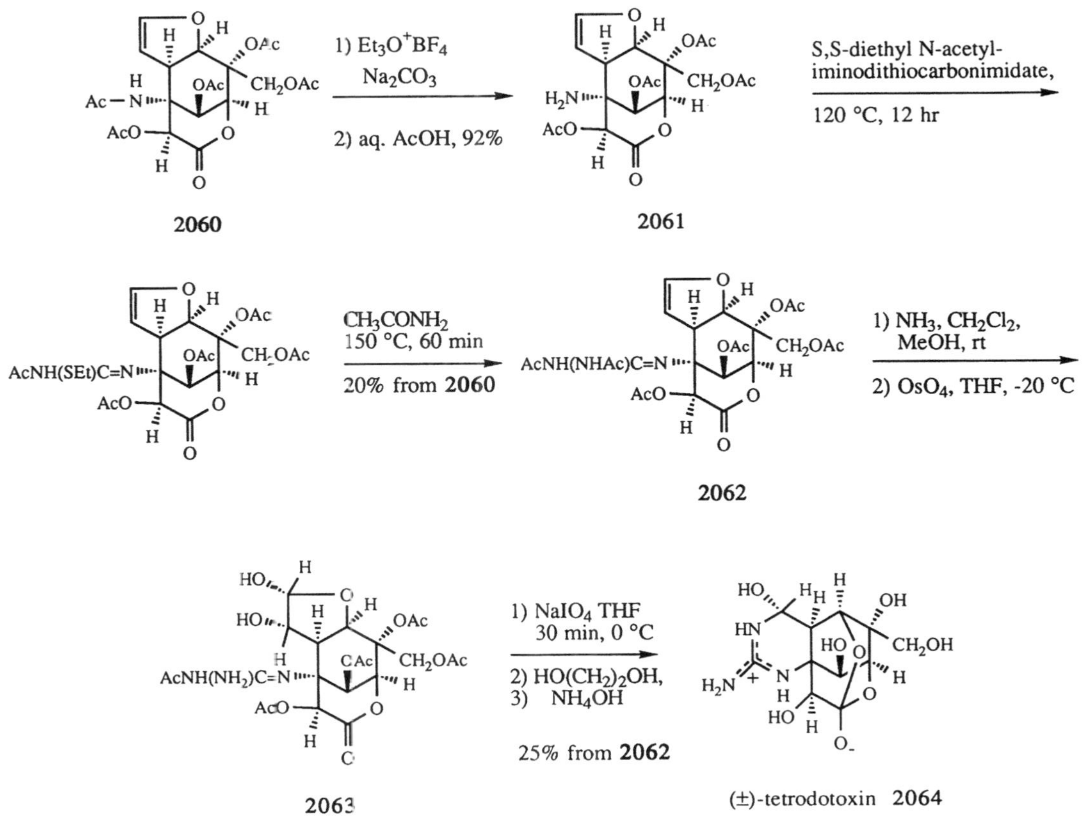

Scheme 364. Kishi Total Synthesis of (±)-Tetrodotoxin.

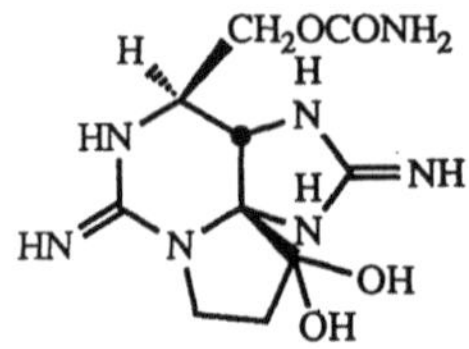

Saxitoxin **2075**

(*Saxidomus giganteus*), toxic mussels (*Mytilus californianus*), and axenic cultures of *Gonyaulax catenella* [552]. It has also been found in aged extracts of scallops collected during a *G. tamarensis* bloom and in soft shell clams (*Mya arenaria*) collected during red tide blooms on the New England coast [553]. Although initially purified [554] in 1957, its molecular structure was not completely elucidated until 1975 when it was subjected to X-ray analysis [555]. The first synthesis of **2075** is that of Kishi [556] (Scheme 365). Conversion of thioamide **2066** to vinylogous carbamate **2067** occurs upon treatment with $CH_3COCHBrCO_2CH_3$ followed by elimination. Condensation of **2067** with benzyloxyacetaldehyde and silicon tetraisothiocyanate yields gives thiourea ester **2068**. Ester **2068** is converted to urea **2069**, the ketal exchanged for the more acid stable thioketal group, and the saxitoxin ring system obtained by warming **2070** with acetic acid to obtain cyclic urea **2071**. At this point, three tactical transformations must be achieved to obtain saxitoxin. The thiourea unit must be converted to a guanidine moiety, the 1,3-dithiane must be removed to give the carbonyl hydrate, and the benzyloxy group must be transformed into the carbamoyl functionality. The diguanidine **2072** is obtained by sequential treatment of **2071** with $Et_3O^+BF_3^-$ and ammonium propionate. Removal of the benzyl and 1,3-dithiane protecting groups gives hydrate **2073**. Addition of chlorosulfonyl isocyanate to **2074** followed by workup with hot water and isolation via ion exchange chromatography yields saxitoxin (**2075**) in 17 steps and about 6.5% overall yield.

Jacobi's synthesis [557] of saxitoxin is outlined in Scheme 366. Hydrazide **2076** (derived from 2-imidazolone) upon treatment with methyl glyoxylate hemiacetal gives pyrazolidine derivative **2078** via kinetically controlled 1,3-dipolar cycloaddition of the intermediate azomethine imine **2077**. Epimerization of **2078** and sodium borohydride reduction of the ester provides intermediate **2079**, containing all of the stereocenters of the product with the correct relative configurations. Removal of the amide carbonyl is achieved by treatment with borane dimethylsulfide complex to give pyrazolidine **2080**. Ring expansion of **2080** is achieved by conversion of **2080** to thiourea **2082**. This occurs via intermediate carbamate **2081** by treatment with sodium in liquid ammonia. Acylation of the free hydroxyl and completion of the synthesis using the same methodology as Kishi (Scheme 365) provides saxitoxin (**2075**) in ten steps and about 13% yield from compound **2076**.

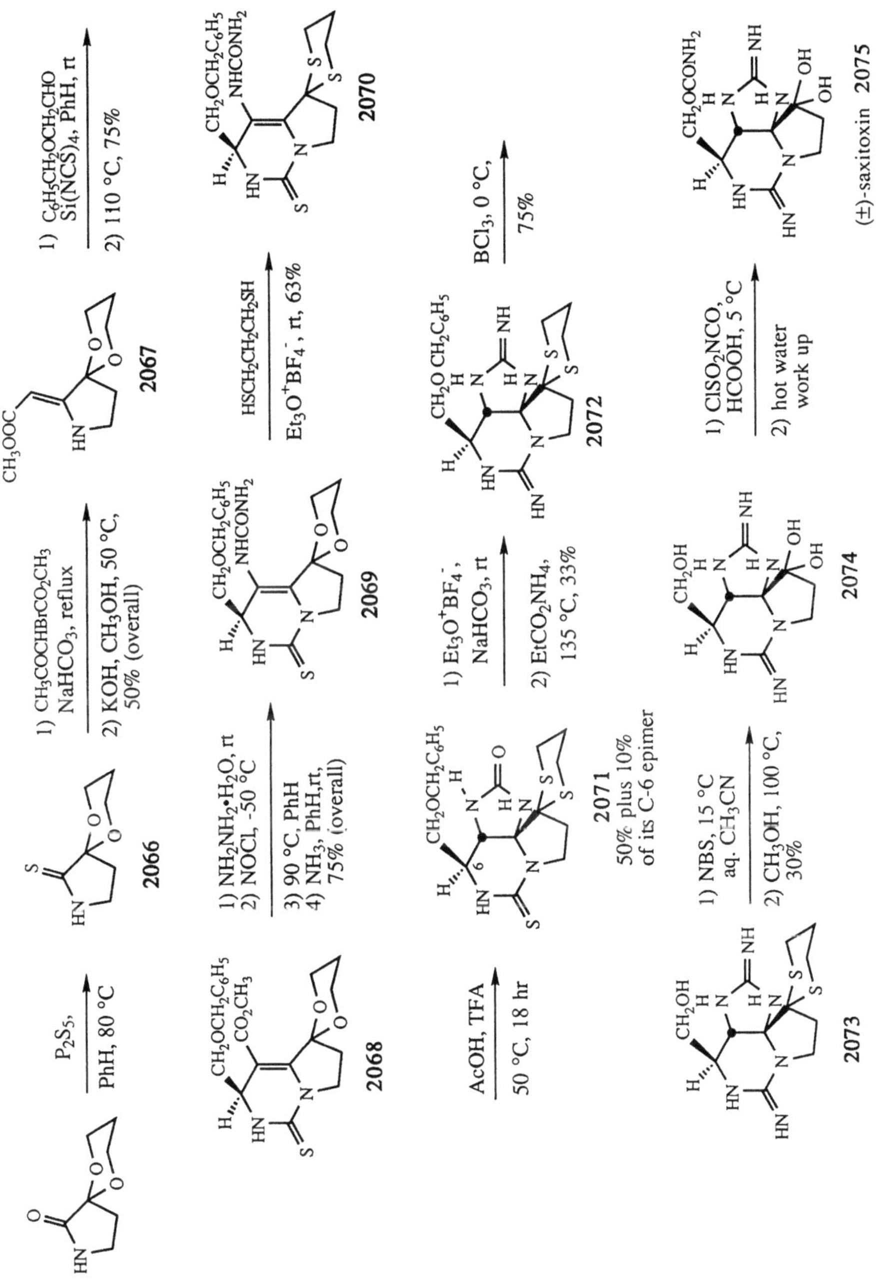

Scheme 365. Kishi Synthesis of (±)-Saxitoxin

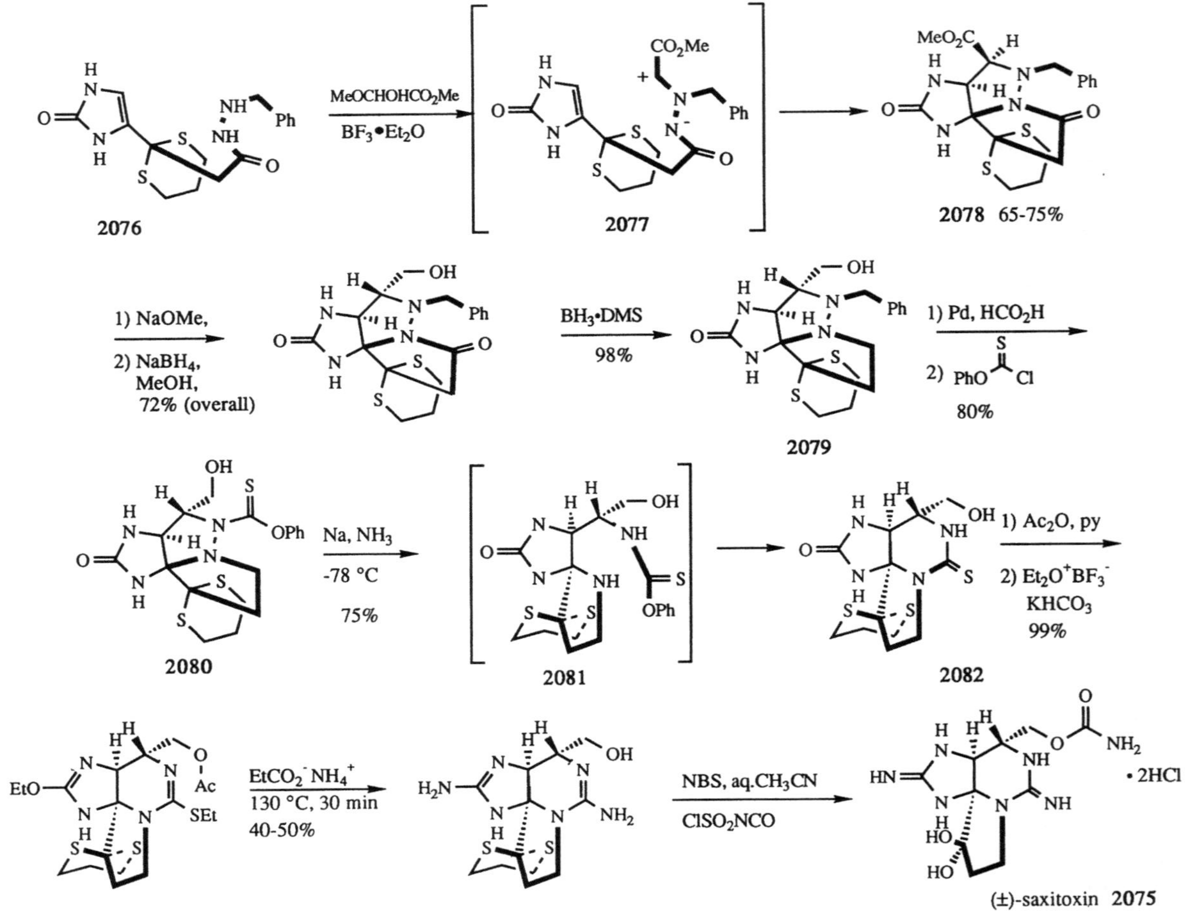

Scheme 366. Jacobi Synthesis of (±)-Saxitoxin

5.4.3 Ptilocaulin

Ptilocaulin **2087** is a guanidine-containing marine natural product isolated [558] from the rope-like orange sponge *Ptilocaulis aff. P. spiculifer*. It exhibits significant antileukemic and antimicrobial properties and, not surprisingly, several syntheses have been reported. Snider described the first synthesis of this substance, proceeding through the bicyclic enone **2086**, a biomimetic strategy which was subsequently adopted by Hassner, and Uyehara, but not by Roush. After a preliminary report [559] by Snider, a full account, including an improved version of the synthesis was given [560]. The first synthesis is shown in Scheme 367. Sequential alkylation and aldol reaction of *t*-butyl acetoacetate produced a keto-aldehyde as a mixture of isomers. Aldol cyclization produced enone **2085**, still as a mixture. Conjugate addition of a terminal butenyl group proceeded with good stereoselectivity with respect to the methyl group producing the required *trans* arrangement. Ozonolysis and aldol cyclization produced the key enone **2086**, still as a 1:1 mixture at the butyl-bearing α-carbon. Treatment with guanidine for a day at reflux produced ptilocaulin with all four stereocenters intact. The groundwork had now been laid for successive work. Once the relative stereochemistry at C5a and C7 had been established, the thermodynamically more stable arrangement at C3a and C8b was assured by this strategy. The improved Snider synthesis utilized the cuprate from compound **2088** thus circumventing an ozonolysis since the adduct of **2088** and enone **2085** could be directly cyclized to bicyclic enone **2086**. Optically pure (−)-**2087** was also produced via a similar route by this group. This served to establish the absolute stereochemistry of naturally-occurring (+)-ptilocaulin.

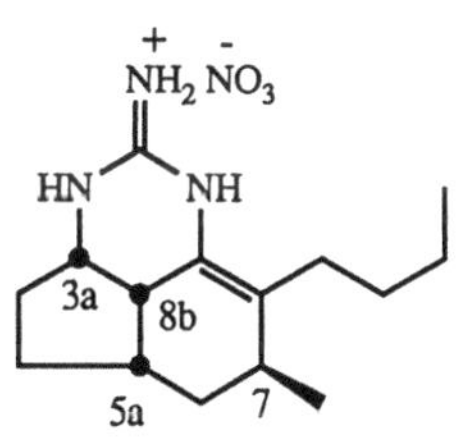

Ptilocaulin 2087

Hassner [561] has described a route to **2087** based on intramolecular nitrile-oxide cycloaddition (Scheme 368). Ketone **2090** was added to the oxime dianion of hexanal to produce **2091**. Treatment of the oxime with sodium hypochlorite produced the oximidoylhalide, the precursor to the nitrile-oxide which underwent cycloaddition to produce adduct **2092** stereochemically homogeneous at the ring fusion and C3a. Elimination of the alcohol and reductive cleavage of the isoxazoline gave β-hydroxy ketone **2093**. Birch reduction gave rise to **2094** possessing all of the stereocenters in the correct relative orientation. β-Elimination of the alcohol produced the key enone **2086** which underwent guanidination as before to produce ptilocaulin.

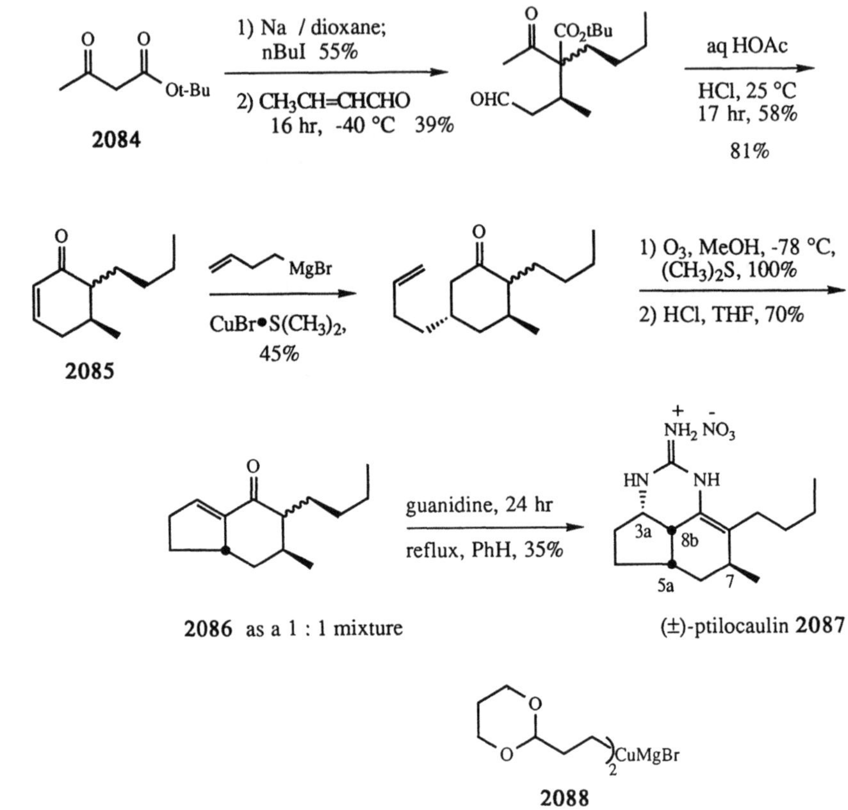

Scheme 367. Snider Synthesis of (±)-Ptilocaulin

The Uyehara racemic synthesis [562] began with the known conversion of tropolone to the bridged bicyclic compound **2095** (Scheme 369). Alkylation processes led to **2096**. Acid-promoted rearrangement led to ketone **2097**, which was irradiated to induce 1,3-rearrangement to **2098**. Refunctionalization of this substance gave way to the familiar enone **2086** from which racemic ptilocaulin was produced as before.

Roush [563] opted for strict control of stereochemistry at all four stereocenters in a synthesis of (−)-**2087** (Scheme 370). (+)-3-Methylcyclohexanone was converted by a standard series of nine operations to olefinic aldehyde **2100**. Using a similar key process to that of Hassner but on a different substrate, compound **2100** was treated with *N*-benzylhydroxylamine with heating to produce an intermediate nitrone which underwent intramolecular cycloaddition to give the cycloadduct **2101** with the correct relative stereochemistry at all 4 centers. Conversion to the β-aminoketone **2102** took 3 steps, which was followed by *trans*-guanidination to produce (−)-ptilocaulin.

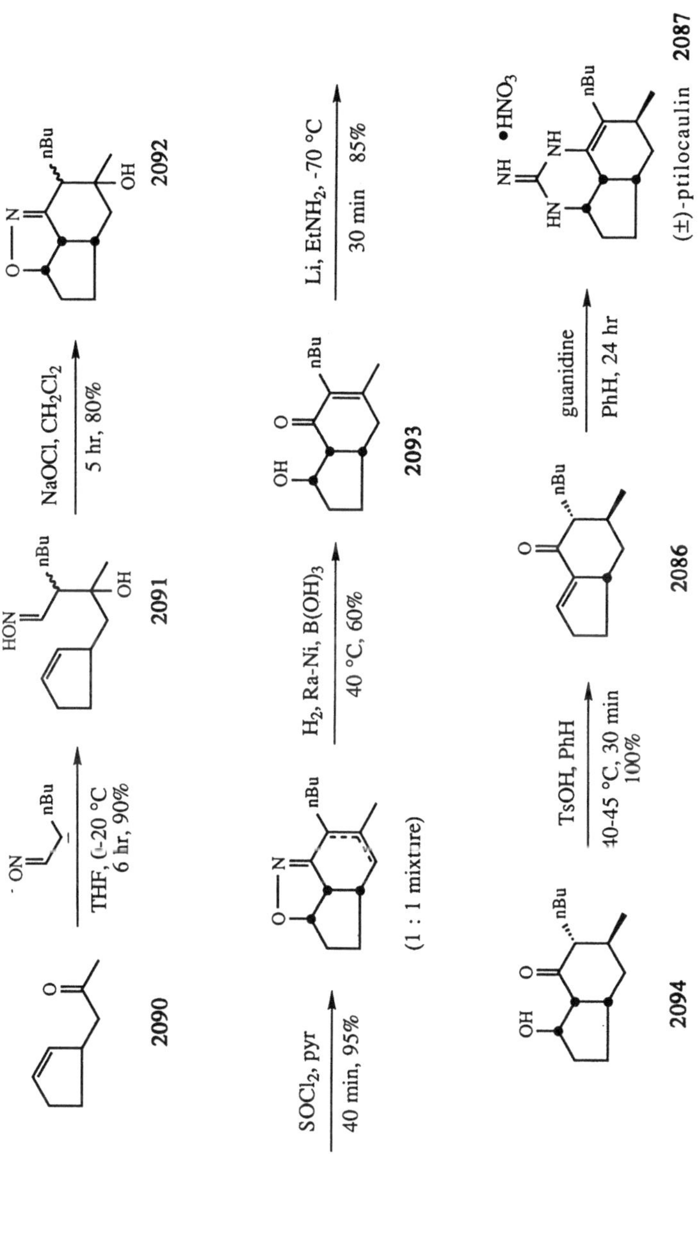

Scheme 368. Hassner Synthesis of (±)-Ptilocaulin

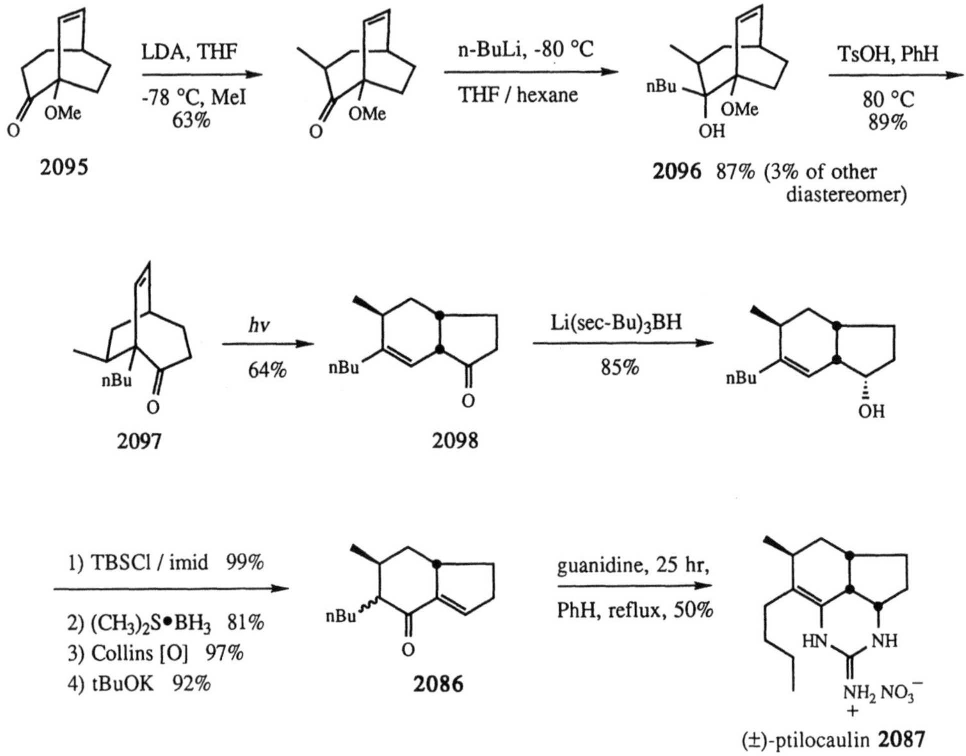

Scheme 369. Uyehara Synthesis of (±)-Ptilocaulin

5.4.4 Oroidin

A series of interesting multi-nitrogenous metabolites have been isolated from a variety of sponges in the genus *Agelas*. A central substance in this family is oroidin, originally misassigned by Minale [564], but corrected by Garcia [565]. Ahond and Poupat [566] have produced oroidin by two closely related pathways, one of which is shown in Scheme 371. 4-Hydroxymethylimidazole (**2103**) was *N*-tritylated and oxidized to aldehyde **2104**. Chain extension to the protected amine **2105** occurred in a standard fashion. Deprotonation between the nitrogens and addition to phenyl azide produced the azo compound **2106**.

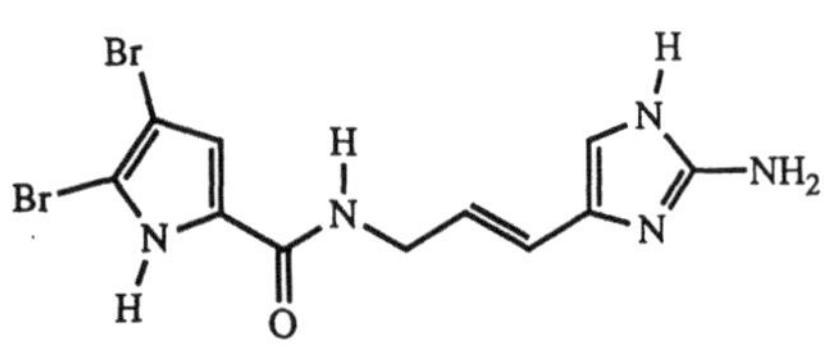

Oroidin

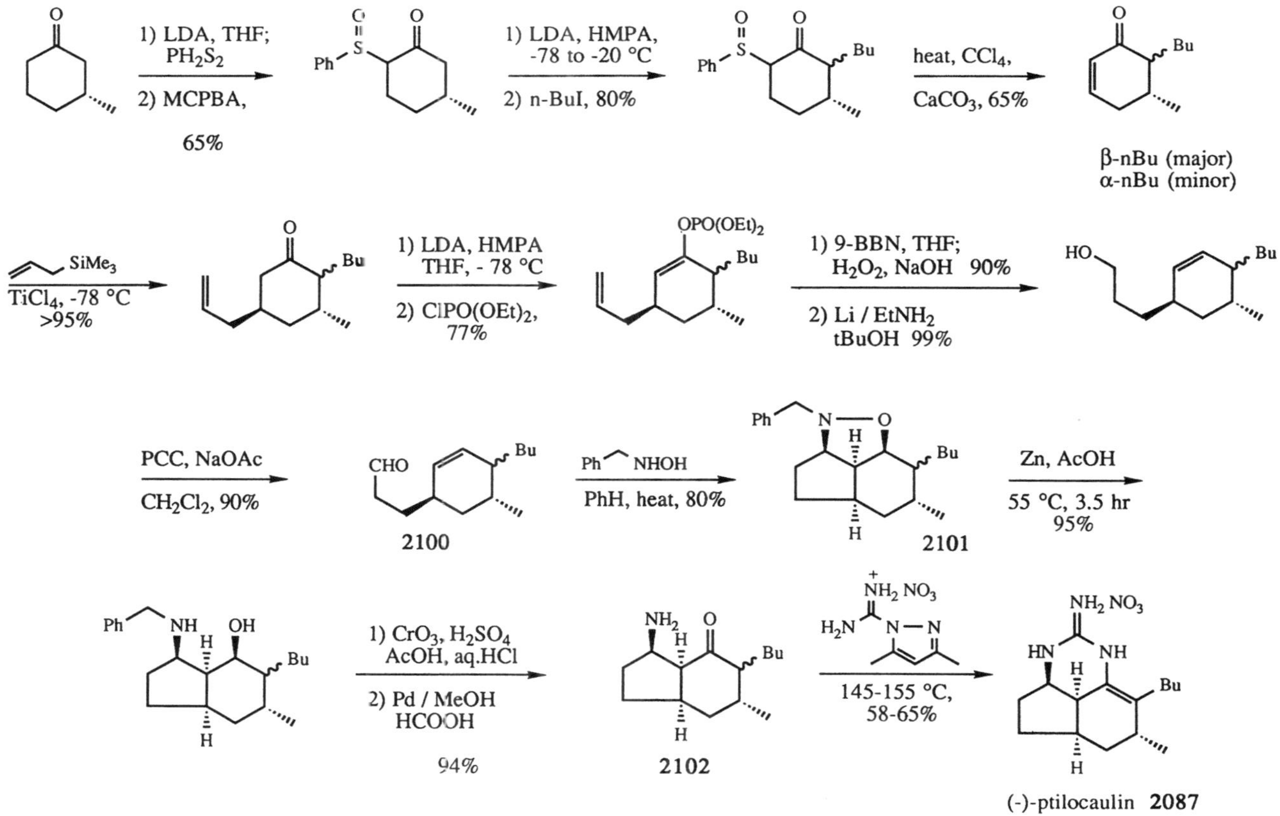

Scheme 370. Rousch Synthesis of (−)-Ptilocaulin

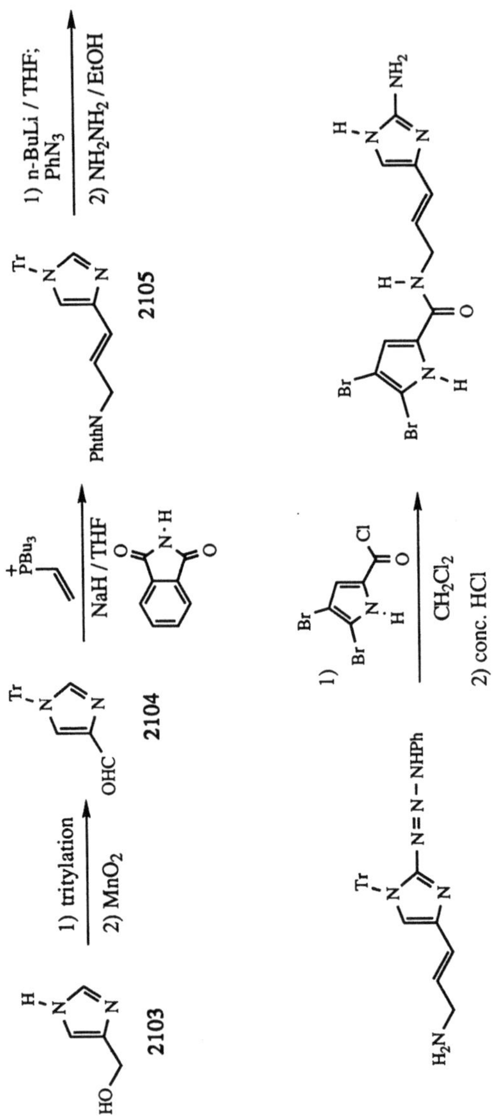

Scheme 371. Ahond Synthesis of Oroidin

The amine was freed and acylated with the appropriate pyrrole on the pathway to the final product. Their second synthesis simply used a different series of protecting groups.

5.4.5 Dibromophakellin

A more complex substance in this family is dibromophakellin, which has been produced in an efficient biomimetic route by Büchi [567]. Scheme 372 shows the conversion of (+)-citrulline to the aminoimidazole **2108** in four steps in good yield. Treatment with base followed by the acylating agent **2109** produced dihydrooroidin (**2110**). Treatment with bromine yields a compound which may be **2111**, but in any case, on treatment with *t*-BuOK dibromophakellin is produced quantitatively.

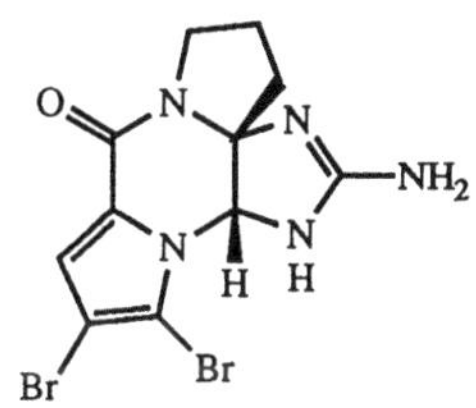

Dibromophakellin

5.4.6 Girolline

The unrelated aminoimidazole girolline from the sponge *Pseudaxinyssa canthar-ella* has been synthesized by Ahond (Scheme 373) [568]. The tritylated imidazole aldehyde **2104** produced in the earlier oroidin synthesis was treated with vinylmagnesium bromide and the alcohol was silylated to give **2112**. Vicinal oxyamination provided **2113** as a mixture of diastereomers. The chloride **2114** was produced from **2113** presumably by an inversion process, still leaving a mixture of isomers. Removal of the protecting groups left **2115** which was aminated between the imidazole nitrogens to give a mixture of final products. The *erythro* isomer was found to be identical to the natural product.

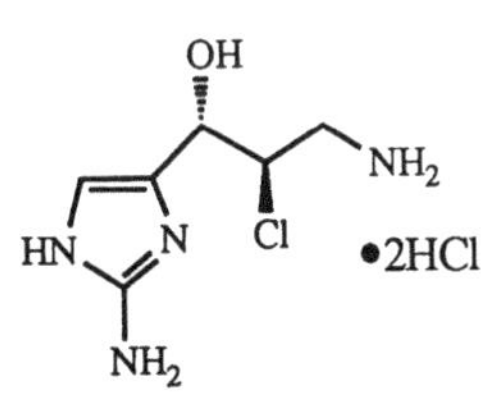

Girolline

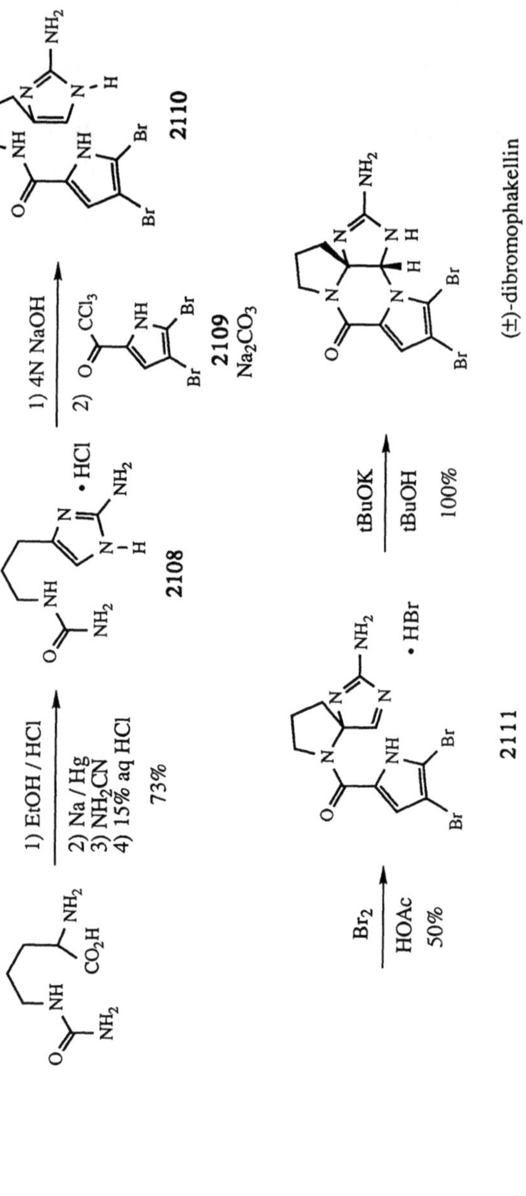

Scheme 372. Buchi Synthesis of (±)-Dibromophakellin

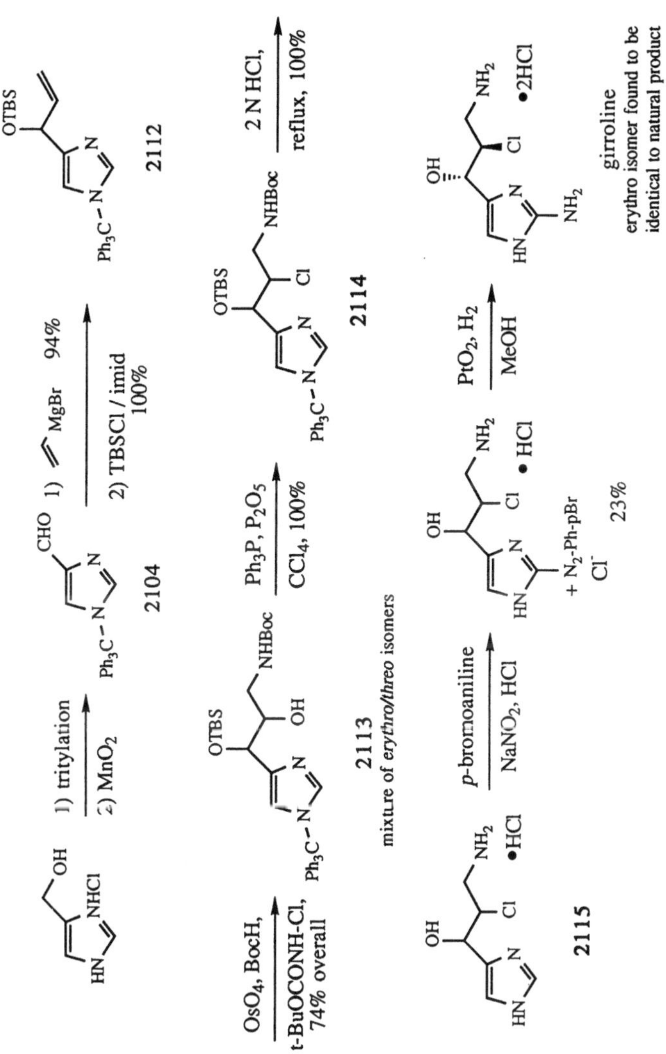

Scheme 373. Ahond Synthesis of Girolline

5.5 Nucleosides and Related Substances

5.5.1 Mycalisine A

The lipophilic extracts of a marine sponge *Mycale* sp., collected in the Gulf of Sagami, Japan, inhibit the cell division of fertilized starfish eggs. One of the active components was found to be mycalisine A, a pyrrolo[2,3-*d*]pyrimidine nucleoside with unsaturation between the 4′ and 5′-positions of the ribose moiety [569]. Townsend's synthesis of mycalisine A is outlined in Scheme 374

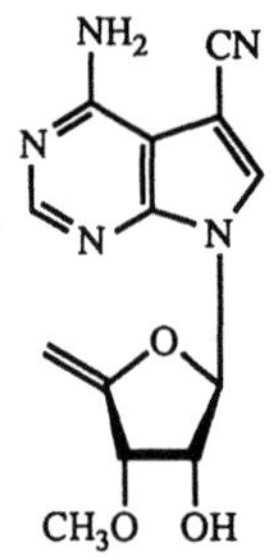

Mycalisine A

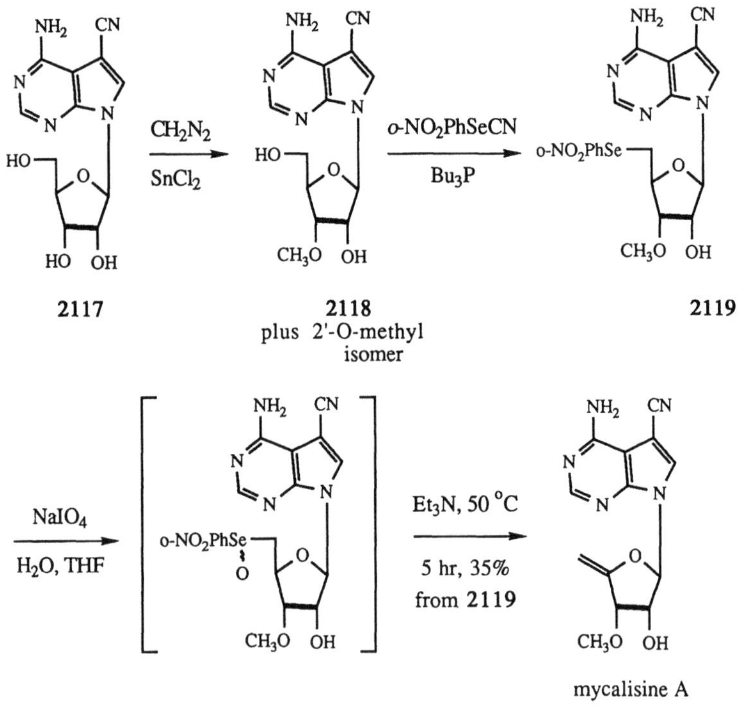

Scheme 374. Townsend Synthesis of Mycalisine A

[570]. Methylation of nucleoside antibiotic toyocamycin (**2117**) yields a mixture of the 2′-*O*-methyl and 3′-*O*-methyl isomers. The two isomers were separated by acetylation, recrystallization and deacylation to give **2118** as a single isomer. Phenylselenylation of the primary hydroxyl followed by selenoxide elimination provides mycalisine A in 35% yield from **2119**.

5.5.2 Phidolopin

Phidolopin (**2123**) is a xanthine derivative isolated from the "lacey" bryozoan *Phidolopora pacifica*, collected in Barkley Sound, British Columbia [571]. It shows in vivo antifungal and antialgal activity [572] and is of interest because it is of animal rather than plant origin and it contains a nitro group which is relatively rare in natural products. Hirota's synthesis [573] of **2123** is shown in Scheme 375. Bromination of the protected cresol **2120** affords the benzyl

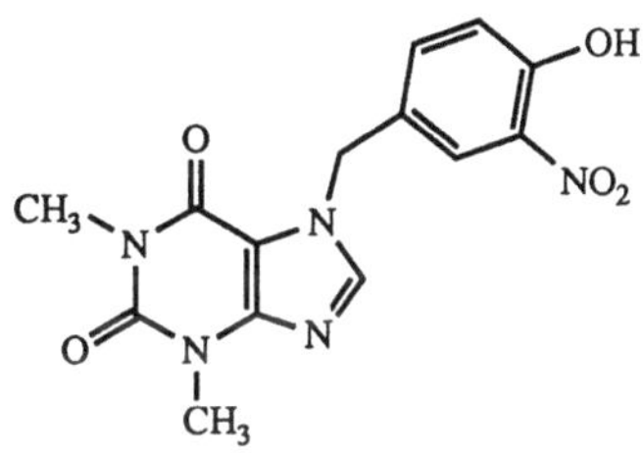

Phidolopin **2123**

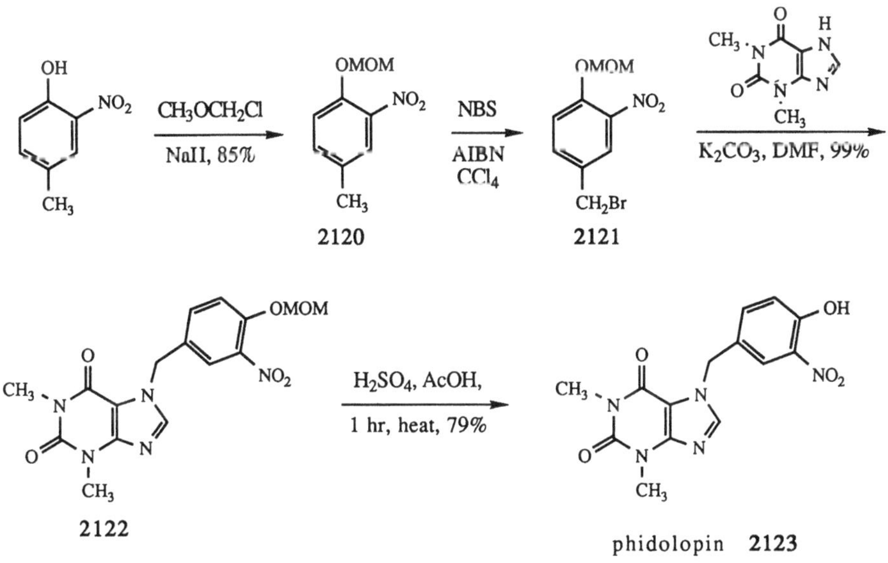

Scheme 375. Hirota Synthesis of Phidolopin

bromide **2121**. Coupling of **2121** with theophylline in the presence of potassium carbonate affords **2122**, which upon treatment with acid gives phidolopin (**2123**), in a total of four steps.

5.5.3 6-Imino-1,9-Dimethyl-8-Oxopurine

The adenine derivative 6-amino-1,9-dimethyl-8-oxopurine (**2128**), has been isolated from the English Channel sponge *Hymeniacidon sanguinea* Grant in the form of its acetyl derivative [574]. The synthesis of the parent base, by Fujii, [575] is outlined in Scheme 376. Bromination of 1-methyladenine (**2125**)

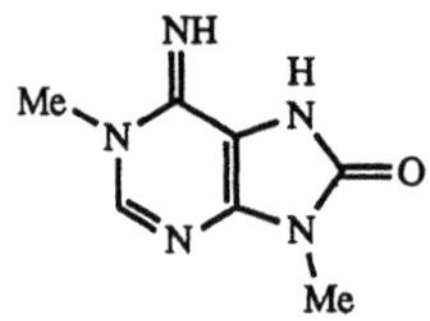

6-Imino-1,9-dimethyl-8-oxopurine **2128**

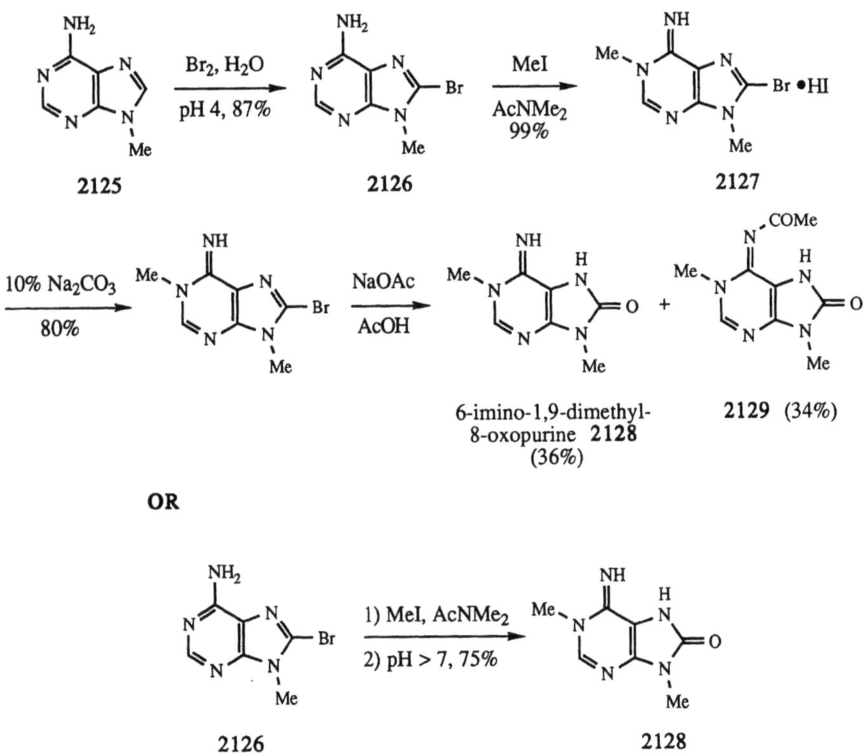

Scheme 376. Fuji Synthesis of 6-Imino-1,9-Dimethyl-8-Oxopurine

followed by methylation gives the methylated bromide **2127** as its hydroiodide. Hydrolysis of **2127** gives a 36% yield of **2128** (25% overall) along with 34% of the N_6-acetyl derivative **2129**. Alternatively, methylation of **2126** and hydrolysis under basic conditions yields **2128** directly in two steps and 65% yield.

5.5.4 1-Methylisoguanosine (Doridosine)

The aqueous ethanolic extracts of the sponge *Tedania digitata* Schmidt, collected off Newport Reef, Sydney, Australia, exhibit muscle relaxant, anti-inflammatory and other pharmacological activities. The active component has been found to be a new methylated purine nucleoside, 1-methylisoguanosine (**2134**) [576]. The preparation of **2134** by Cook [577] begins with the acetylation of 5-amino-4-carbamoyl-1β-D-ribofuranosylimidazole (**2130**) to afford triacetate **2131** (Scheme 377). Conversion of **2131** to the nitrile **2132** and reaction with

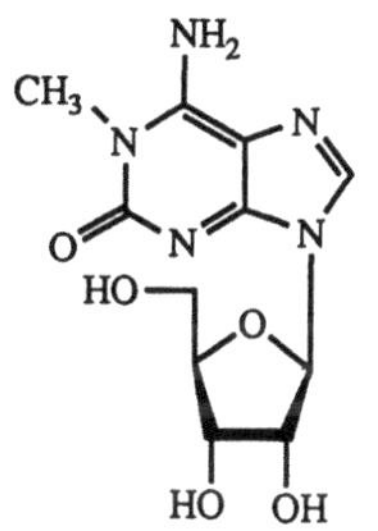

1-Methylisoguanosine 2134

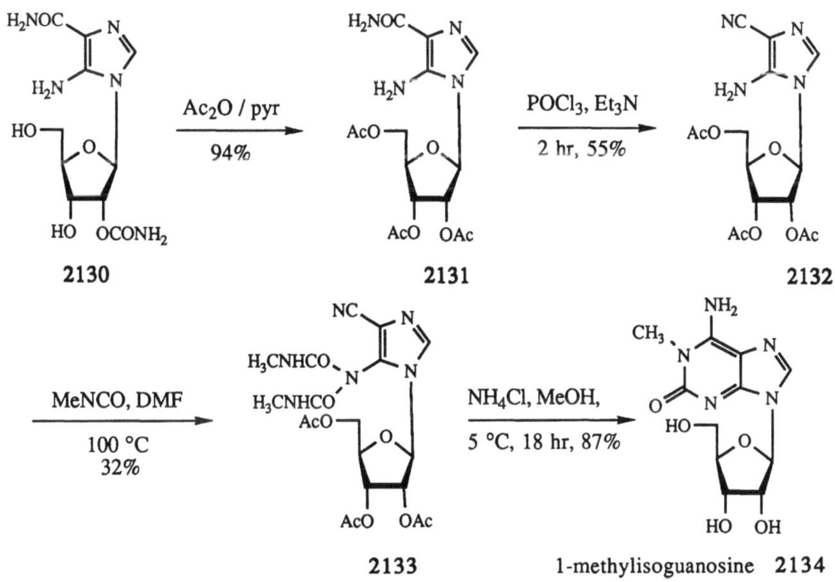

Scheme 377. Cook Synthesis of 1-Methylisoguanosine

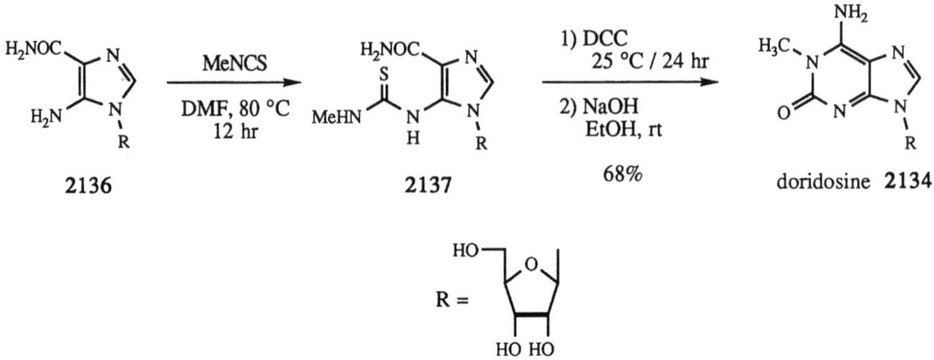

2136 2137 doridosine **2134**

Scheme 378. Townsend Synthesis of 1-Methylisoguanosine (doridosine)

methyl isocyanate produces the bis(methylcarbamoyl) adduct **2133**. Treatment of **2133** with aqueous ammonium chloride induces cyclization to give 1-methylisoguanosine (**2134**) in four steps and 14% overall yield. Alternatively, **2133** can be cyclized without isolation to give **2134** in 22% overall yield.

Using a similar starting material, Townsend [578] also prepared **2134** using a shorter sequence consisting of two steps (Scheme 378). Treatment of **2136** with methylisothiocyanate gave the thiourea **2137**. DCC mediated cyclization and saponification gave **2134** in good overall yield.

5.6 *Reniera* Alkaloids

5.6.1 Mimosamycin

Mimosamycin was first isolated from the culture filtrate of *Streptomyces lavendulae* No. 314. It exhibits antibiotic activity towards mycobacteria, including streptomycin resistant strains of human tubercle bacilli [579]. It is the first example of a naturally occurring "isoquinoline quinone" to be reported. Mimosamycin has also been isolated as a minor metabolite from a bright blue sponge *Reniera* sp. [580].

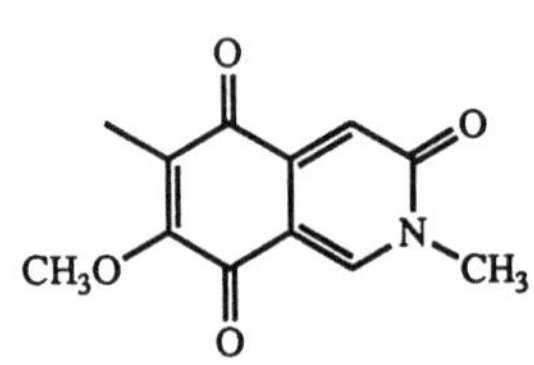

Mimosamycin **2141**

The structure elucidation and first synthesis of mimosamycin (**2141**), by Fukumi [581], appeared in 1977 (Scheme 379). The synthesis begins with the conversion of benzaldehyde **2138** to isoquinoline **2139** in eight steps using standard transformations. Air oxidation of **2139** in the presence of morpholine and copper (II) acetate affords morpholino-substituted *o*-quinone **2140**, containing all the functionality needed to complete the synthesis. Hydrolysis and methylation completes the right hand portion of the molecule. Reduction of the newly formed *p*-quinone and protection as the diacetate allows *N*-methylation of the isoquinoline and hydrolysis to the corresponding *N*-methyl pyridone. Oxidative methanolysis provides mimosamycin **2141** in 14 steps and 5.2% overall yield.

Parker's synthesis [582] of mimosamycin (**2141**) begins with the highly functionalized benzonitrile **2143** (Scheme 380). Allylation of the phenol, Claisen rearrangement, and methylation provides hexasubstituted benzene **2144**. Permanganate oxidation reveals the latent acetic acid moiety in **2145** which is esterified to give ester **2146**. Reduction of the nitrile in the presence of acetic formic anhydride provides amide **2147** which cyclizes to lactam **2148** upon treatment with diborane. Oxidative demethylation of **2148** with silver oxide yields mimosamycin **2141**. Difficulties in reduction of the nitrile and problems with isolation of the final product lessen the yield considerably. A final yield of less than 3% is obtained over nine steps.

A very short synthesis of mimosamycin, which is applicable to the preparation of a number of derivatives, has been developed by McKillop [583] (Scheme 381). Hetero Diels-Alder reaction of quinone **2149** with 2-aza-1,3-bis(*tert*-butyldimethylsilyloxy)-1,3-butadiene affords intermediate **2150** which yields lactam **2151** upon treatment with HCl. The use of concentrated solutions of the diene is crucial to the success of the cycloaddition step. Methylation of **2151** under phase-transfer conditions provides mimosamycin in three steps and 54% overall yield.

5.6.2 *Reniera* Isoindole

Along with renierone (**2179**) and mimosamycin (**2141**), the bright-blue sponge *Reniera* sp., found near Isla Grande, Mexico, provides a wide variety of antibiotic metabolites. One of these, 2,5-dimethyl-6-methoxy-4,7-dihydroisoindole-4,7-dione (**2156**), the first example of a naturally occurring isoindole, has been isolated and synthesized by Faulkner [584] (Scheme 382). The

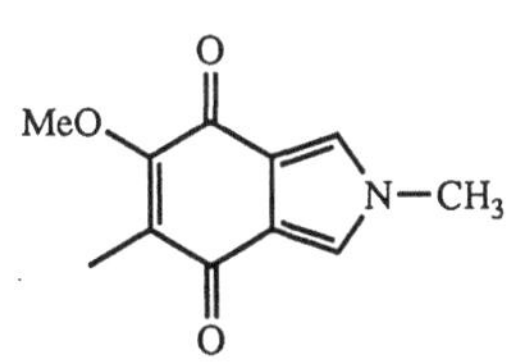

Reniera isoindole **2156**

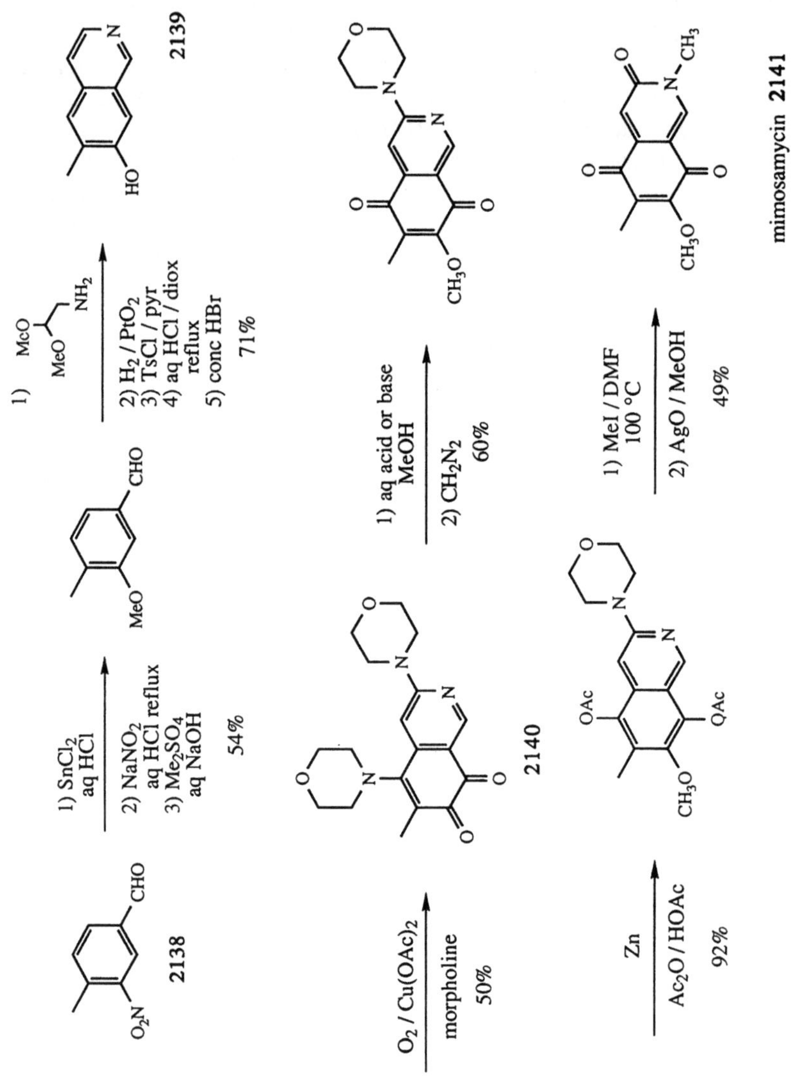

Scheme 379. Fukumi Synthesis of Mimosamycin

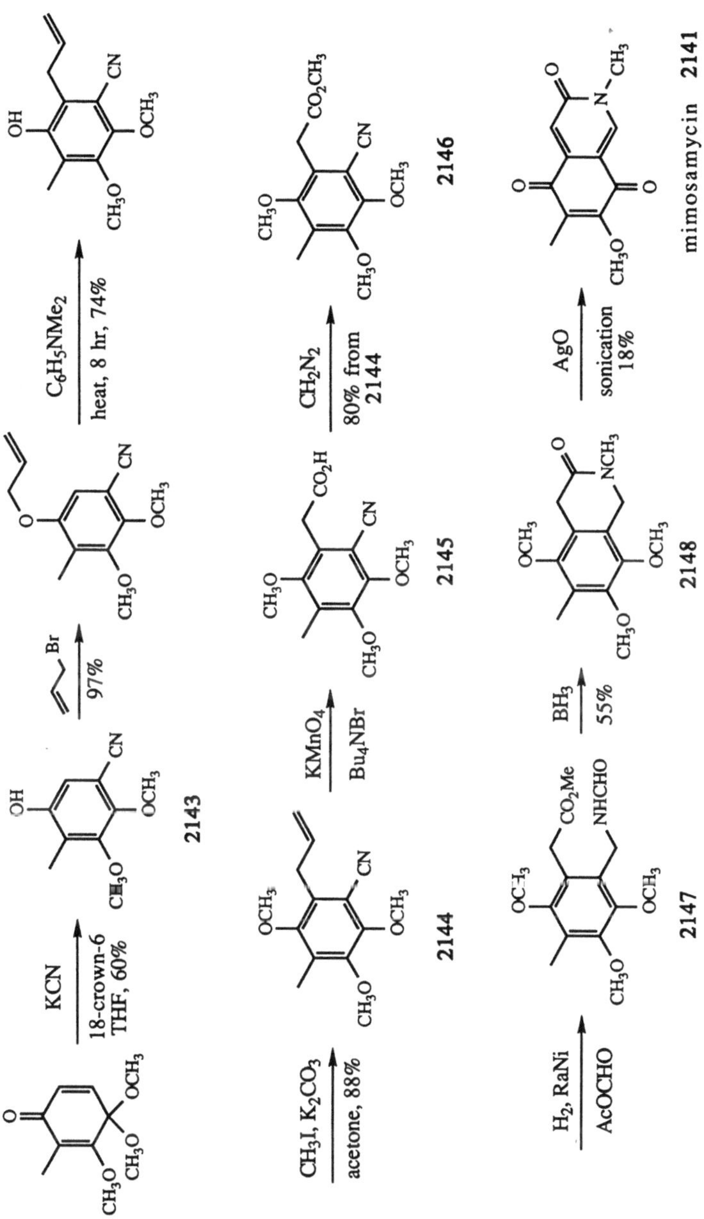

Scheme 380. Parker Synthesis of Mimosamycin

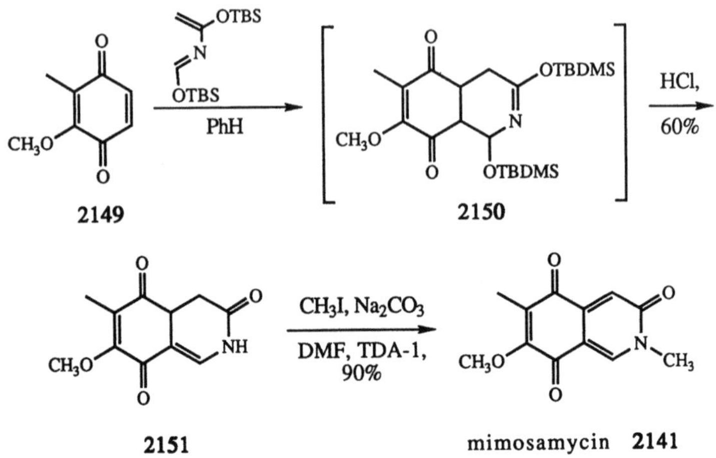

Scheme 381. McKillop Synthesis of Mimosamyscin

synthesis involves building the quinone portion onto an *N*-methylated pyrrole. Addition of one equivalent of 2-lithio-2-ethyl-1,3-dithiane to 3,4-dicarbomethoxy-1-methylpyrrole (**2153**) gives ketone **2154** in 27% yield. Hydrolysis of the dithiane, under conditions that do not preferentially attack the pyrrole, gives α-diketone **2155**. Annulation and methylation with diazomethane provides **2156**, thus confirming the structure of the natural product. A yield of 3% of **2156** is obtained over four steps.

Padwa and Parker's synthesis [585] of isoindole **2156** (Scheme 383) involves annulation of the pyrrole moiety onto a preformed quinone ring. Electrochemical oxidation of 1,2,4-trimethoxy-3-methylbenzene (**2157**) in methanol yields quinone monoacetal **2158** which upon hydrolysis provides quinone **2159**. Treatment of aminosilane **2160** with five equivalents of silver fluoride generates an unstable azomethine ylide which undergoes cycloaddition with quinone **2159** giving isoindole **2156** in 68% yield.

5.6.3 7-Methoxy-1,6-Dimethyl-5,8-Dihydroisoquinoline-5,8-Dione and *N*-Formyl-1,2-Dihydrorenierone (2171) (2164)

Two other *Reniera* metabolites, 7-methoxy-1,6-dimethyl-5,8-dihydroisoquinoline-5,8-dione (**2164**) and (±)-*N*-formyl-1,2-dihydrorenierone (**2171**), have been synthesized by Kubo [586] (Schemes 384 and 385). The dihydroisoquinoline **2164** was prepared via two different pathways. Tosylation of carbinol **2162** and reduction gives 1-methylisoquinoline **2163**. Ceric ammonium nitrate oxidative demethylation of **2163** provides the natural product in 30% yield along with 42% of the ortho-quinone. An alternative pathway begins with nitro

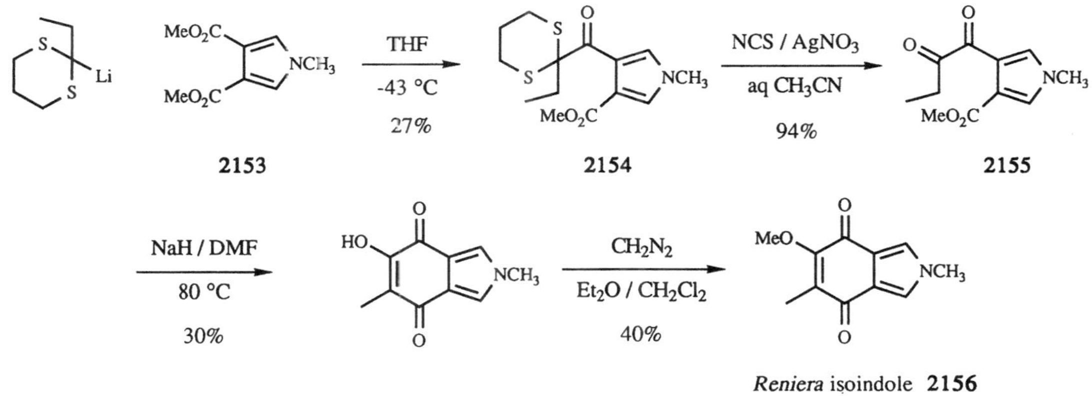

Scheme 382. Faulkner Synthesis of *Reniera* Isoindole **2156**

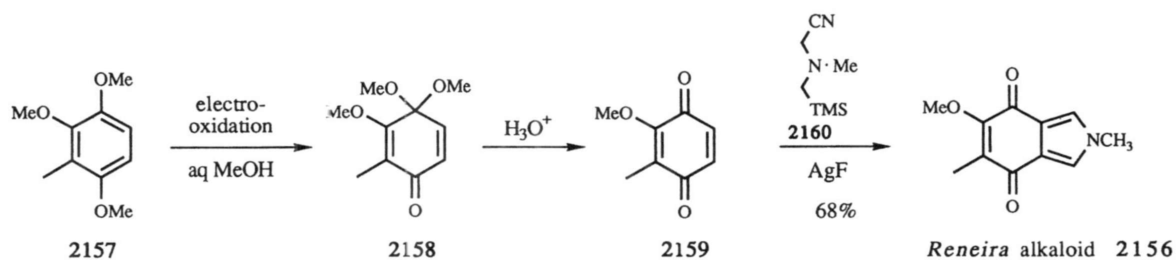

Scheme 383. Padwa/Parker Synthesis of *Reniera* Isoindole **2156**

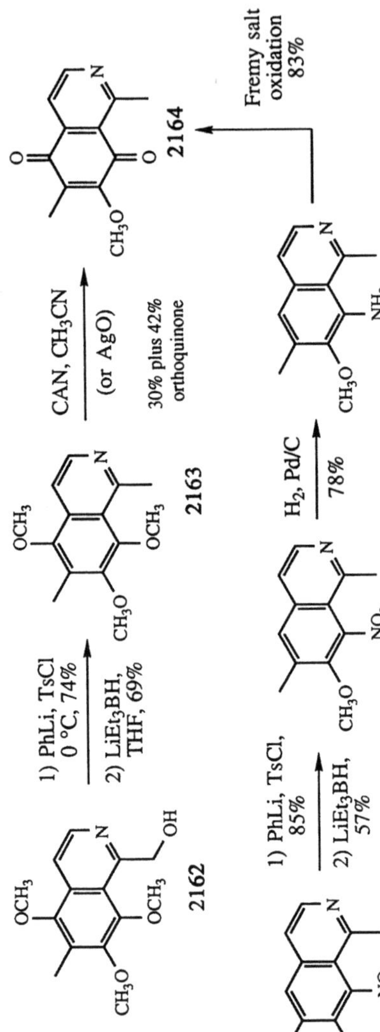

Scheme 384. Kubo Synthesis of *Reniera* Metabolite **2164**

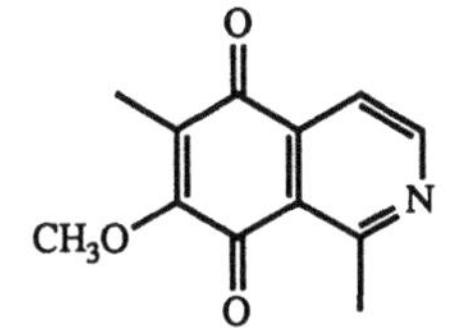

2164

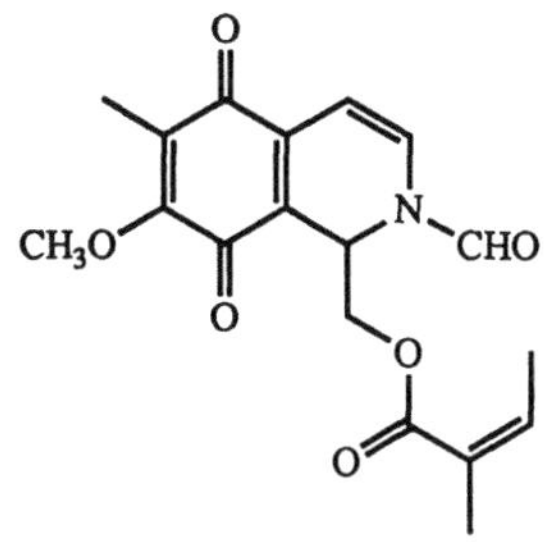

2171

substituted carbinol **2165**. Tosylation and reduction provides the nitro substituted 1-methylisoquinoline **2166**. Catalytic reduction of **2166** and oxidation with Fremy's salt provides **2164** in 37% overall yield.

The synthesis of (±)-*N*-formyl-1,2-dihydrorenierone (**2171**), as shown in Scheme 385, also begins from isoquinoline carbinol **2162**. Simultaneous reduction and formylation of **2162** provides *N*-formyltetrahydroisoquinoline **2167**. Acylation of **2167** with acid chloride **2168** leads to compound **2169** which is converted to quinone **2170** by oxidation with ceric ammonium nitrate. Dehydrogenation with palladium on carbon gives the natural product in four steps from **2162** and 12% overall yield.

5.6.4 Renierone

There have been two syntheses of renierone (**2179**). The first of these, by Danishefsky [587], begins with the chloromethylation of compound **2173** (Scheme 386). Homologation of **2173** with cyanide, reduction to the amine and protection as the carbamate affords **2174**. Formation of the tetrahydroisoquinoline **2175** and aromatization gives isoquinoline **2176**. Reduction of **2176** to the alcohol and acylation with angelic acid gives **2177** which is oxidized to a mixture of the ortho-quinone **2178** and renierone (**2179**). Conversion of the ortho-quinone to renierone increases the overall yield to approximately 1.1% over 14 steps.

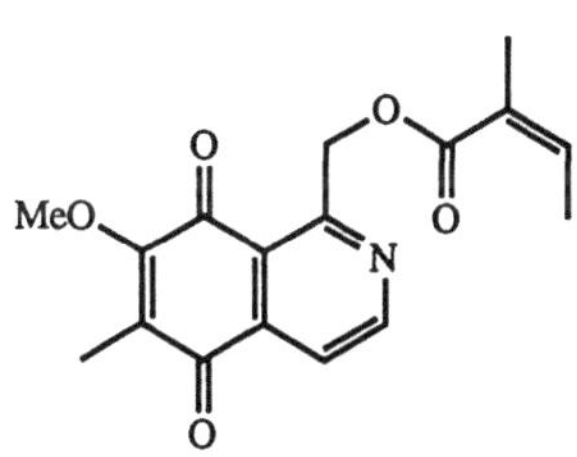

Renieronc **2179**

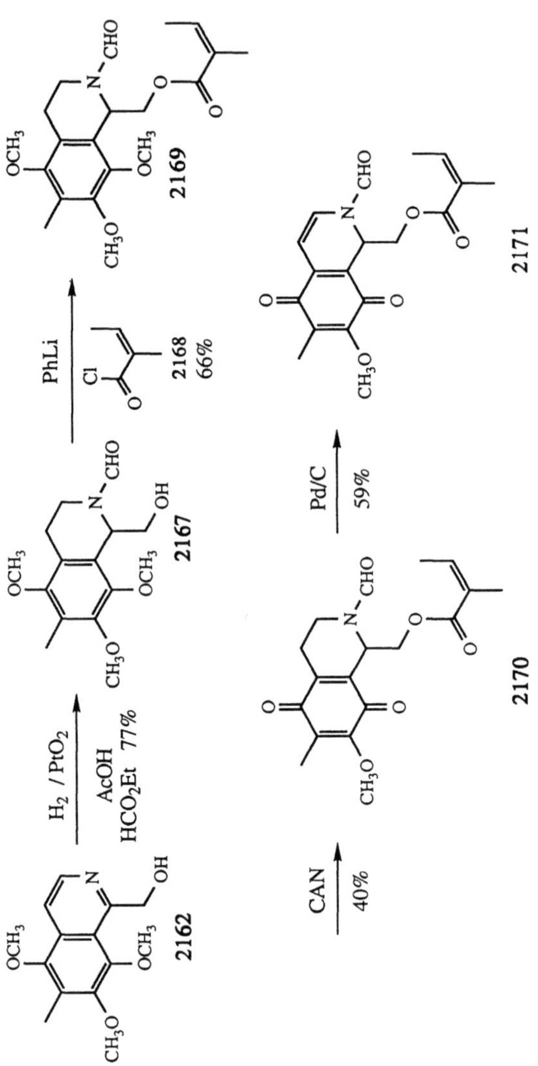

Scheme 385. Kubo Synthesis of *Reniera* Metabolite **2171**

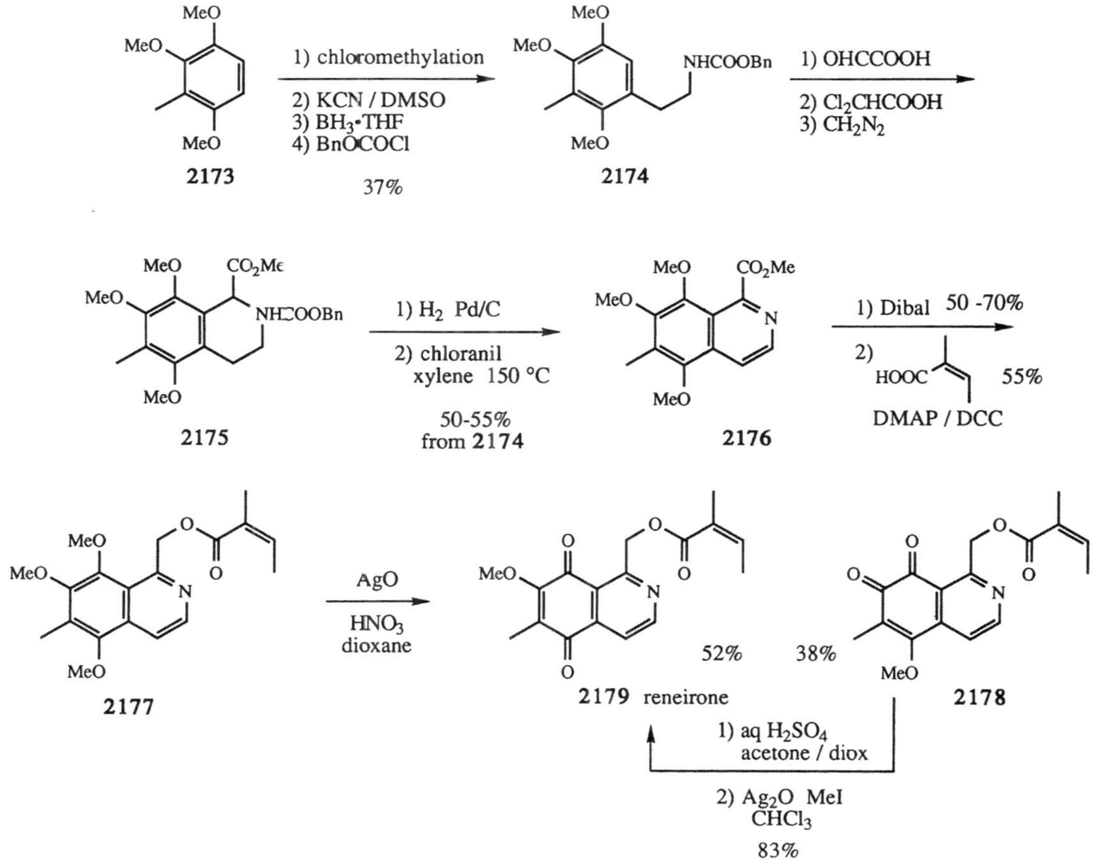

Scheme 386. Danishefsky Synthesis of Reneirone

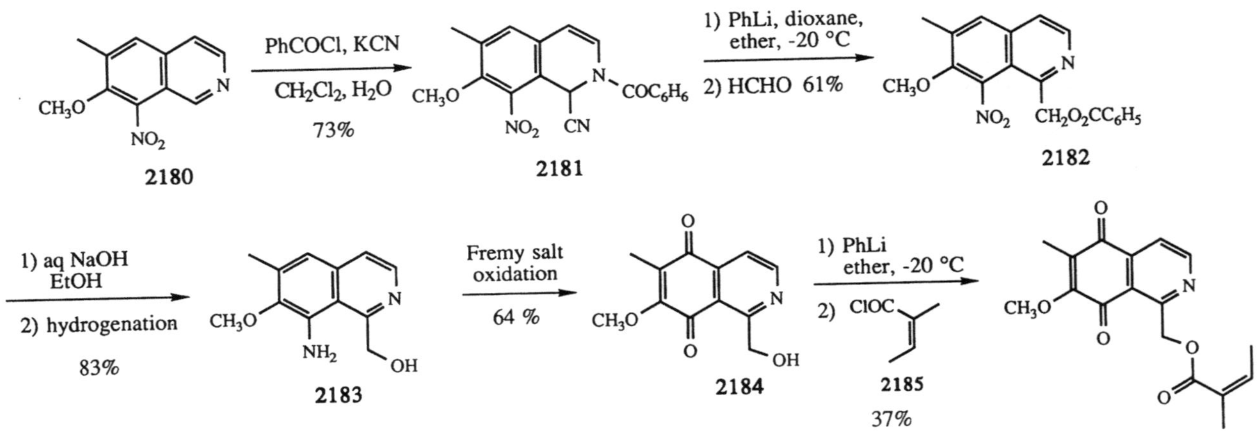

Scheme 387. Kubo Synthesis of Renierone

Kubo's synthesis [588] of renierone is outlined in Scheme 387. Addition of benzoyl chloride and potassium cyanide to isoquinoline **2180** provides Reissert compound **2181**. Lithiation of **2181** and treatment with gaseous formaldehyde yields benzoate **2182** which is saponified and hydrogenated to give amine **2183**. Fremy salt oxidation of **2183** gives quinone **2184** which is acylated with **2185** to provide renierone in six steps and 8.6% overall yield.

5.7 Zoanthoxanthins

Büchi has described a biomimetic route to parazoanthoxanthin A (**2189**) (Scheme 388) and pseudozoanthoxanthin A, which is equally short and efficient [589]. These metabolites are fluorescent pigments produced by marine anthozoans and are dimers of a presumed C_5N_3 biogenetic precursor [590]. Parazoanthoxanthin A was produced in a two-step process from 2-aminobutyrolactone (**2187**). Treatment with cyanamide and sodium amalgam in aqueous ethanol led to the aminoimidazole **2188** in 64% yield. Treatment of this with aqueous acid at elevated temperature gave way to **2189**, probably involving dimerization of an imidazolium cation. The route to pseudozoanthoxanthin A was somewhat longer (Scheme 389) beginning with the oxadiazole **2190** which

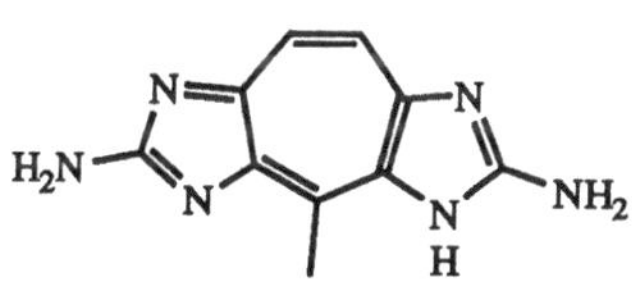

Parazoanthoxanthin A **2189**

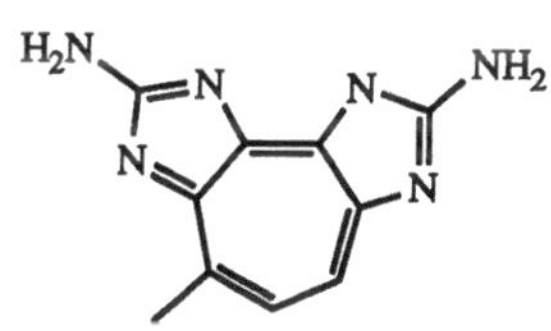

Pseudozoanthoxanthin A

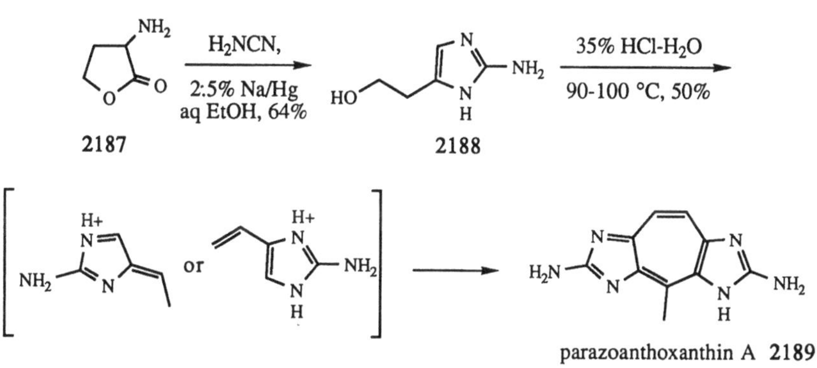

Scheme 388. Buchi Synthesis of Parazooanthoxanthin A

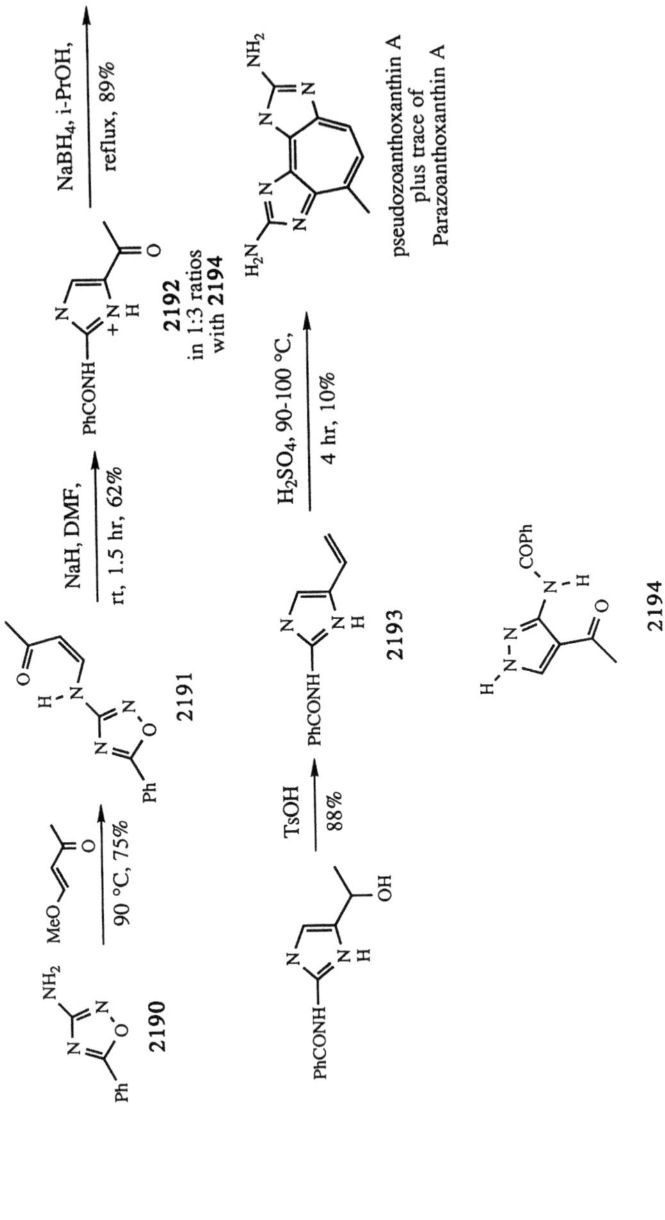

Scheme 389. Buchi Synthesis of Pseudozoanthoxanthin A

was converted to **2191** and rearranged to the imidazole **2192** with base. Conversion of the ketone to the olefin **2193** producing the desired precursor. Acid treatment at 90–100°C provided psudozoanthoxanthin A in 10% yield along with a trace of parazoanthoxanthin A.

5.8 Pyrrole-Containing Alkaloids

5.8.1 *Oscarella Lobularis* Pyrrole Metabolite (3-Octadecyl Pyrrole-2-Carbaldehyde)

The 3-Alkylpyrrole-2-carbonxaldehyde (**2197**) has been isolated from the sponge *Oscarella lobularis* [591], although it has been suggested that the structure of the natural product was misassigned [592]. Muchowski's synthesis of **2197** [593] begins with the bromination of silyl-protected pyrrole **2195** (Scheme 390). Transmetallation and alkylation of **2195** affords compound **2196**. Desilylation and formylation provides **2197** four steps and 38% overall yield. The natural product is obtained as an inseparable mixture with the 5-carboxaldehyde isomer.

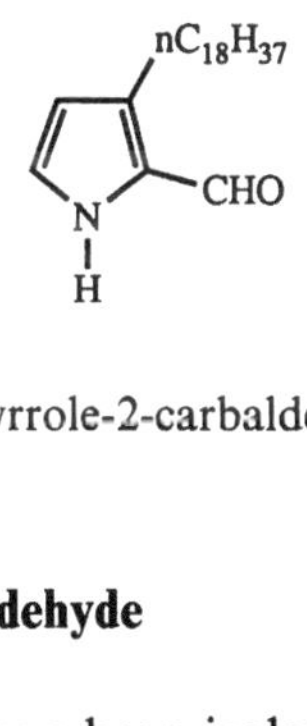

3-Alkylpyrrole-2-carbaldehyde **2197**

5.8.2 5-Nonylpyrrole-2-Carbaldehyde

A number of simple pyrroles have been isolated from various marine sources. 5-Nonylpyrrole-2-carbaldehyde was synthesized by Coll [594] to confirm the structure of a soft coral metabolite (Scheme 391). Acylation of pyrrole-magnesium iodide with nonanoyl chloride gave **2198** which was reduced to **2199**. Standard formylation gave rise to a substance identical to the natural product which was thus assigned as **2200**.

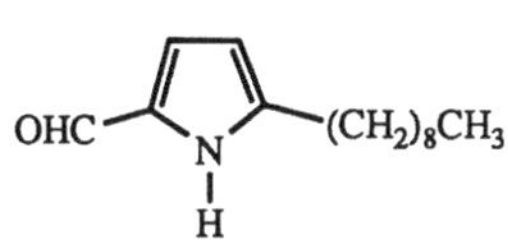

5-Nonylpyrrole-2-carbaldehyde 2200

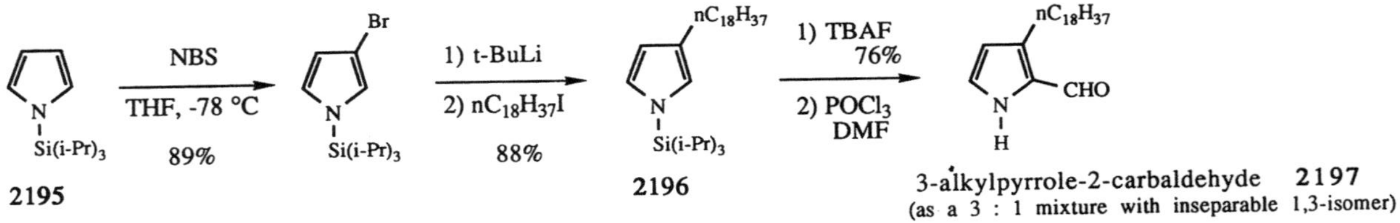

Scheme 390. Muchowski Synthesis of Pyrrole Metabolite **2197** from *Oscarella lobularis*

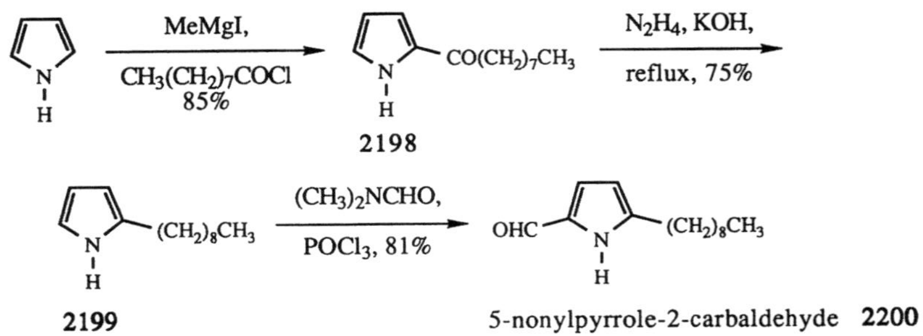

Scheme 391. Coll Synthesis of 5-Nonylpyrrole-2-carbaldehyde

5.8.3 Pentabromopseudilin

Pentabromopseudilin (**2206**) is a highly brominated cytotoxic phenylpyrrole that has been isolated from the marine bacterium *Alteromonas luteo-violaceus.* It exhibits antibiotic and enzyme-inhibitory properties. Laatsch [595] has prepared pentabromopseudilin in five steps as shown in Scheme 392. Substituted benzaldehyde **2202** is converted to ketone **2203** and then cyclized to pyrrole **2204** by treatment with ammonium acetate in refluxing acetic acid. Bromination of the pyrrole ring with pyidinium hydrobromide perbromide affords compound

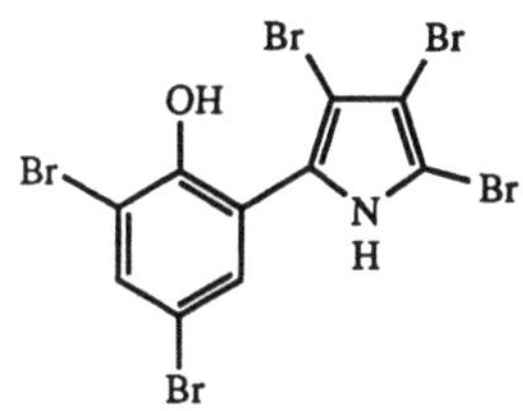

Pentabromopseudilin **2206**

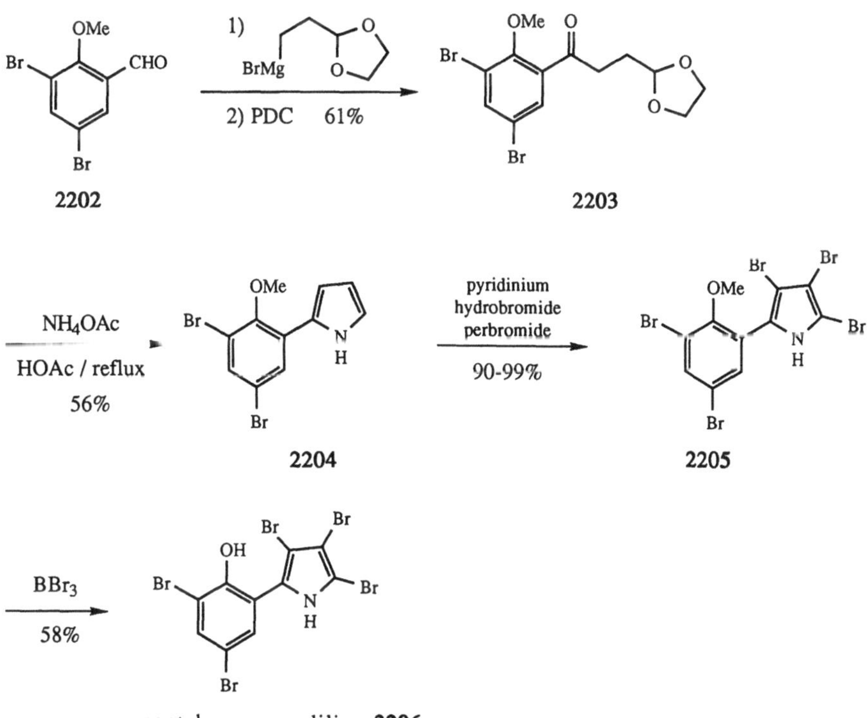

Scheme 392. Laatsch Synthesis of Pentabromopseudilin

2205 which is converted to pentabromopseudilin by demethylation with BBr_3. The natural product is obtained in approximately 19% overall yield.

5.8.4 Bonellin, Methyl Ester

Bonellin is a green pigment isolated from a marine echurian worm, *Bonellia viridis*, native to the sea bottom of the Gulf of Pozzuoli near Naples [596]. Sexually undifferentiated larvae of *B. viridis* that come into contact with bonellin develop into males. Those that avoid contact develop into female worms [597]. In addition, bonellin exhibits anti-tumor activity [598]. The synthetic strategy developed by Battersby [599] for the synthesis of bonellin methyl ester (**2220**) is convergent. The right and left halves of the molecule are constructed separately (Scheme 393), joined via a condensation reaction of a pyrrole and an aldehyde, and then cyclized photochemically (Scheme 394). Preparation of the right hand fragment begins with the oxidation and subsequent reduction of highly functionalized pyrrole **2207** to obtain pyrrolone **2208**. Condensation of **2208** with aldehyde **2209** provides the carbon skeleton of the right-hand fragment **2210**. Decarboxylation and formylation of **2210** with benzoyl chloride and DMF, followed by methylation yields aldehyde **2211**. Construction of the left-hand fragment begins with the transformation of pyrrole **2212** to nitro alkene **2213**. Conjugate addition of the nitropyrrole to the known enone **2214**, catalyzed by TBAF, provides compound **2215**. Reduction of **2215** with $TiCl_3$ provides the desired imine **2216**. Acid-catalyzed condensation of imine **2216** and aldehyde **2211** provides the *seco* system as a mixture of two isomers **2217** and **2218**. Photochemical cyclization of the mixture occurs slowly in the presence of proton sponge to give nitrile **2219**, which is converted to the methyl ester of the natural product by hydrolysis to the amide. Bonellin methyl ester (**2220**) is obtained in 11 steps and 8.6% overall yield.

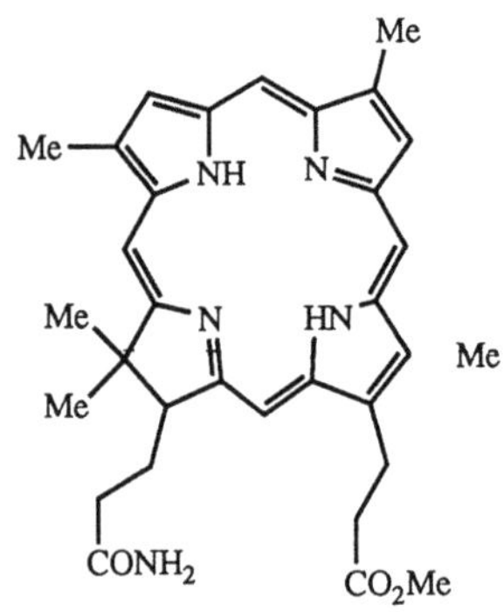

Bonellin methyl ester 2220

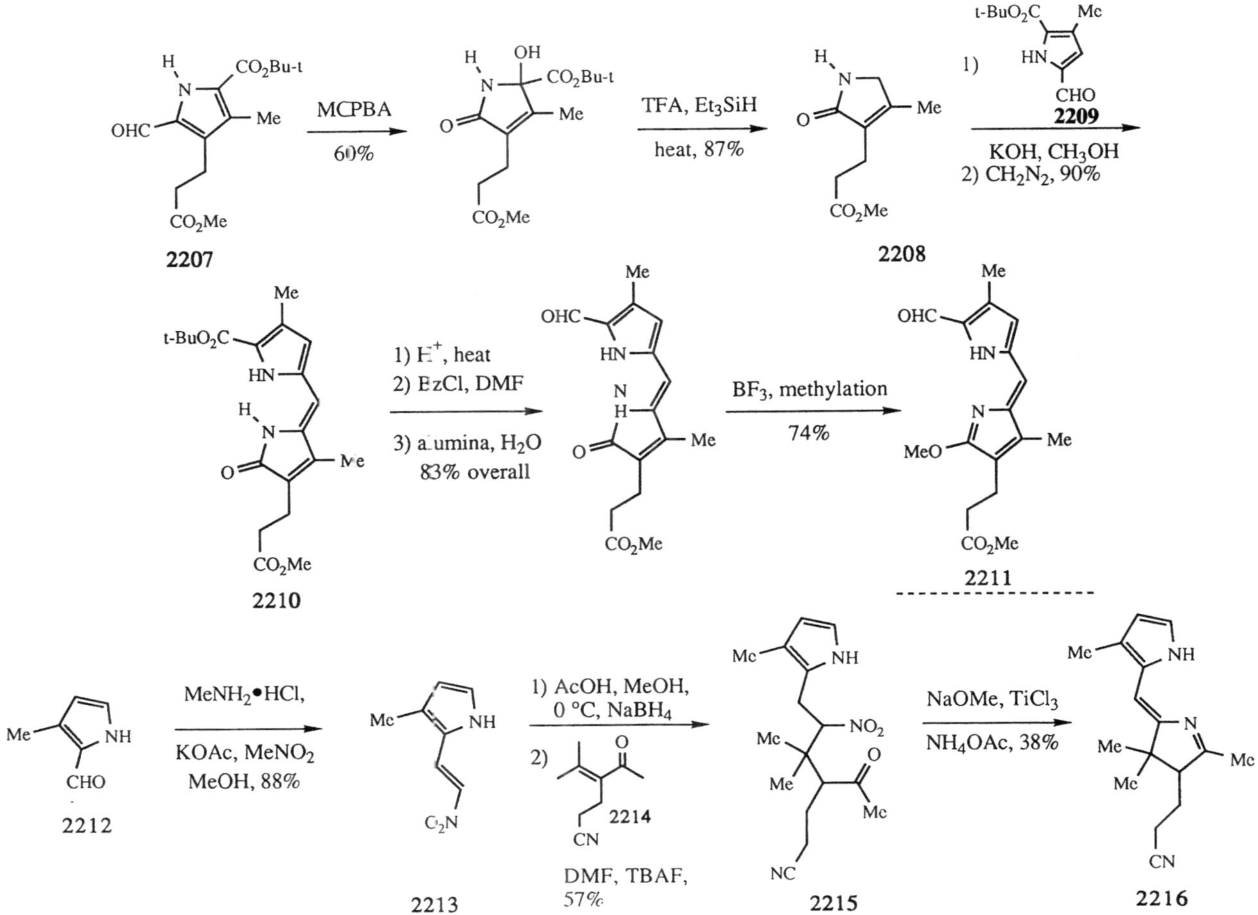

Scheme 393. Battersby Synthesis of (±)-Bonellin Dimethyl Ester

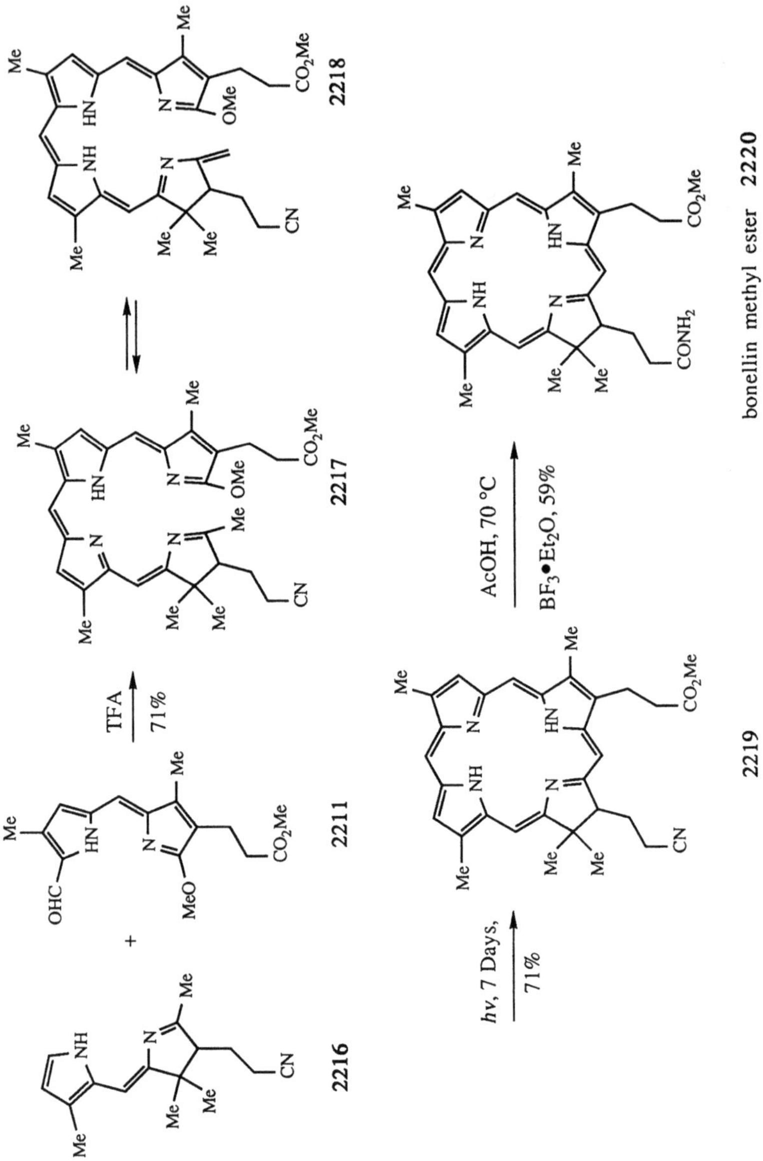

Scheme 394. Joining of Bonellin Dimethyl Ester Fragments (Battersby)

6 Miscellaneous Metabolites

6.1 Metabolites Related to Citric Acid

6.1.1 Delesserine

Delesserine (**2226**) is a secondary metabolite isolated from the alga *Delesseria sanguinea* (Lamouroux). Its relative configuration has been determined by X-ray crystal structure analysis [600]. Although aqueous extracts of *D. sanguinea* collected on the European Atlantic coast are powerful anticoagulants for human blood [601], delesserine does not exhibit this property. The first synthesis of **2226**, by Seebach [602], is shown in Scheme 395. Condensation of ester **2222** with aldehyde **2223** (available in four steps from diethyl tartrate) gives a mixture of four diastereomers **2224**. Oxidation of **2224** and treatment with TsOH in methanol provides lactone **2225** in 32% yield along with 59% of its diastereomer. Removal of the benzyl protecting groups affords (+)-delesserine in four steps (eight steps from diethyl tartrate) and 18% overall yield.

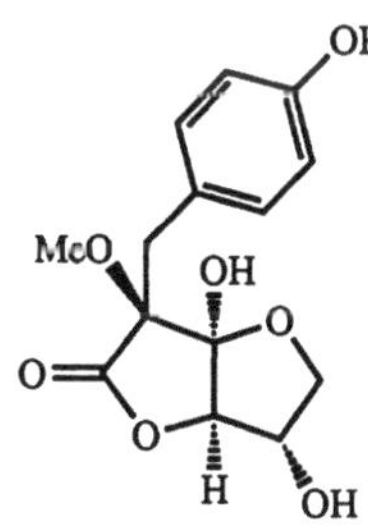

(+)-Delesserine **2226**

Poss utilizes ascorbic acid as a chiral template to provide a short synthesis [603] (Scheme 396) of (+)-delesserine (**2226**) along with the structurally related brominated metabolites (+)-rhodomelol (**2230**) and (+)-methylrhodomelol (**2232**) (**2230** and **2232** have been isolated [604] from the red alga *Polysiphonia lanosa*). Methylation of the dianion of ascorbic acid provides monomethyl derivative **2227** [605]. Reaction of **2227** with *p*-methoxybenzyl alcohol in water at 50 °C gives (+)-delesserine in 80% yield. The reaction is thought to proceed

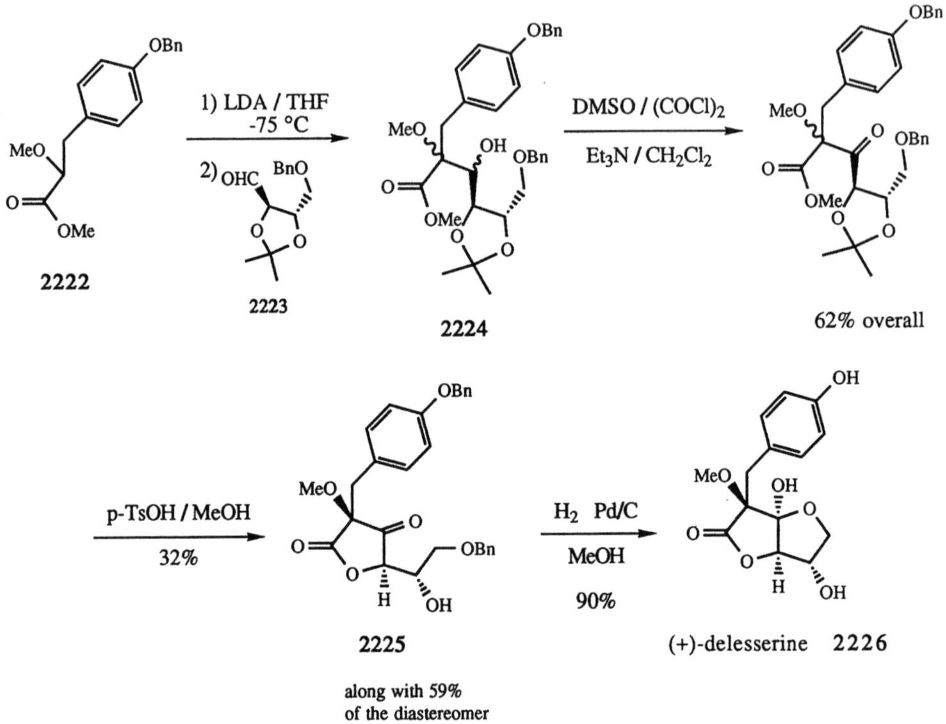

Scheme 395. Seebach Synthesis of (+)-Delesserine

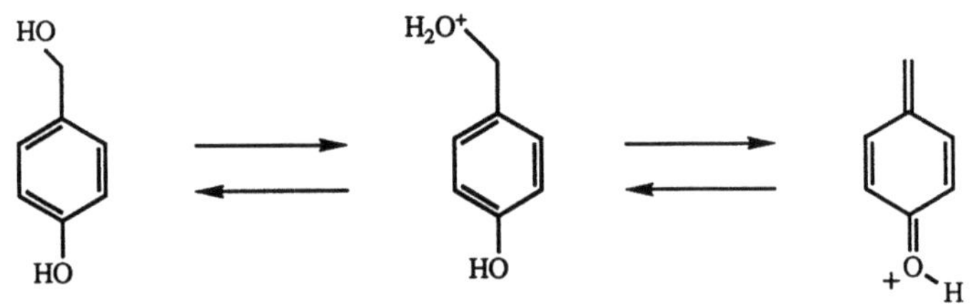

2228

via *p*-quinone methide **2228** as diagrammed followed by C-alkylation of the ascorbic acid to give **2226**. No reaction is observed with either *o*- or *m*-hydroxybenzyl alcohol. In a similar manner, rhodomelol and methylrhodomelol are obtained from brominated phenols **2229** and **2231**.

6.1.2 Leptosphaerin

In the course of studies of chemical constituents of higher marine fungi, leptosphaerin (**2239**) was isolated from laboratory cultures of the ascomycete

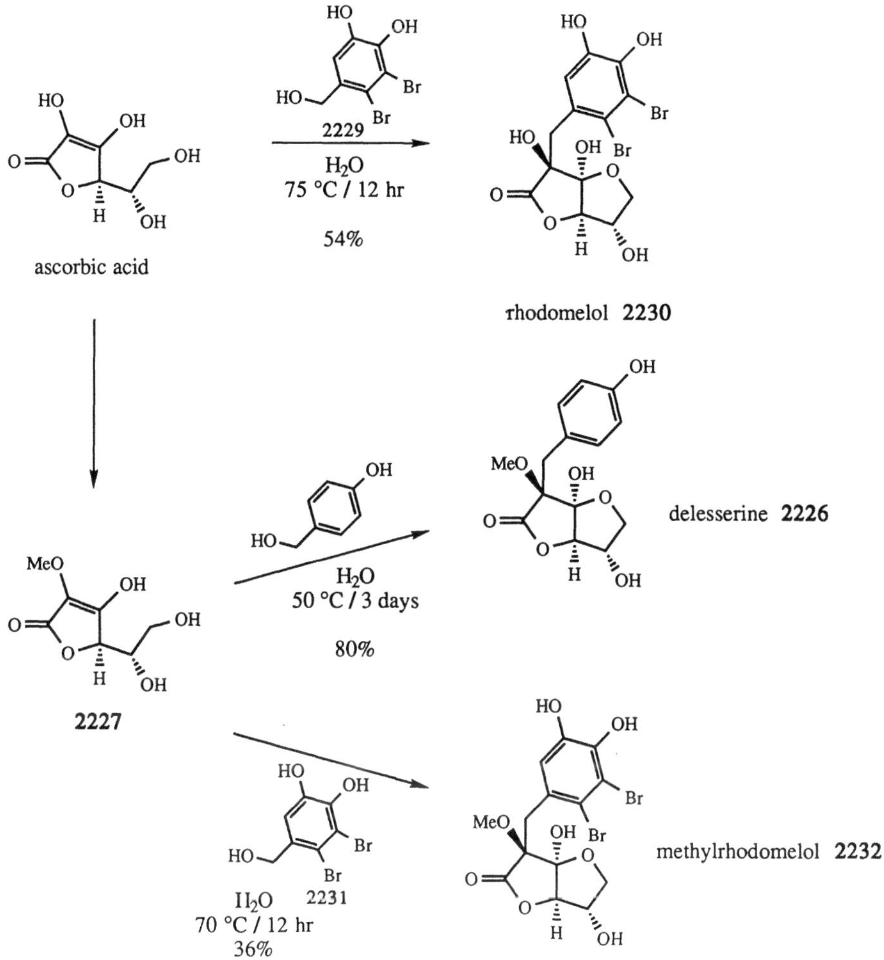

Scheme 396. Poss Synthesis of (+)-Delsserine, (+)-Rhodomelol and (+)-Methylrhodomelol

Leptosphaeria oraemaris (Linder) [606]. Spectroscopic and X-ray crystallographic analyses initially led to an incorrect assignment of the structure of leptosphaerin [607]. This was corrected following the synthesis of **2232**. The

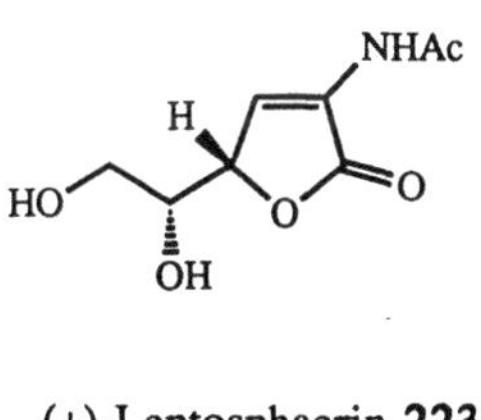

(+)-Leptosphaerin **2239**

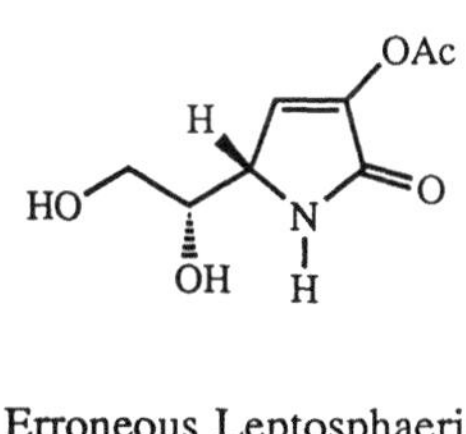

Erroneous Leptosphaerin Structure **2232**

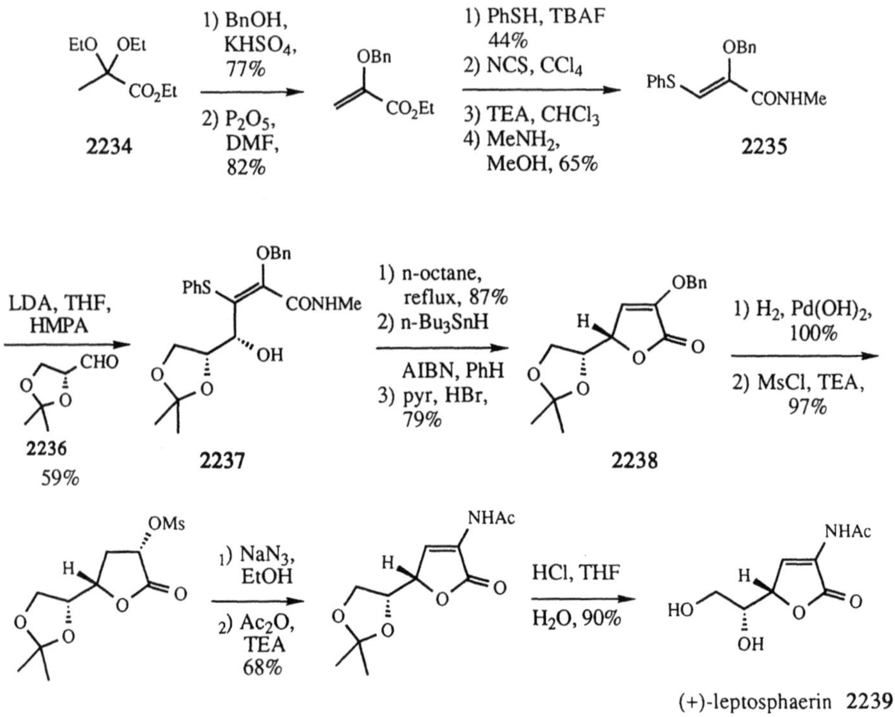

Scheme 397. White Synthesis of (+)-Leptosphaerin

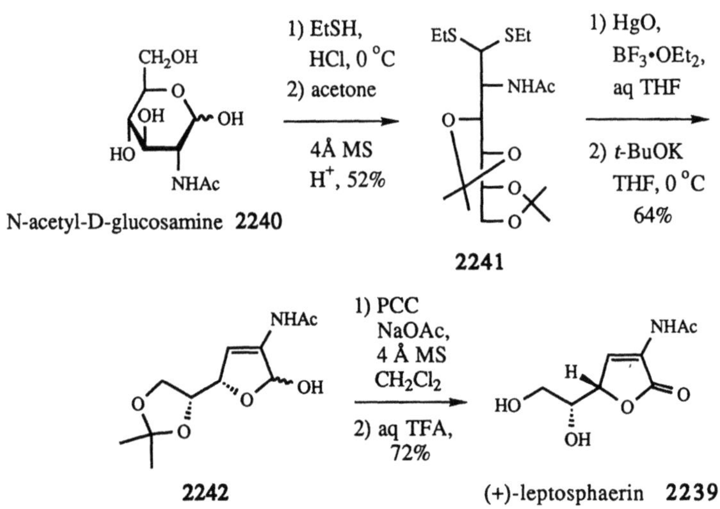

Scheme 398. Rollin Synthesis of (+)-Leptosphaerin

synthesis of leptosphaerin (**2239**) by White [608] is illustrated in Scheme 397. Transformation of the diethyl ketal of ethyl pyruvate (**2234**) to amide **2235** occurs in six steps and 18% yield. Addition of the lithio dianion of **2235** to aldehyde **2236** provides alcohol **2237** as a crystalline product in 59% yield. Closure to the γ-lactone and removal of the phenylthio moiety with *n*-Bu_3SnH gives benzyl enol ether **2238**. Completion of the synthesis requires five steps to replace the benzyloxy substituent with an acetamido group and give (+)-leptosphaerin (**2239**) in a total of 15 steps and an overall yield of 4.3%.

In the following year, a second synthesis of (+)-leptosphaerin by Rollin appeared [609] (Scheme 398). *N*-Acetyl-D-glucosamine (**2240**) is ring-opened and fully protected to give **2241**. Unmasking of the aldehyde followed by α,β-elimination gives the previously known lactol **2242** as a mixture of anomers. Oxidation and hydrolysis gives leptosphaerin in six steps and 24% overall yield.

6.2 Brominated Phenolic Ethers

A variety of highly brominated diphenyl ethers that act as self-defense substances have been isolated from *Dysidea herbacea* and *Ptychodera flava laysanica* [610]. Many of these exhibit antibacterial activity against Gram-positive and -negative bacteria. The synthesis of a highly brominated *P. flava* metabolite by Yamamura [611] is shown in Scheme 399. Anodic oxidation of brominated phenol **2244** affords trimer **2245**. Reduction of **2245** with zinc in acetic acid provides biphenyl ether **2246** in 26% yield along with 43% of byphenyl **2247**. Demethylation of **2246** by treatment with BBr_3 gives **2248** in three steps and 26% overall yield.

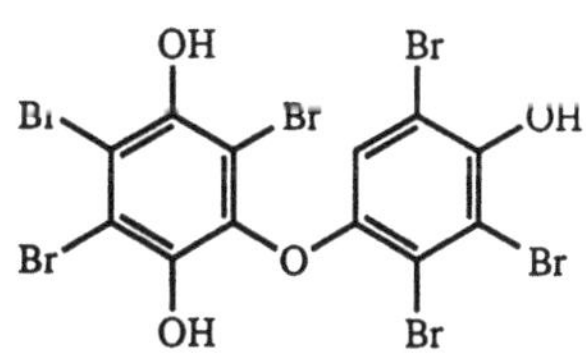

2248

A series of 2-phenoxy substituted brominated phenols has been isolated from sponges of the *Dysidea* sp. [612] and one member of the *Callyspongiidae* family [613]. Ghisalberti has synthesized [614] five of these, which are shown in Scheme 400. Coupling of potassium phenoxide **2250** with nitro-activated phenyl bromide **2249** provides diphenyl ether **2251**. Control of temperature is crucial at this point, as higher temperatures lead to the formation of dioxins. Reduction of the nitro group affords the amine **2252** which is subjected to substitutive

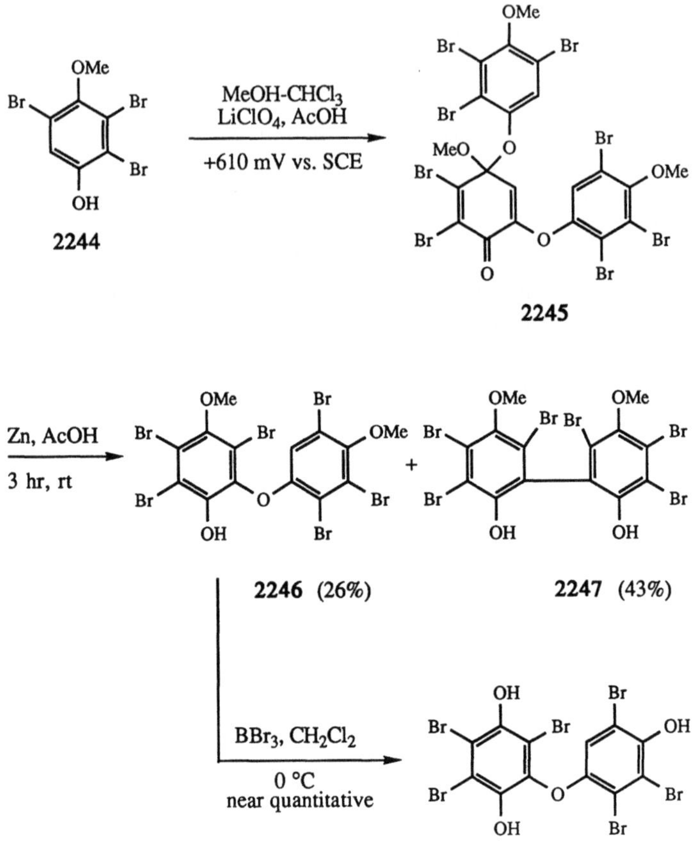

2244 2245 2246 (26%) 2247 (43%) 2248

Scheme 399. Yamamura Synthesis of Polybrominated Diphenyl Ethers from *Ptychodera flava*

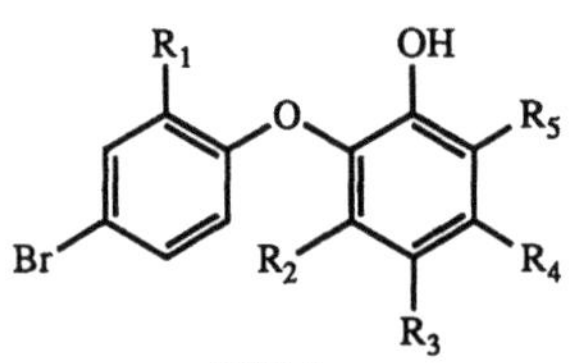

2254

	R_1	R_2	R_3	R_4	R_5
a	Br	Br	Br	Br	H
b	Br	Br	H	Br	H
c	H	Br	H	Br	H
d	H	Br	H	H	H
e	Br	H	Br	H	Br

deamination according to the procedure of Doyle [615] to give methoxy compound **2253**. Demethylation yields the natural products **2254**.

As part of the structure elucidation process, Sharma and Vig synthesized the antibacterial pentabromodiphenyl ether **2258** as shown in Scheme 401 [616]. The phenol **2256** was tribrominated and coupled to 2,4-dinitro-chlorobenzene

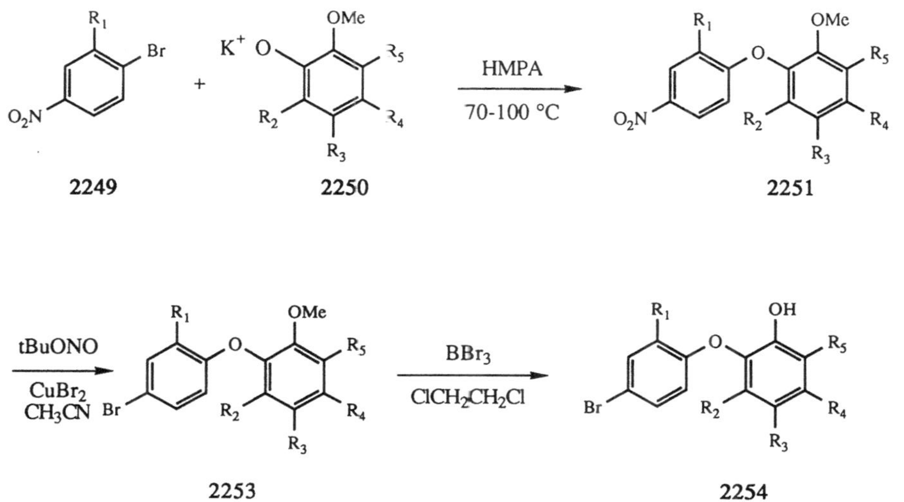

	R_1	R_2	R_3	R_4	R_5
a	Br	Br	Br	Br	H
b	Br	Br	H	Br	H
c	H	Br	H	Br	H
d	H	Br	H	H	H
e	Br	H	Br	H	Br

Scheme 400. Ghisalberti-Francesconi Synthesis of Polybrominated Diphenyl Ethers from Sponges

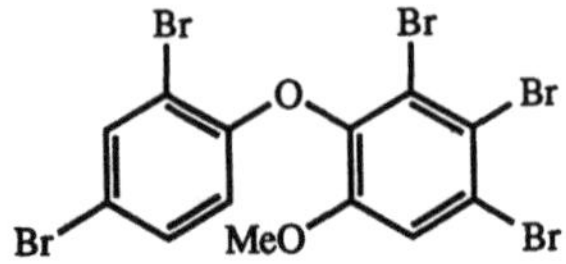

2258

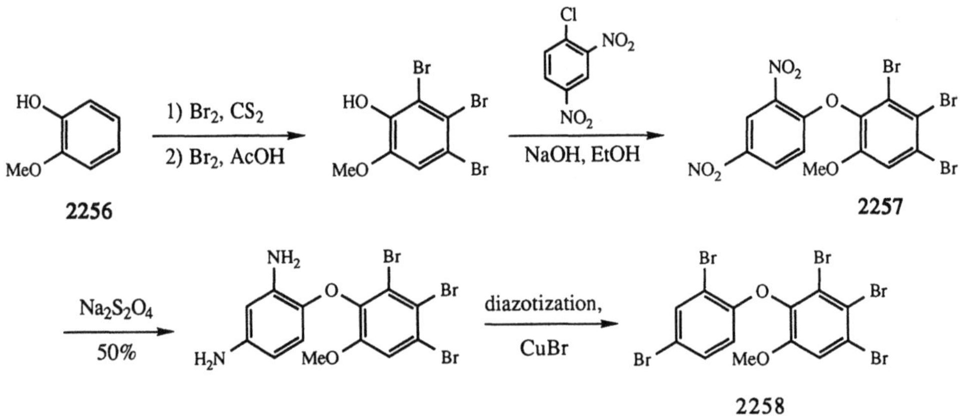

Scheme 401. Sharma Synthesis of Brominated Diphenyl Ether from *Dysidea herbacea*

via an addition-elimination process to give **2257**. Reduction of the nitro groups to the amines followed by Sandmeyer reaction led to **2258** which was identical to the natural product.

6.3 Others

6.3.1 Metabolites of *Delisea fimbriata*

Red algae of the Bonnemaisoniaceae are sources of large amounts of a number of polyhalogenated substances. Sims has synthesized several halogenated 1-octen-3-ones as part of the structure elucidation process [617] (Scheme 402).

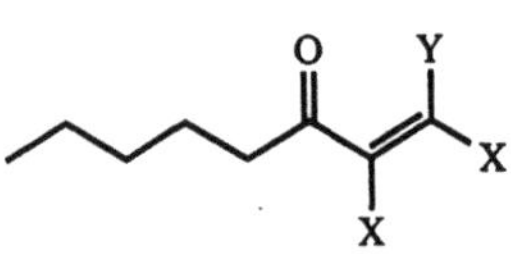

2262 X, Y = Br
2263 X = Br, Y = I

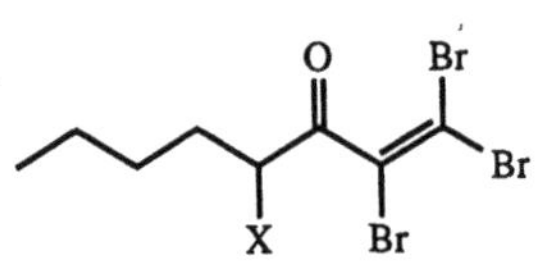

2264 X=Br
2267 X=Cl

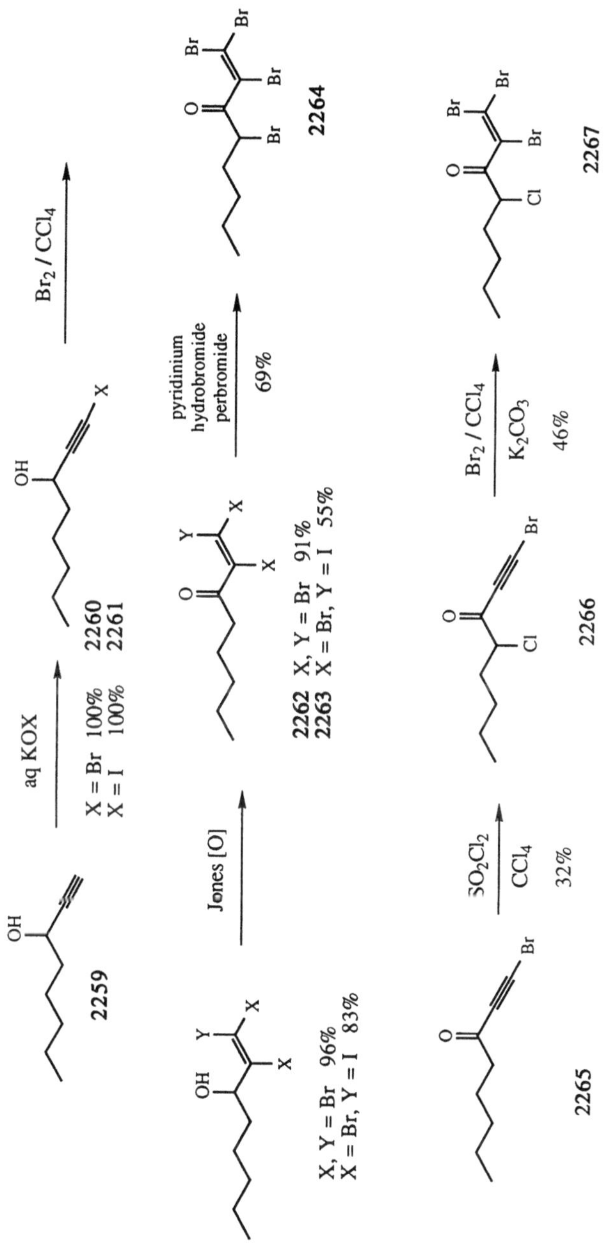

Scheme 402. Sims Synthesis of Haloketones from *Delisia fimbriata*

The known octynol **2259** was terminally halogenated to produce the iodide **2261** and bromide **2260**. The bromoalkyne was further brominated and oxidized to metabolite **2262**. Controlled α-monobromination of **2262** produced the tetrabromo metabolite **2264**. The iodoalkyne **2261** was treated similarly to produce metabolite **2263**. Metabolite **2267** was produced by α-chlorination of the alkynone **2265** to give **2266** which was then brominated. It is thought that natural products in *Asparagopsis* and *Bonnemaisonia* arise via a bromoperoxidase mechanism. Support for this idea was found by Hager [618] who produced a number of naturally occurring simple haloketones by halogenation catalyzed by an algal-derived enzyme extract.

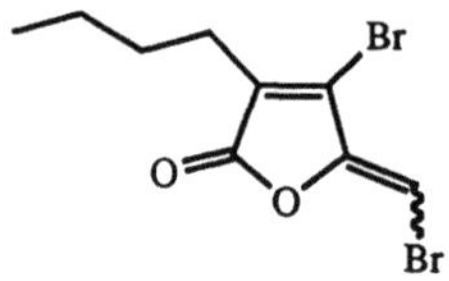

Fimbrolides
(*E* and *Z*)

Delisea fimbriata produces a number of halogenated lactones, including the fimbrolides (Scheme 403) [619]. Sims [620] has also synthesized two members of this family, confirming their structures. The keto acid was produced using standard chemistry. Symmetrical bromination to **2269** occurred on reaction with bromine/$CHCl_3$ to give a dibromide which was not separated, but directly cyclized with accompanying oxidation using 100% H_2SO_4 to a mixture of the isomeric fimbrolides shown.

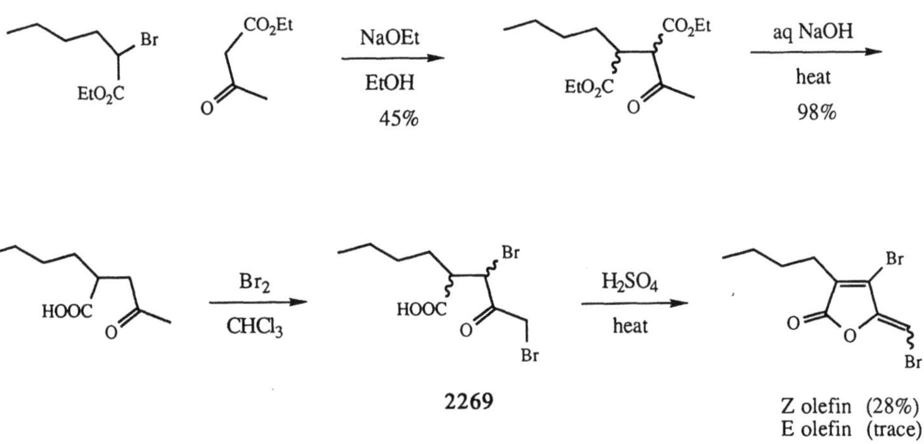

2269

Z olefin (28%)
E olefin (trace)

Scheme 403. Sims Synthesis of Fimbrolides from *Delisea fimbriata*

6.3.2 Kjellmanianone

Kjellmanianone is a member of the cyclopentanoid class of antibiotics. It was first isolated in 1980 from the brown alga *Sargassum kjellmanianum* and was shown to have moderate activity against Gram-positive bacteria. Its absolute configuration has been determined by single crystal X-ray analysis [621]. Smith's synthesis [622] of kjellmanianone (**2272**) is illustrated in Scheme 404. Oxidation of the enolate of **2271** with MCPBA provides **2272** in two steps and 32% yield from ketone **2270**. The synthesis can be made stereoselective, if an (+)-*N*-aryl-camphorsulfonyloxaziridene replaces MCPBA as the oxidizing agent. The natural enantiomer (+)-kjellmanianone is obtained in 36.5% enantiomeric excess.

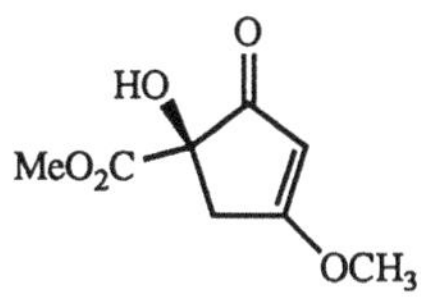

(+)-Kjellmanianone 2272

6.3.3 Pukeleimide A

Along with the known irritants debromoaplysiatoxin and lyngbyatoxin, the blue-green alga *Lyngbya majuscula* also produces [623] the pukeleimides. The first synthetic entry into this class of 5-ylidenepyrrole-2-ones is the preparation of pukeleimide A (**2277**) by Pattenden (Scheme 405) [624]. Treatment of *N*-methylimide **2274** with five equivalents of carboethoxymethylenetriphenylphosphorane regio- and stereo-selectively affords ester **2275** in low yield. Reaction of

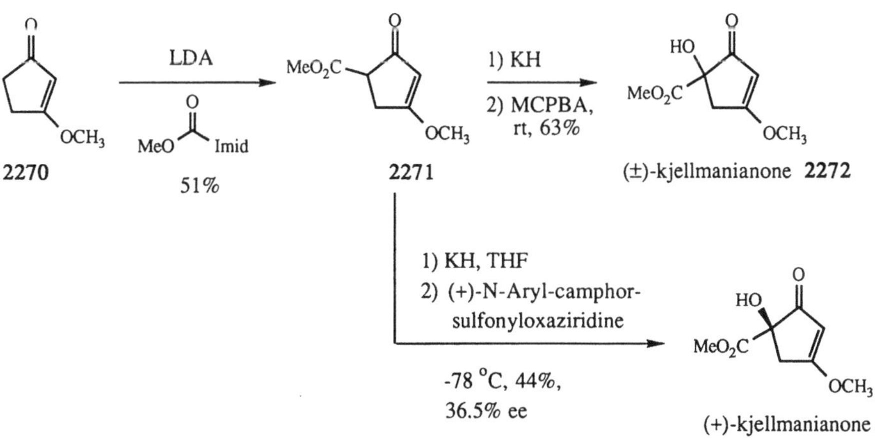

Scheme 404. Smith Syntheses of (±)- and (+)-Kjellmanianone

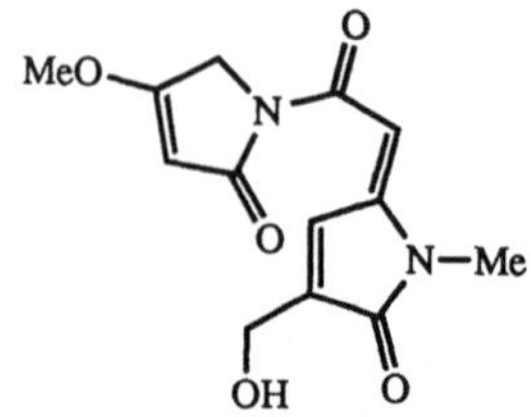

Pukeleimide A **2277**

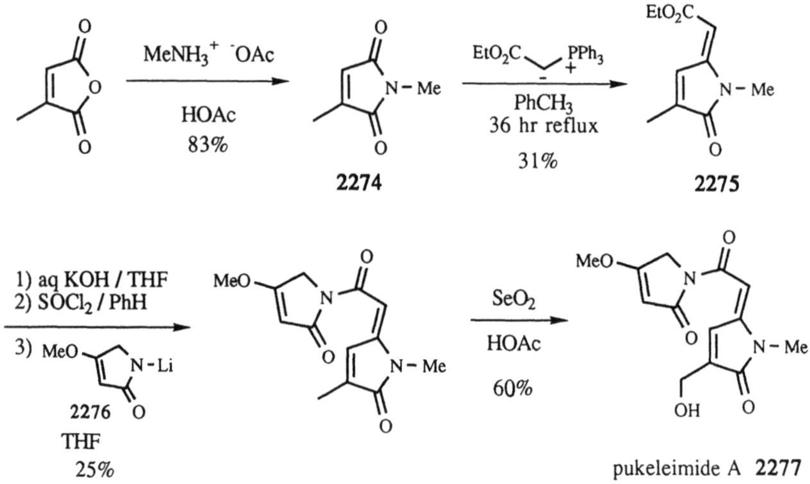

Scheme 405. Pattenden Synthesis of Pukeleimide A

the acid chloride of **2275** with lithiated amide **2276** followed by allylic oxidation provides pukeleimide A in six steps with an overall yield of 3.9%.

6.3.4 Latrunculin B

Latrunculin B (**2288**) and the related latrunculin A are novel macrocyclic toxins isolated from the Red Sea sponge *Latrunculia magnifica* (Keller), found in the Gulf of Eilat. When disturbed, the sponge emits a reddish fluid that causes any nearby fish to leave the vicinity immediately [625]. The latrunculins also exhibit the interesting property of inducing reversible reorganization of cytoskeletal proteins. The structure of latrunculin A was determined through spectroscopic, degradative and X-ray crystal analysis. The structure of latrunculin B was determined by spectroscopic comparison with latrunculin A. Smith's synthesis [626] of (+)-latrunculin B, shown in Scheme 407, is convergent. The general

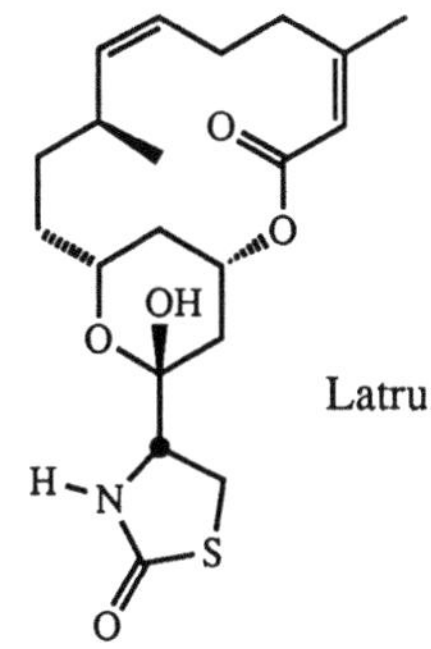

Latrunculin B **2288**

strategy involves aldol coupling of ketone **2283** with aldehyde **2281** to afford compound **2284**. Rearrangement of **2284**, homologation and macrolactonization completes the synthesis. Fragments **2281** and **2283** are prepared as shown in Scheme 406. Baeyer-Villiger oxidation of ketone **2278** and methylation affords a 1:1 mixture of diastereomeric lactones **2279**. Conversion to the

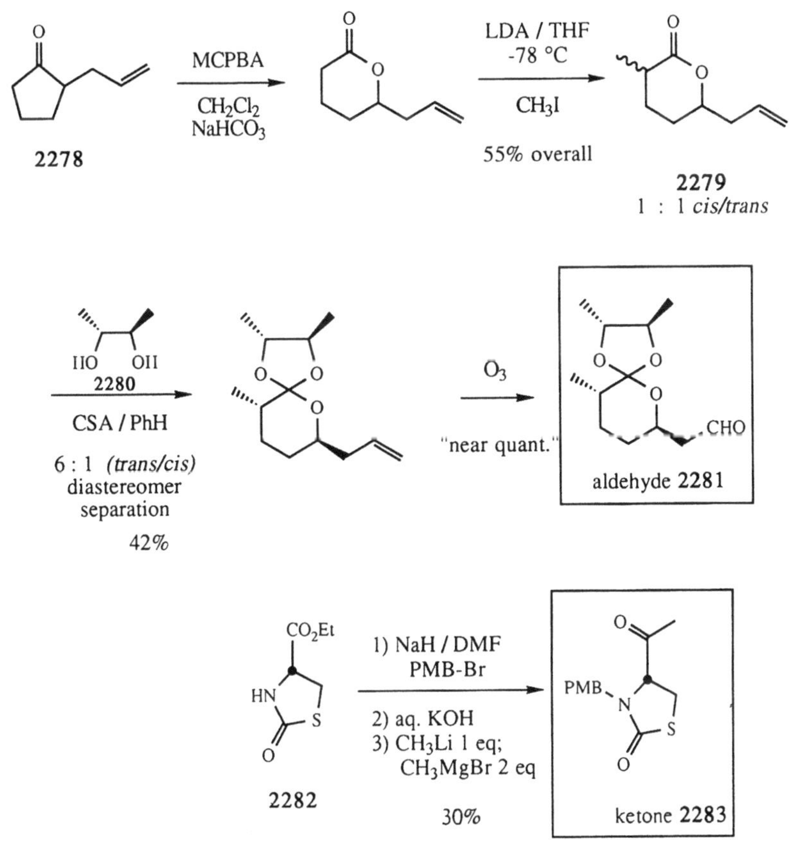

Scheme 406. Synthesis of Latrunculin B Fragments

orthoesters with chiral diol **2280** equilibrates the diastereomers to give a 6:1 *trans* to *cis* ratio and allows separation, to give aldehyde **2281** in 23% yield after ozonolysis. Ketone **2283** is obtained from ester **2282** [627] in three steps and 30% yield. Aldol condensation between **2281** and **2283** proceeds to give diastereomeric mixture **2284** (Scheme 407). Treatment of **2284** with catalytic acid in methanol results in hydrolysis of the orthoester and formation of the mixed methyl ketal **2285**. Conversion of **2285** to aldehyde **2286** and Wittig olefination provides lactone precursor **2287**. Lactonization of **2287** and deprotection yields (+)-latrunculin B in 13 steps and 0.68% overall yield.

6.3.5 Bisucaberin

Siderophores are iron-chelators which are produced by various organisms to acquire exogenous iron(III). The naturally occurring siderophore bisucaberin has been isolated from the marine bacterium *Alteromonas haloplanktis* and possesses a symmetrical hydroxyamide structure [628]. Bergeron [629] has produced this substance in short order, naturally taking advantage of this symmetry (Scheme 408). *O*-Benzylhydroxylamine was protected and alkylated to give **2290**, which was converted to both fragments **2291** and **2292**. These were coupled classically to give **2293** and converted to the *seco* compound **2294**. Macrocyclization occurred in 43% yield and debenzylation gave the natural siderophore **2295**.

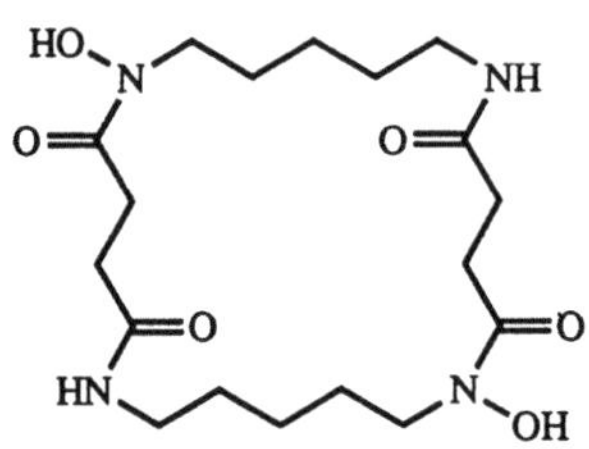

Bisucaberin **2295**

6.3.6 Hormothamnione

Hormothamnione (**2300**) was isolated [630] by Gerwick from the blue-green alga *Hormothamnion enteromorphoides* in 1986. It is the first example of a naturally occurring styrylchromone. Hormothamnione exhibits potent cytotoxicity vs. P388 lymphocytic leukemia and HL-60 human promyelocytic leukemia cell lines in vitro and appears to be a selective inhibitor of RNA synthesis. The first syntheses of **2300** appeared in 1988. Brossi [631] prepared **2300** in a 14-step process (Scheme 409). Conversion of 2,3,4-trimethoxybenzaldehyde (**2296**) occurs in seven steps to afford the key intermediate pentamethoxybenzene **2297** in 54% yield. Friedel-Crafts acylation of **2297**

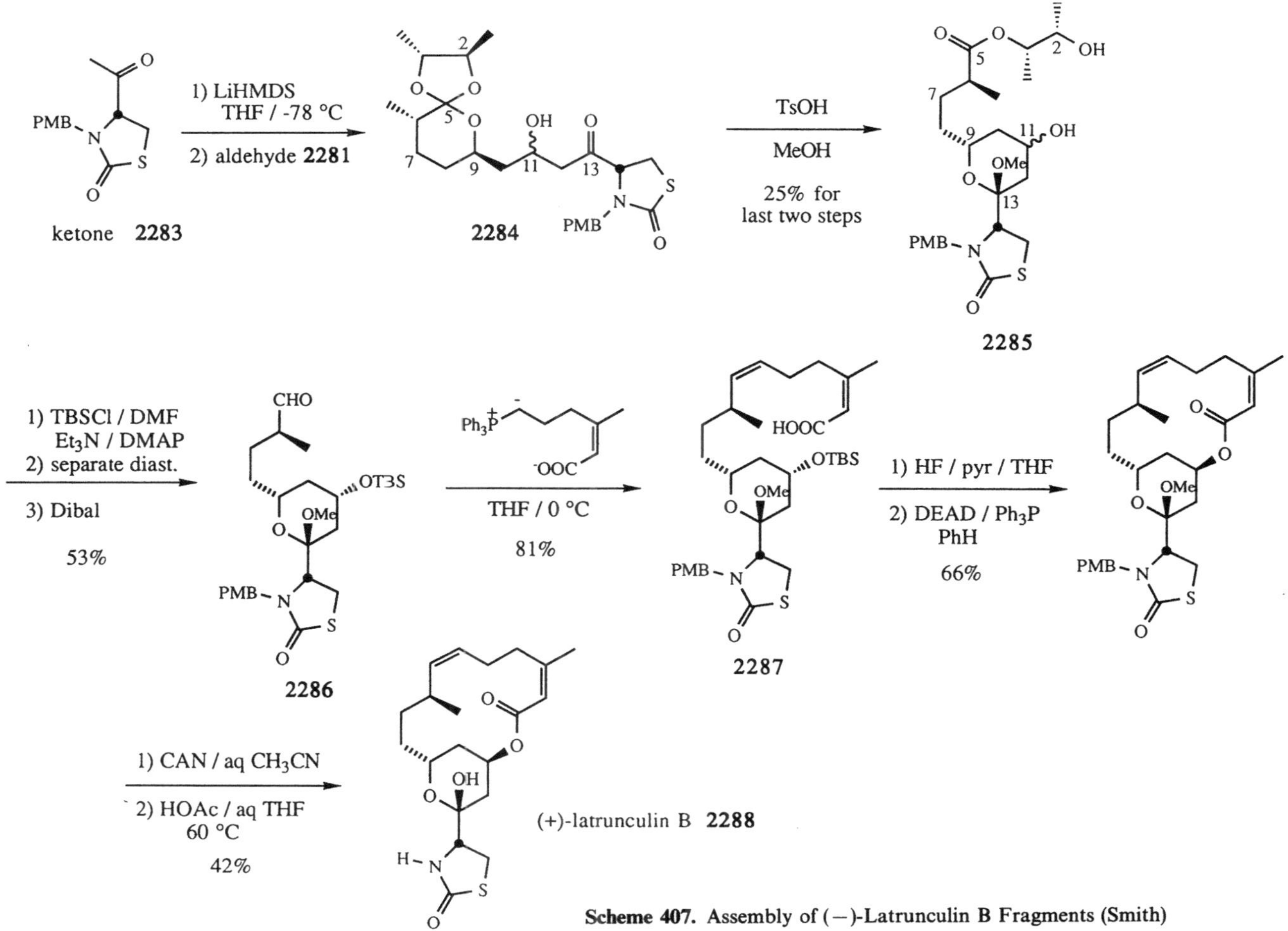

Scheme 407. Assembly of (−)-Latrunculin **B** Fragments (Smith)

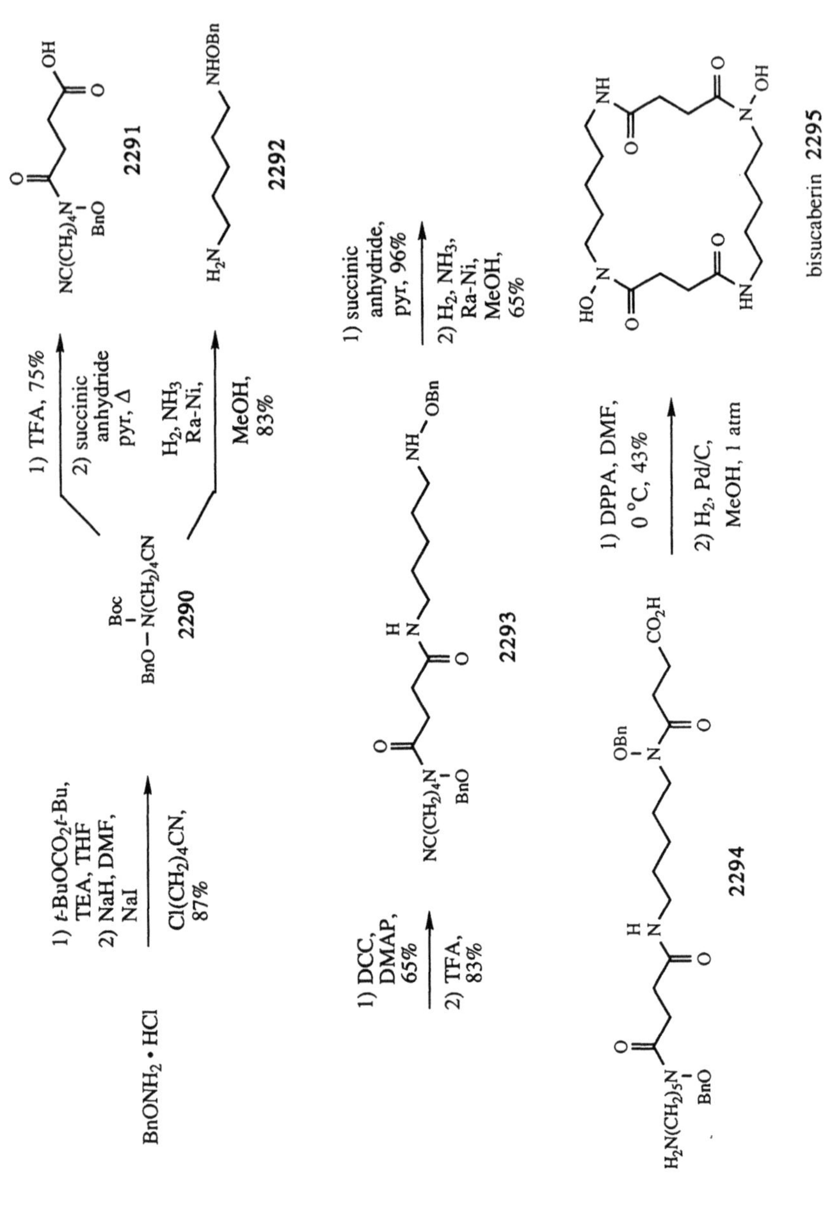

Scheme 408. Bergeron Synthesis of Bisucaberin

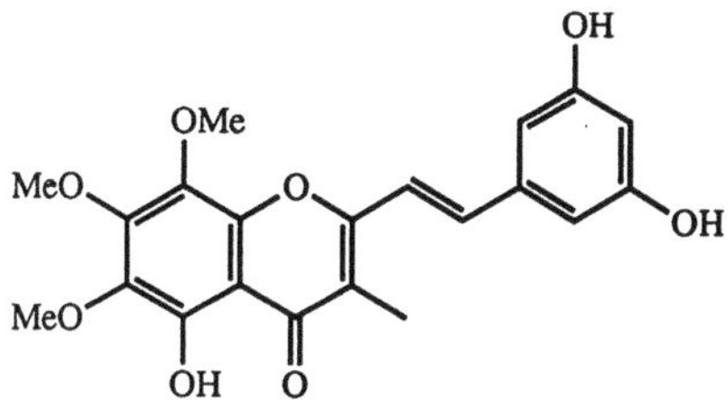

Hormothamnione **2300**

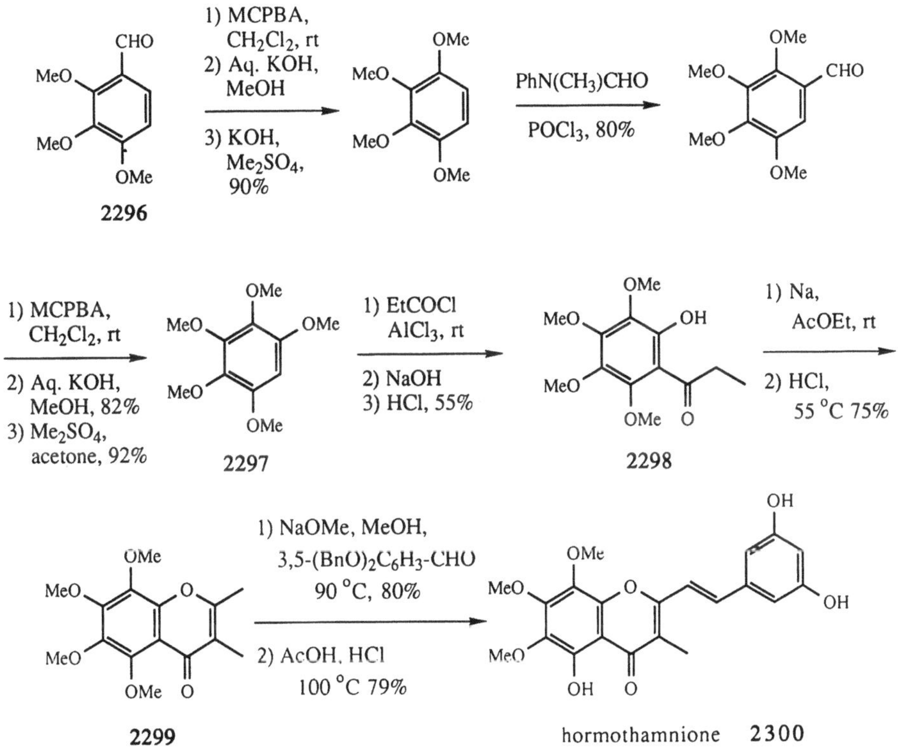

Scheme 409. Brossi Synthesis of Horomothamnione.

with propionyl chloride occurs with monodemethylation of an ortho methoxy group to give **2298**. Claisen condensation of ketone **2298** with ethyl acetate followed by cyclization gives chromone **2299**. Condensation of **2299** with 3,5-dibenzyloxybenzaldehyde followed by debenzylation and selective demethylation of the adduct provides hormothamnione **2300** in 14% overall yield.

A second synthesis of hormothamnione by Ayyangar [632] appeared later (Scheme 410). Monomethylation of ketone **2302** produces the same intermediate

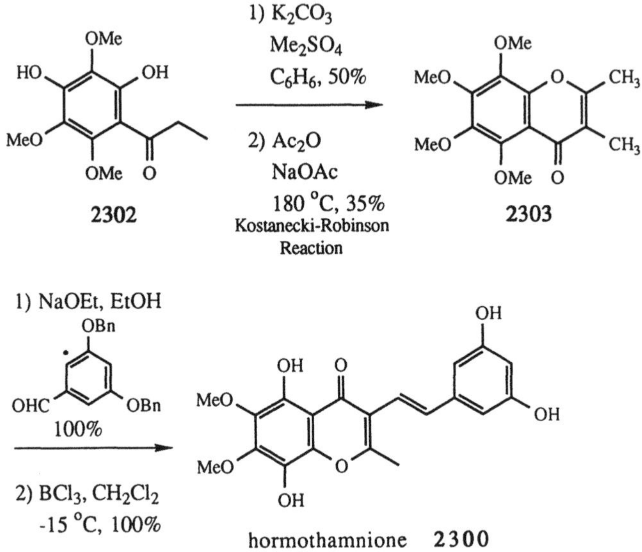

Scheme 410. Ayyangar Synthesis of Hormothamnione

as that utilized by Brossi (Scheme 409). Application of the Kostanecki-Robinson reaction to this tetramethoxypropiophenone yields chromone **2203**. Condensation of **2203** with 3,5-dibenzyloxybenzaldehyde, via the dienolate and monodemethylation affords hormothamnione (**2300**) in 35% yield from the common intermediate.

6.3.7 Bissetone

Bissetone (**2306**) is a metabolite of the soft coral *Briareum polyanthes*, isolated [633] in 1987. Lichtenthaler's synthesis [634] of **2306** is outlined in Scheme 411. The key starting material for Lichtenthaler's preparation of bissetone is the dihydropyranone **2304**, available form D-glucose in two one-flask conversions in 67% yield. Stereospecific addition of methylallyl titanium isopropoxide followed by saponification gives diol **2305**. Use of the lithium enolate of acetone gives **2306** and its epimer in a 5:1 ratio. Ozonolysis of **2305** provides bissetone in three steps and 59% overall yield (four steps and 47% yield from D-glucose).

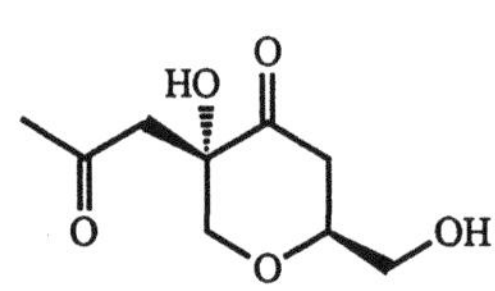

Bissetone 2306

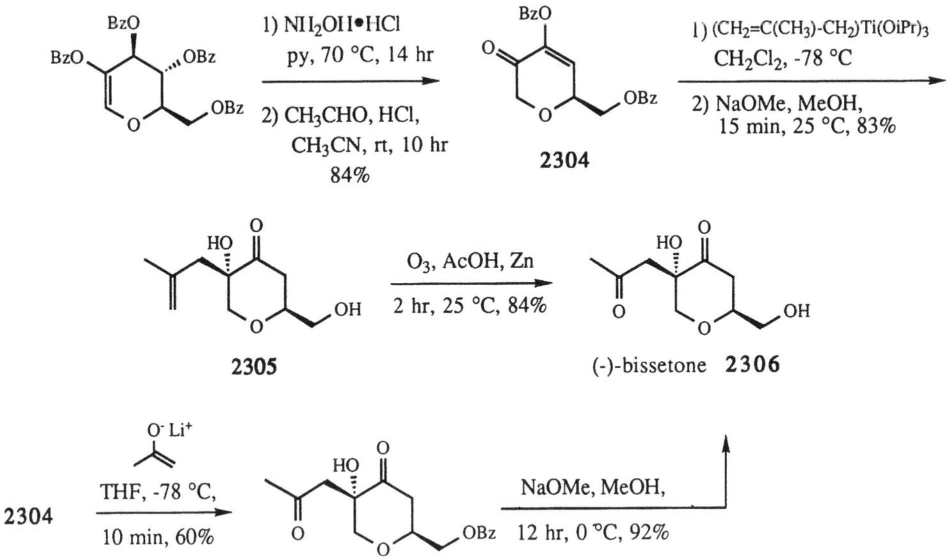

Scheme 411. Lichtenthaler Synthesis of (−)-Bissetone

6.3.8 *(S,S)*-Palythazine

Along with the very poisonous palytoxin and a number of cyclohexanoid iminium salts, *Palythoa tuberculosa* produces [635] two heterocycles containing the dipyranopyrazine skeleton **2310** and **2311**. One of these, (*S*,*S*)-palythazine (**2310**), is clearly related to bissetone and has also been prepared by Lichtenthaler [636] (Scheme 412). Formation of the oxime **2308** (readily available from D-glucose) followed by saponification gives pyranone **2309**. Hydrogenation to afford the amine and air oxidation gives the natural product **2310** in three steps and 49% yield from **2308**. Interestingly, the physical properties of **2310** suggest that the original structural assignments of palythazine and isopalythazine may be reversed.

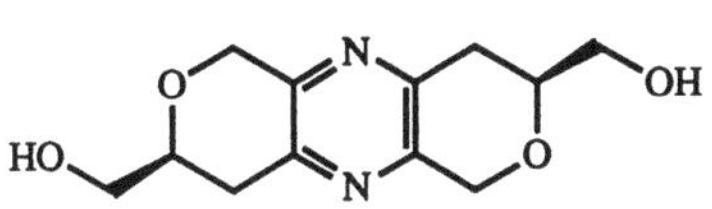

Palythazine **2310**

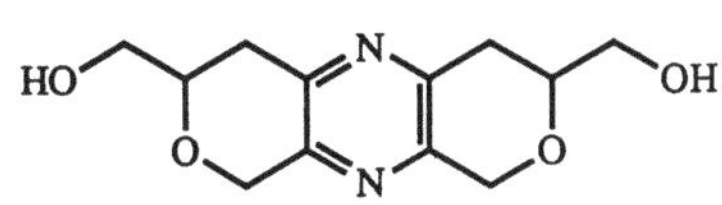

Isopalythazine **2311**

6.3.9 Dysidin

Williard's synthesis [637] of dysidin [638] appeared in 1984 (Scheme 413). The general strategy involves the independent preparation of two fragments (**2314**

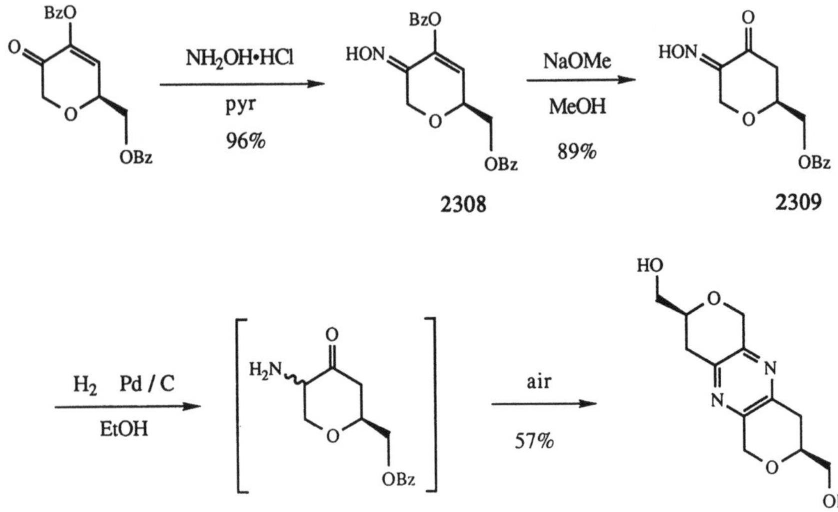

(S, S)-palythazine 2310

Scheme 412. Lichtenthaler Synthesis of (*S*, *S*)-Palythazine

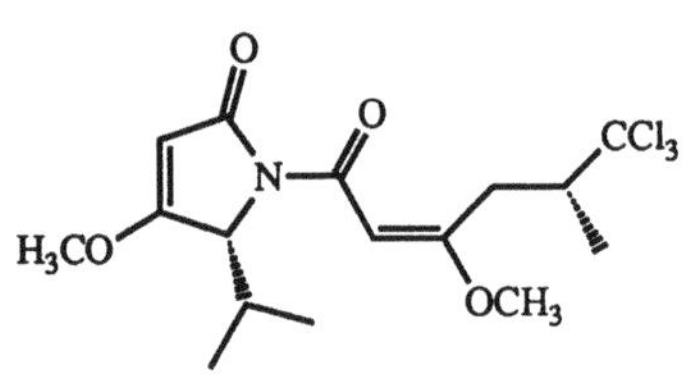

Dysidin **2317**

and **2316**) which are coupled in the last step. Because of its relative inertness, the trichloromethyl functionality was introduced early in the synthesis. Radical chain addition of bromotrichloromethane to crotonic acid affords acid **2312** as a mixture of diastereomers. Reduction of **2312**, conversion to the acid chloride and formation of the adduct with Meldrum's acid allows easy preparation of ester **2313**. Ester **2313** was chosen due to its ease of hydrolysis after the formation of methyl enol ether fragment **2314**. Fragment **2316** is formed from the *N*-phthalimide substituted acid chloride **2315** which is converted to the *O*-methyltetramic acid in three steps and 15% overall yield. Activation of **2314** by treatment with thionyl chloride and addition of **2316** gives a 1:1 mixture of (±)-dysidin along with its diastereomer. Dysidin (**2317**) is obtained in eight steps and 3.4% overall yield.

A second and similar synthesis was reported by Gerlach [639] in that same year which indicated that the absolute configuration assigned to dysidenin and isodysidenin should be revised. The Gerlach synthesis of (−)-dysidin (Scheme 414) involved the coupling of an optically active trichloromethyl hexenoic acid

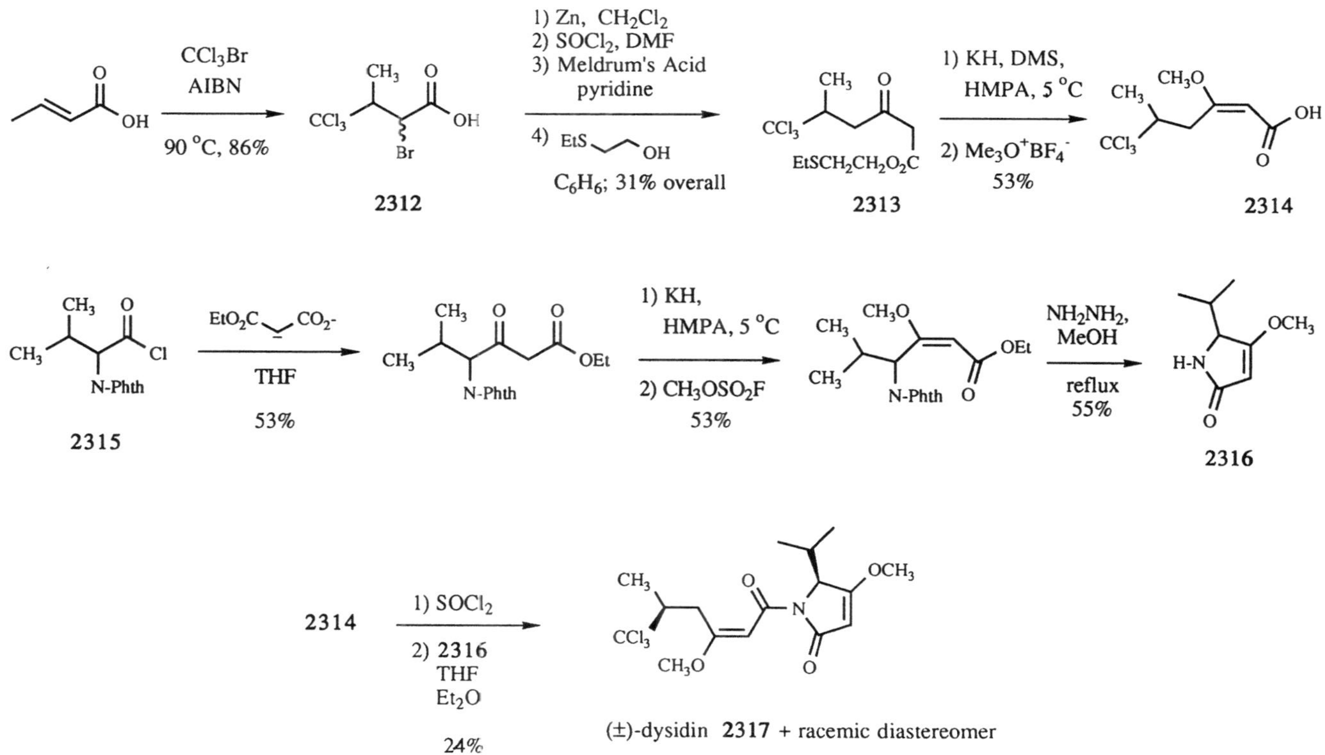

Scheme 413. Williard Synthesis of (±)-Dysidin.

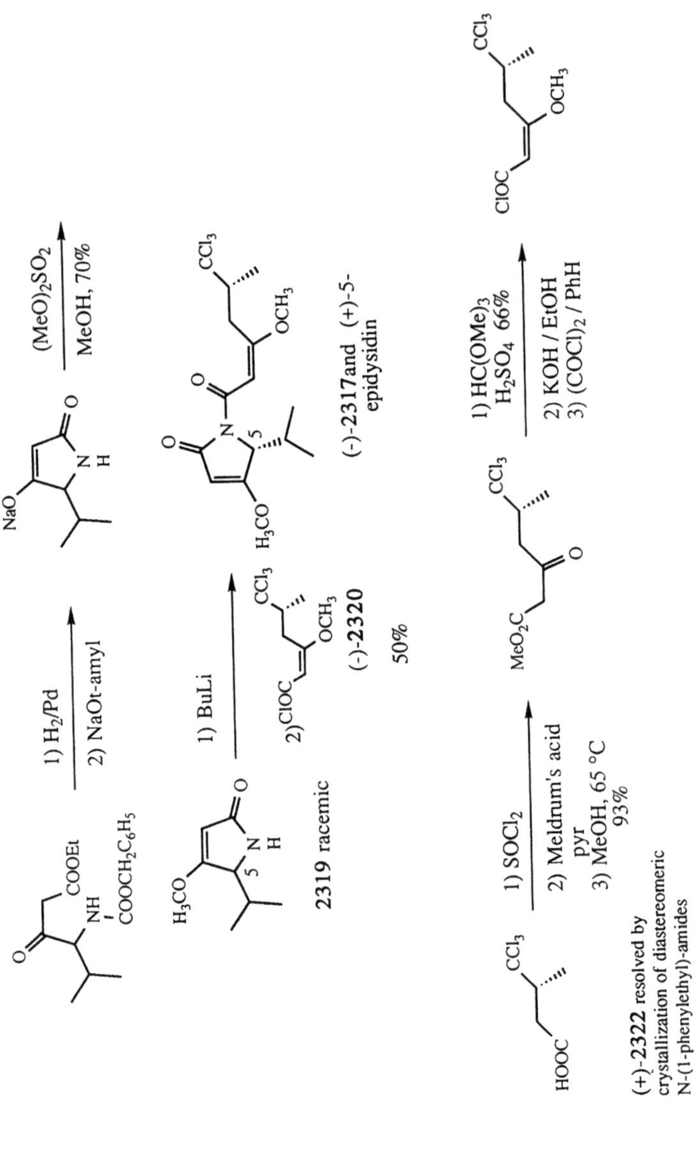

Scheme 414. Gerlach Synthesis of Dysidin

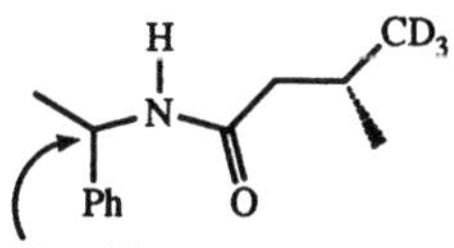

2323

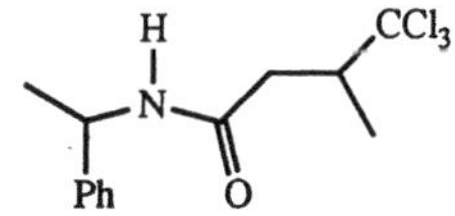

2321

piece with a racemic pyrrolinone piece. The racemic carboxylic acid **2322** was resolved via crystallization of diastereomeric amides to give the *R* isomer. This configuration was assigned by comparison of the NMR spectra of the isomers of **2322** with the corresponding trideuterio-compounds of known absolute configuration. Therefore, this structural assignment is only as good as the spectral analogy. In any case, the (−)-acid chloride **2320** was coupled with the lithium salt of the racemic pyrrolinone **2319** giving (−)-dysidin and its 5-*epi*-isomer.

6.3.10 *Grateloupia Filicina* Metabolite and Related Compounds

The pyrogallol derivative **2327** along with 3,4,5-trihydroxybenzyl methyl ether are found in the red alga *Grateloupia filicina* [640]. Both compounds show moderate antibacterial activity against *Bacillus subtilis.* An attempted synthesis of **2327** by Nakayama [641] that confirmed its structure is shown in Scheme 415. Ester **2324** is converted to sulfide **2325** in 70% yield. Reduction and oxidation produces sulfone **2326** in 4 steps and 21% overall yield. Attempts to demethylate the phenolic hydroxyls were unsuccessful. However, methylation of the natural product and comparison with sulfone **2327** confirmed the identity of **2327**.

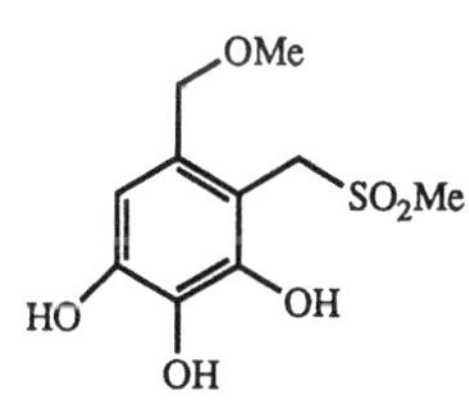

Grateloupia filicina Metabolite **2327**

During an investigation of the constituents of marine red algae, three brominated hydroxy dibenzyl ethers were isolated from the red alga *Symphyocladia latiuscla* [642]. Amiya's [643] syntheses of the methyl ethers of these compounds are illustrated in Scheme 416. The starting material, 3,4-dimethoxybenzyl acetate (**2329**), is obtained via standard methods from vanillin. Bromination of **2329** affords the tribromobenzyl bromide **2330**, which can be hydrolyzed to benzyl alcohol **2334**. Williamson ether coupling of **2330** with ethanol, methanol and substituted benzyl alcohol **2334** produces the methyl ethers **2332**, **2333** and **2331** of the natural products. Confirmation of the structures of the

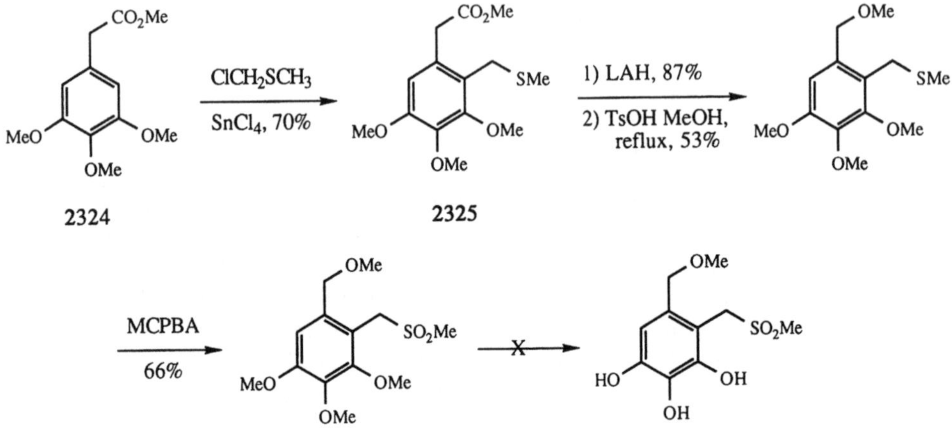

2324 2325 2326 2327

Demethylation step unsuccessful; the natural product (2326) was methylated to establish the identity of (2327)

Scheme 415. Nakayama Attempted Synthesis of *Grateloupia filicina* Metabolite **2327**

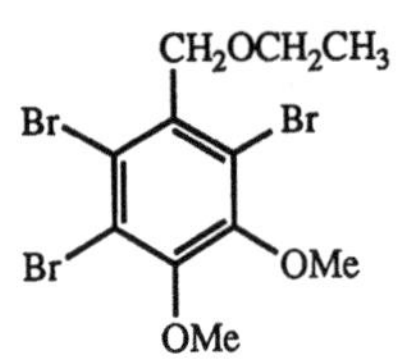

2331

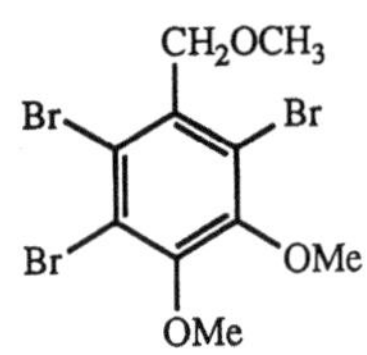

2333

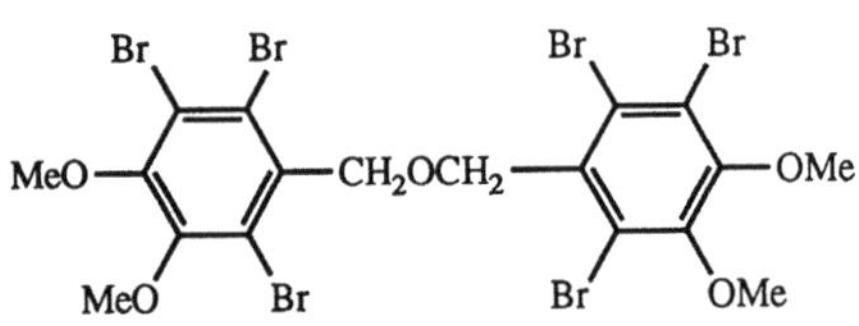

2332

Halogenated *Symphyocladia* Metabolites

methyl ethers was obtained by methylation of authentic samples of the natural materials.

6.3.11 Didemnenones A and B

The didemnid tunicate *Trididemnum cyanophorum*, collected on the seagrass beds off Shroud Cay, Bahama Islands, produces (+)-didemnenones A (**2342**)

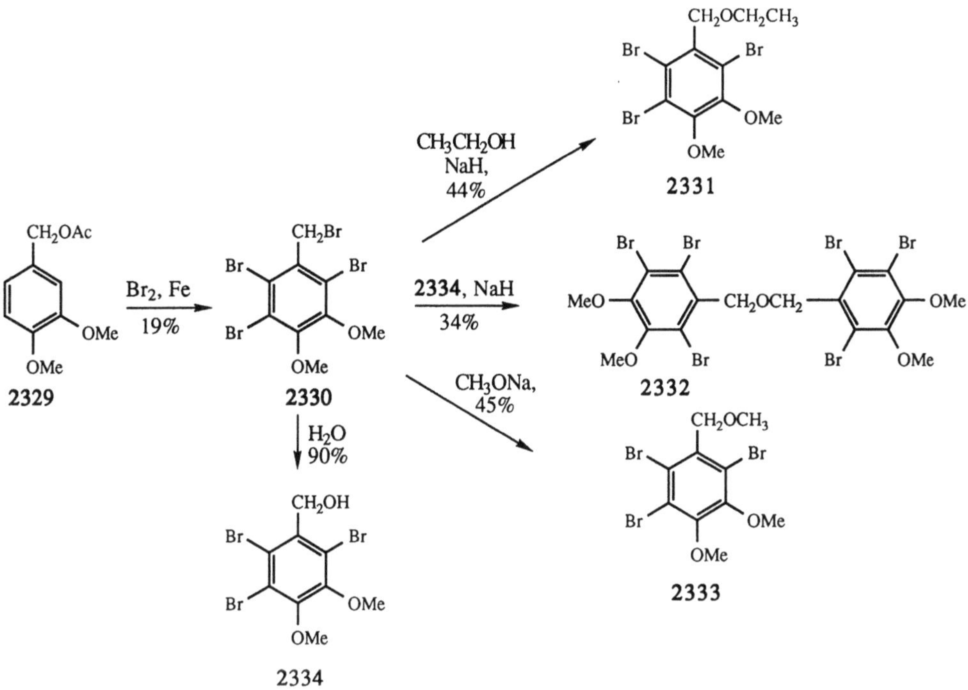

Scheme 416. Synthesis of Brominated Phenolic Compounds of the Rhodomelaceae

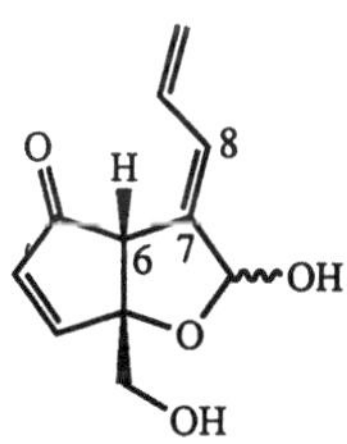

Didemnenone A **2342** (α-OH)
Didemnenone B **2343** (β-OH)

and B (**2343**). The relative stereostructures of A and B were determined by X-ray diffraction analysis followed by chemical and spectral correlations [644]. Didemnenones A and B show antibacterial activity and antifungal activity vs. the pathogenic marine fungus *Lagenidium callinectes*. Scheme 417 illustrates Clardy's synthesis of **2342** and **2343** [645]. Addition of hydroxymethyl anion equivalent **2336** to cyclopentenone **2335** (94% ee) affords alcohol **2337** as a 7:1 diastereomeric mixture. Conversion of **2337** to propargyl ether **2338** sets the stage for the key step of the synthesis, i.e., the formation of the C6–C7 bond. Mercuric chloride mediated cyclization of the alkyne onto the silyl enol ether

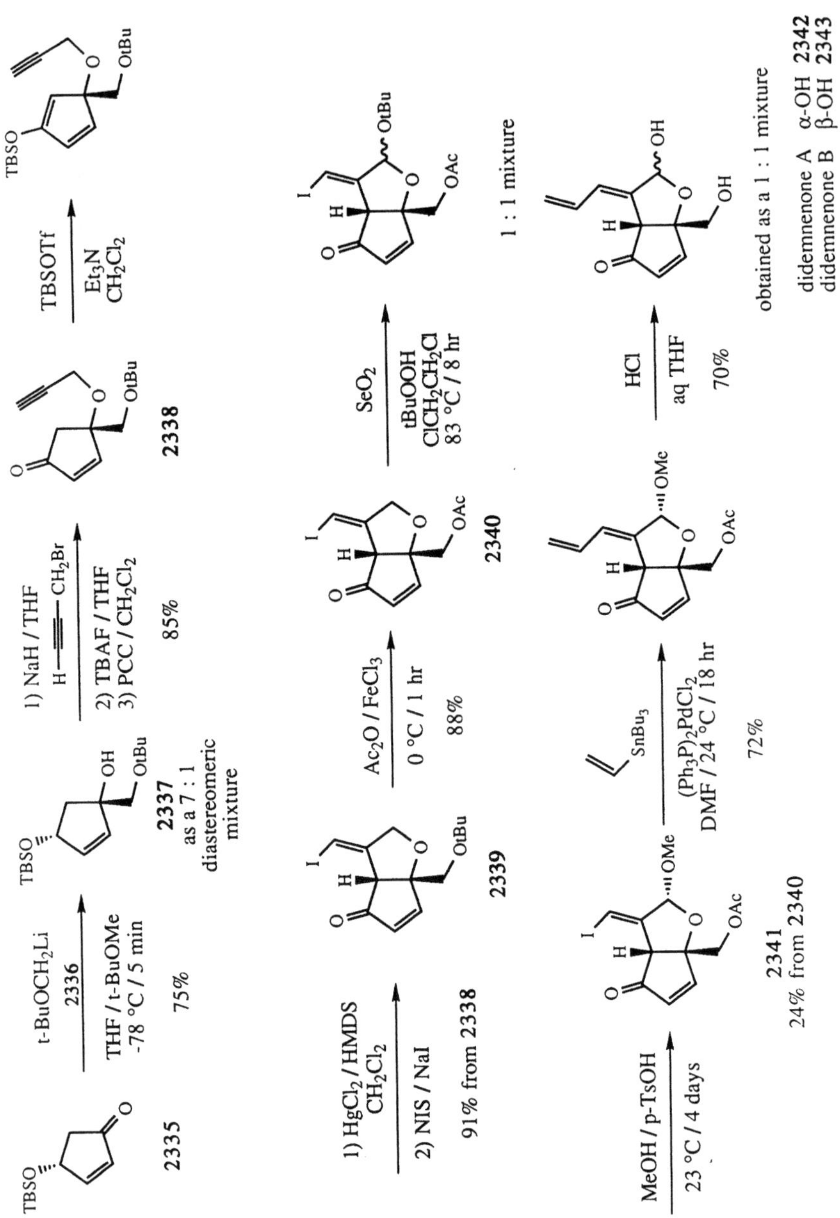

Scheme 417. Clardy Synthesis of (+)-Didemnenones A and B

followed by treatment with NIS gives iodide **2339** with the correct geometry for subsequent formation of the diene system. Iodide **2339** is transformed into compound **2341** in three steps. Palladium-catalyzed coupling of **2341** with tri-*n*-butylvinylstannane and hydrolysis provides didemnenones A and B as a 1:1 mixture. The didemnenones are obtained in only 12 steps and 5.1% overall yield.

6.3.12 *Tridacna Maxima* Metabolite

Arsenic-containing metabolites have been isolated from a variety of marine sources. One of these, the naturally occurring carbohydrate (*R*)-2′,3′-dihydroxy-propyl 5-deoxy-5-dimethylarsinyl-β-D-riboside (**2350**) has been the subject of a total synthesis by Stick [646] (Scheme 418). Metabolite **2350** has been isolated from both the brown kelp *Ecklonia radiata* [647] and the kidney of the giant clam *Tridacna maxima* [648]. Stick's synthesis of **2350** begins with orthoester **2345**, easily obtainable from commercially available 1-*O*-acetyl-2,3,5-tri-*O*-benzoyl-β-D-ribose. Transesterification of **2345** with alcohol **2346** gives orthoester **2347** which is immediately rearranged into glycoside **2348** in 76% yield. Exchange of protecting groups and chlorination by treatment with *N*-dichloromethylene-*N,N*-diethylammonium chloride yields chloride **2349**. Displacement of the chloride with dimethylarsinosodium and oxidation completes the synthesis to provide **2350** in nine steps and 22% overall yield, thus establishing the relative and absolute configuration of **2350**.

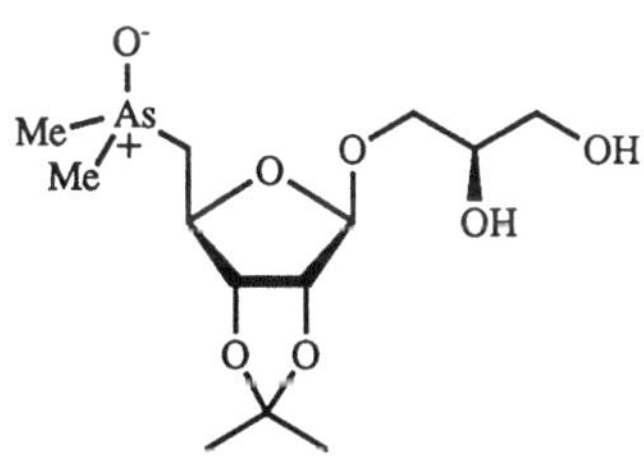

Tridacna maxima Metabolite 2350

6.3.13 Nereistoxin

The 1,2-dithiolane nereistoxin (**2357**) [649] is a neurotoxin produced by the worm *Lumbriconereis heteropoda*. It and related substances have been prepared by Hagiwara [650] from 1,3-dichloro-2-propanol (Scheme 419) in one of the earliest references to marine natural products synthesis. Double chloride displacement on **2351** by sodium phenylsulfide gave **2352** as a protected form of a disulfide. Treatment with $SOCl_2$ gave a mixture of the desired **2353** as well as

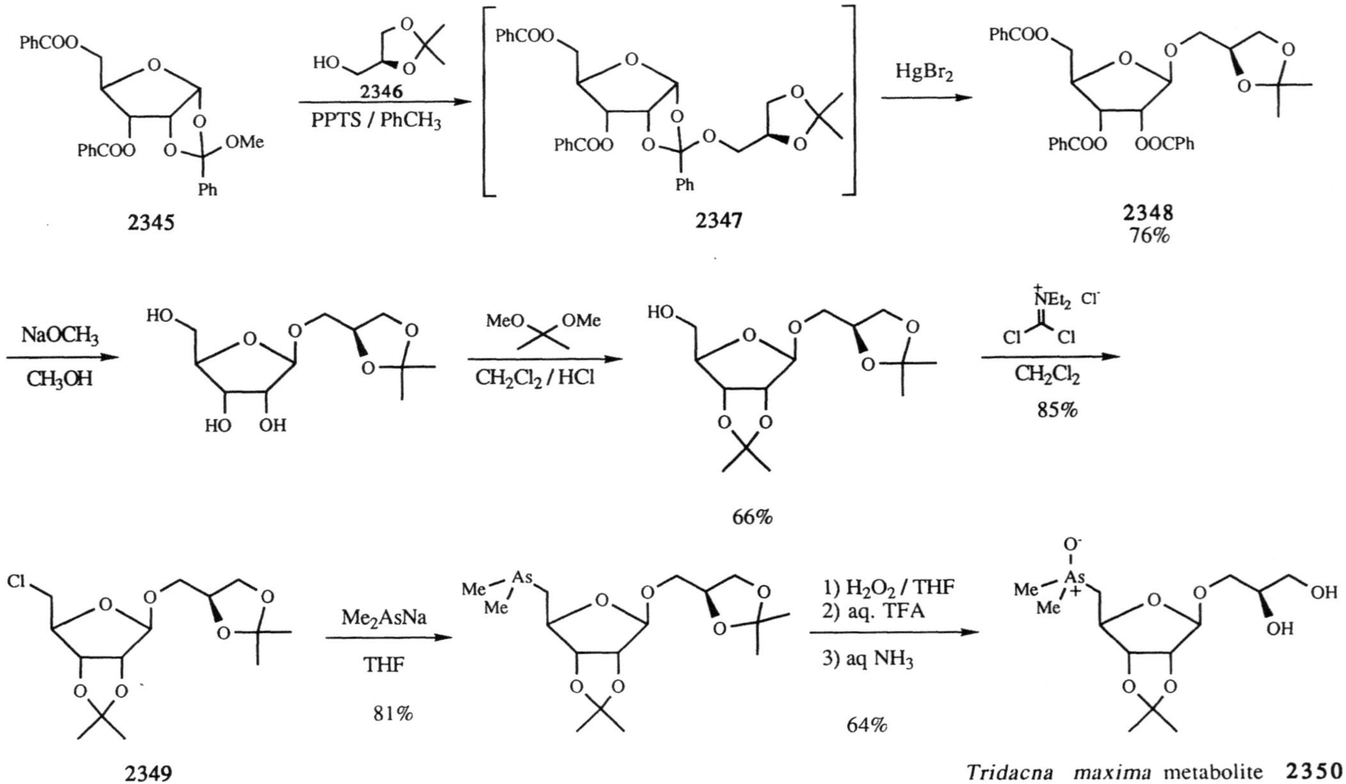

Scheme 418. Stick Synthesis of *Tridacna maxima* Metabolite **2350**

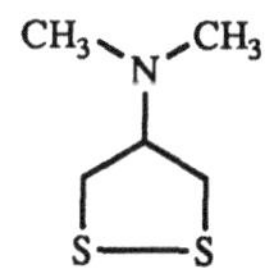

Nereistoxin **2357**

2354, probably derived by cationic rearrangement involving an episulfonium ion intermediate. Chloride displacement by dimethylamine gave **2355** and **2356**. Deprotection and air oxidation of the resulting dithiol gave the naturally-occurring disulfide **2357**.

6.3.14 3-*n*-Hexyl-4,5-dithiacycloheptan-5-one

Another interesting disulfide is the dithiacycloheptanone **2361** isolated from the brown alga *Dictyopteris plagiogramma*. Although this is clearly related to the C11 brown algal pheromones reviewed earlier (Sect. 5), it is included here because of its structural similarity, albeit somewhat vague, to nereistoxin. Moore has prepared this substance as part of the structure elucidation process (Scheme 420) [651]. Bis-Michael addition of thioacetic acid to the dienone **2359** gave rise to **2360**. Methanolysis of the thioacetates and iodine oxidation gave the cyclic disulfide **2361** in short order, confirming the structure of this substance.

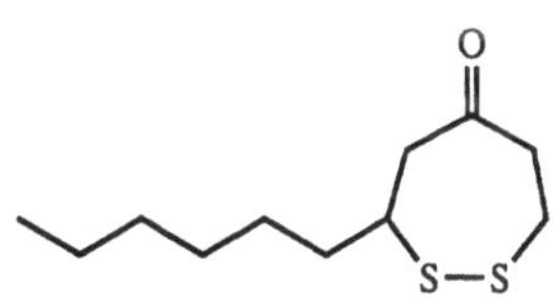

3-*n*-Hexyl-4,5-dithiacycloheptanone **2361**

6.3.15 3-Methylnavenone B

When molested, the blind carnivorous sea-slug *Navanax inermis* releases a photosensitive mixture of trail-breaking alarm pheromones. The components of this bright yellow mixture have been identified as a series of 10-aryldeca-3,5,7,9-tetraen-2-ones along with several minor metabolites [652]. One of these minor metabolites, the yellow tetraenone 3-methylnavenone B (**2364**), has been prepared by Knox [653] in a single step from triene aldehyde **2362** (Scheme 421).

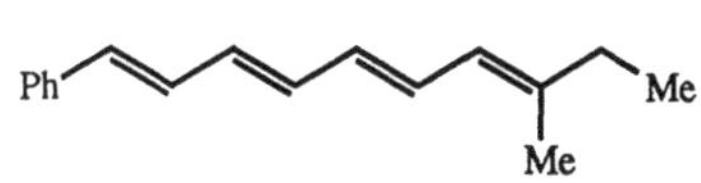

3-Methylnavenone B **2364**

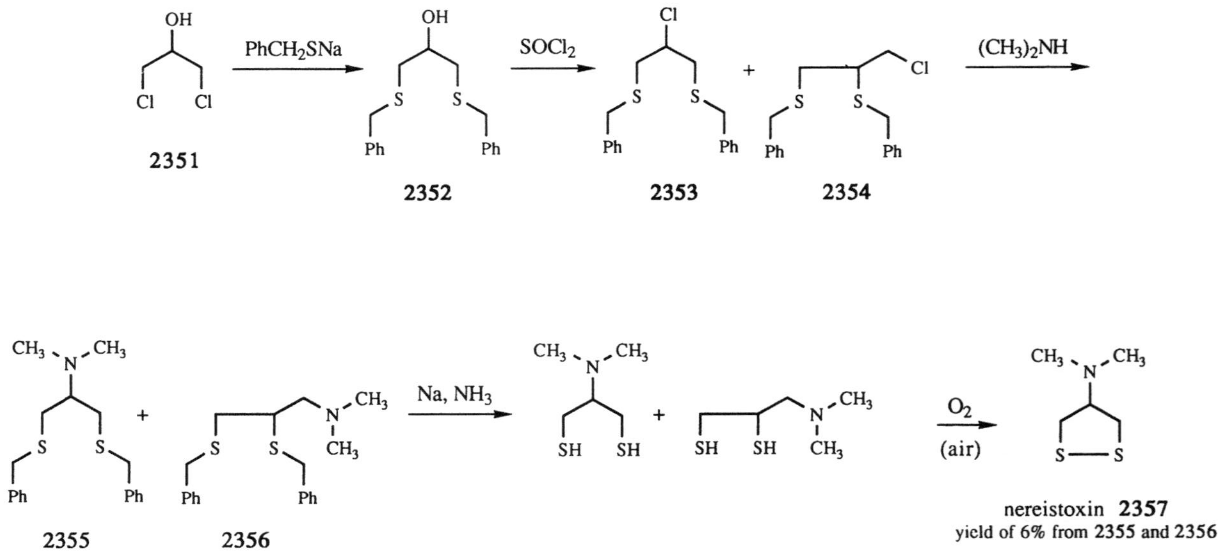

Scheme 419. Hagiwara Synthesis of Nereistoxin

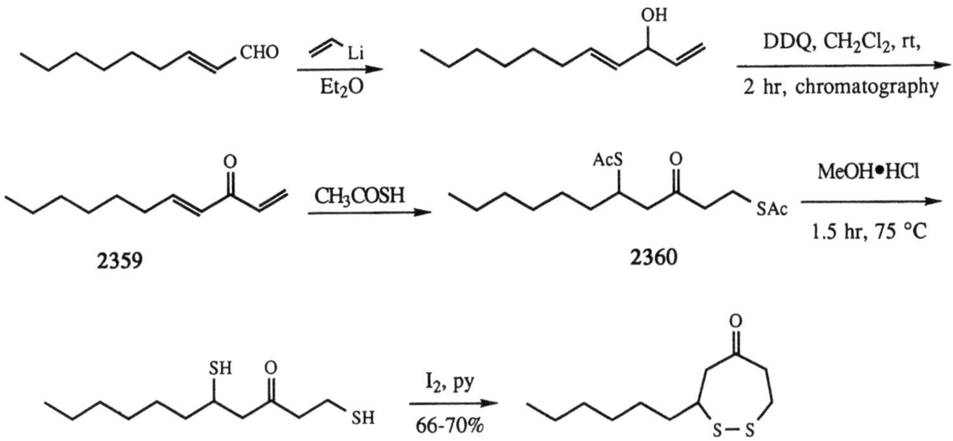

2359 2360

3-n-hexyl-4,5-dithiacycloheptanone **2361**

Scheme 420. Moore Synthesis of 3-*n*-Hexyl-4,5-dithiacycloheptanone

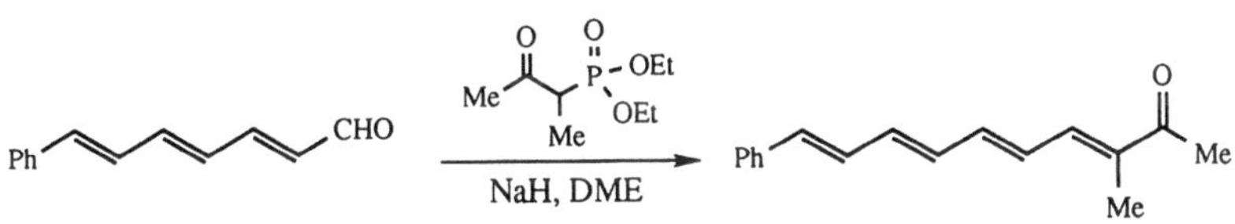

2362 3-methylnavenone B **2364**

Scheme 421. Knox Synthesis of 3-Methylnavenone B

Olefination of **2362** provides the natural product **2364**. The stereochemistry of the natural product was determined unambiguously via formation of its mono- and di-$Fe(CO)_3$ complexes, followed by regeneration of the original tetraene.

6.3.16 Malyngolide

Malyngolide is a six-membered lactone isolated from the lipid extract of the shallow-water variety of the blue-green alga *Lyngbya majuscula* [654]. It

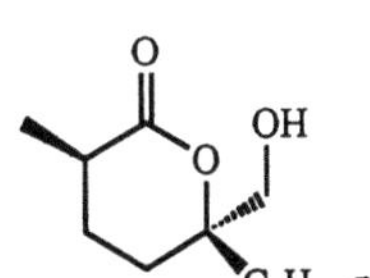

Malyngolide **2370**

exhibits antibiotic activity against *Mycobacterium smegmatis* and *Streptococcus pyogenes.* Due to its structural simplicity, its combination of two chiral centers (one of which is quaternary), and the large number of more complex natural products that possess the 5-substituted δ-lactone moiety, malyngolide has been the target of a large number of syntheses designed to test new methodology.

The first reported synthesis of malyngolide is that of Mukaiyama [655] shown in Scheme 422. The key intermediate in this synthesis is the chiral α-hydroxy aldehyde **2368**, which is obtained in three steps from proline derivative **2366**. Reduction of **2368** and selective silylation of the resulting diol followed by ozonolysis provides lactol **2369**. Oxidation and methylation of **2369** yields malyngolide and 2-epimalyngolide with poor diastereoselectivity. Malyngolide (**2370**) is obtained in 15% yield over nine steps. The yield of (−)-malyngolide can be increased by epimerization of 2-epimalyngolide to a 1:1 mixture of diastereomers followed by separation.

A racemic, though diastereospecific, synthesis of malyngolide by Babler [656] is illustrated in Scheme 423. The Michael reaction between diethyl methylmalonate and 1-dodecen-3-one (**2371**) followed by decarboxylation and Wittig olefination affords δ,ε-unsaturated acid **2372**. MCPBA oxidation of **2372** in a toluene/cyclohexane mixture produces epoxide **2373** which cyclizes in situ to give malyngolide as a single diastereomer. Malyngolide (**2370**) is obtained in six steps and an overall yield of 17%.

Cardillo [657] has developed two syntheses of malyngolide based upon the addition of the dianion of tiglic acid to either an aldehyde or ketone (Scheme 424). Addition of decanal to the tiglic acid dianion proceeds initially at the α-position. Allowing the resulting β-hydroxy carboxylate to warm gives the thermodynamically more stable δ-addition product **2375**. Hydroxy acid **2375** is converted in five steps to the δ,ε-unsaturated acid **2376**. Iodolactonization of **2376** followed by mercuric ion assisted hydrolysis of the resulting iodide gives malyngolide **2370** with almost no stereoselectivity. Malyngolide is obtained in 21% overall yield over eight steps. An alternative procedure for the preparation of **2370** involves the addition of tiglate dianion to THP protected α-hydroxy ketone **2377**. Hydrogenation of **2378** and treatment with 6N HCl provides malyngolide as a 1:1 mixture of diastereomers. After separation malyngolide is obtained in six steps and 45% yield from decanoyl chloride.

A synthesis of malyngolide (**2370**) by Torii [658] also proceeds with an absence of diastereoselectivity (Scheme 425). Electrooxidative cleavage of α-hydroxycyclopentanone **2379** affords the cleavage product **2380** in 93% yield. Wittig olefination and hydrolysis followed by iodolactonization gives iodolactone **2381**. Due to difficulties in the hydrolysis of the iodide, the iodolactone was converted to the epoxy ester **2382** and reclosed with BBr_3 to give a 1:1 mixture of the malyngolide diastereomers. After separation, a 27% yield of malyngolide was obtained in six steps from substituted cyclopentanone **2379**.

Matsuo has developed two approaches towards the synthesis of malyngolide (**2370**). The first of these [659] involves alkylation of β-ketoester **2384** to give **2385** (Scheme 426). Protection of the ketone as its enolate and reduction of the

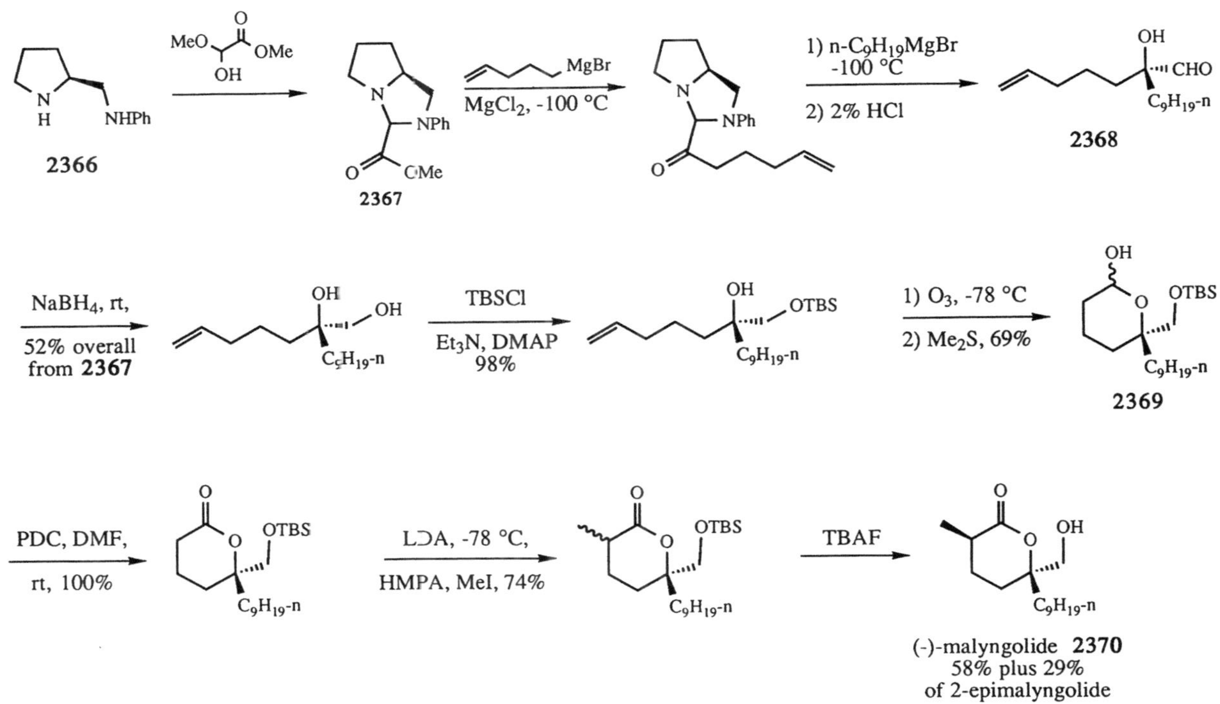

Scheme 422. Mukaiyama Synthesis of (−)-Malyngolide

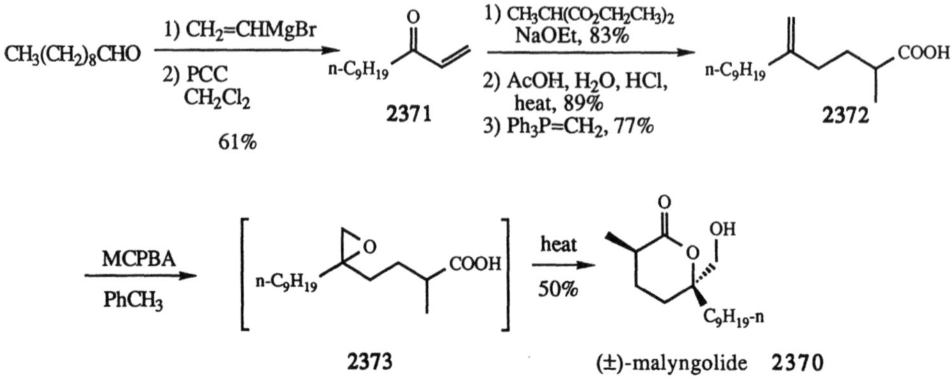

Scheme 423. Babler Synthesis of (±)-Malyngolide

ester affords β-hydroxy ketone **2386**, which is subjected to Baeyer-Villiger oxidation giving lactone **2387**. Protection of the hydroxyl as its THP ether and methylation results in a 5:4 mixture of malyngolide (**2370**) and its C2 epimer. Malyngolide is obtained in six steps and an overall yield of 11%.

The second approach by Matsuo [660] diastereoselectively provides malyngolide (**2370**) in four steps (Scheme 427). Sequential alkylation of β-ketoester **2384** produces a mixture of diastereomers. Protection of the ketone as its enolate and LAH reduction of the ester results in formation of the alcohol and epimerization at C2 to provide **2386** as one diastereomer. Baeyer-Villiger oxidation of **2386** provides malyngolide in 62% overall yield.

The key step in the synthesis of malyngolide by Kozikowski [661] involves the reductive coupling of an organomercurial with an electron deficient olefin (Scheme 428). Alkylation of dianion **2388** with 1-iodooctane and protection of the hydroxyl as its THP ether affords alkene **2389**. Oxymercuration affords organomercurial **2390** which undergoes radical addition to methylacrylonitrile when reduced with NaBH $(OMe)_3$. Treatment of the resulting δ-hydroxy nitrile **2391** with TsOH produces a 1:1 mixture of malyngolide diastereomers. Malyngolide (**2370**) is obtained in five steps and $< 11\%$ yield after separation.

A synthesis of malyngolide that is somewhat similar to the Matsuo synthesis (Scheme 426) is that of Kim [662] (Scheme 429). Baeyer-Villiger oxidation of β-ketoester **2393** followed by methylation and ester hydrolysis gives lactone **2394**. Reduction of the acid via its mixed anhydride yields a 70:30 mixture of malyngolide **2370** along with its C2 epimer. Malyngolide is obtained in 42% yield over five steps. The yield is increased by epimerization of the C2 epimer.

An enantiospecific, though non-diastereoselective, synthesis of malyngolide was developed by Sinay [663] (Scheme 430). The overall strategy involves the use of methyl 4,6-*O*-benzylidene-2-deoxy-α-D-*erythro*-hexopyranosid-3-ulose (**2395**) as a chiral template. Ulose **2395** is available in five steps from the commercially available methyl α-D-glucopyranoside. Stereospecific addition of nonylmagnesium bromide to **2395** followed by hydrolysis and olefination gives

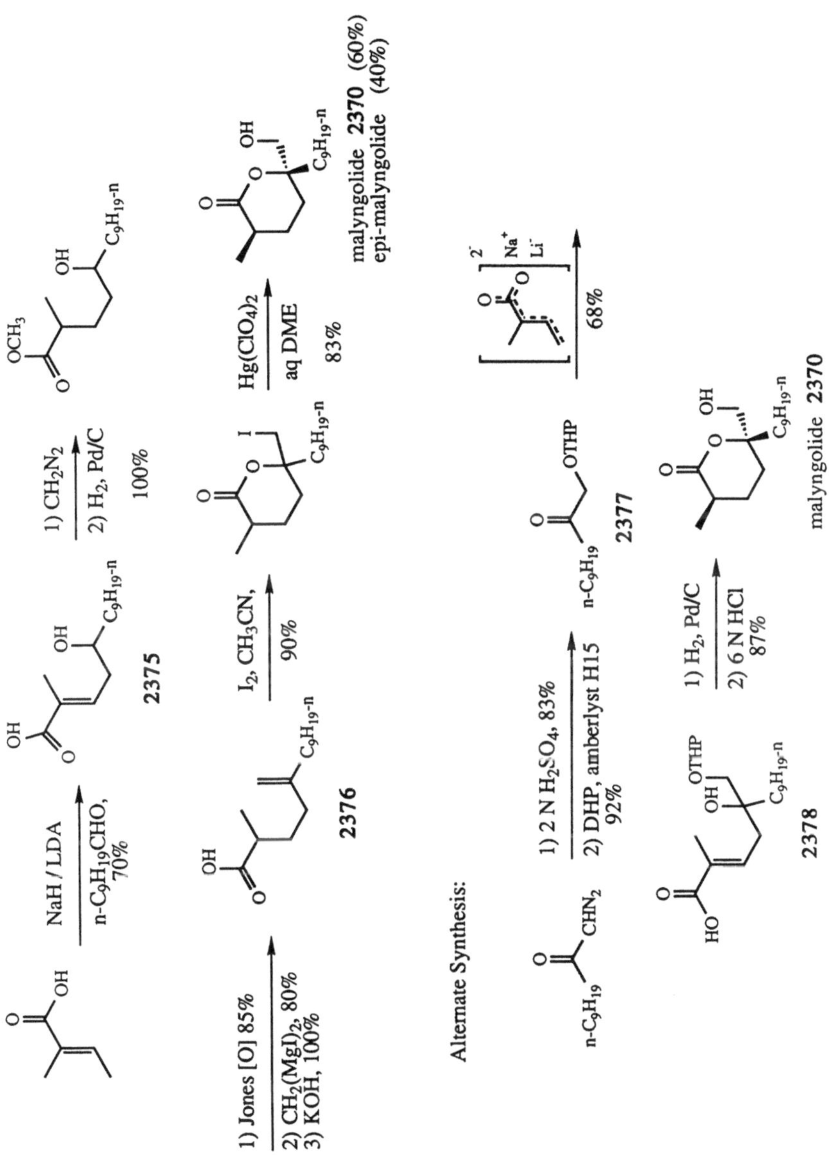

Scheme 424. Cardillo Synthesis of Malyngolide

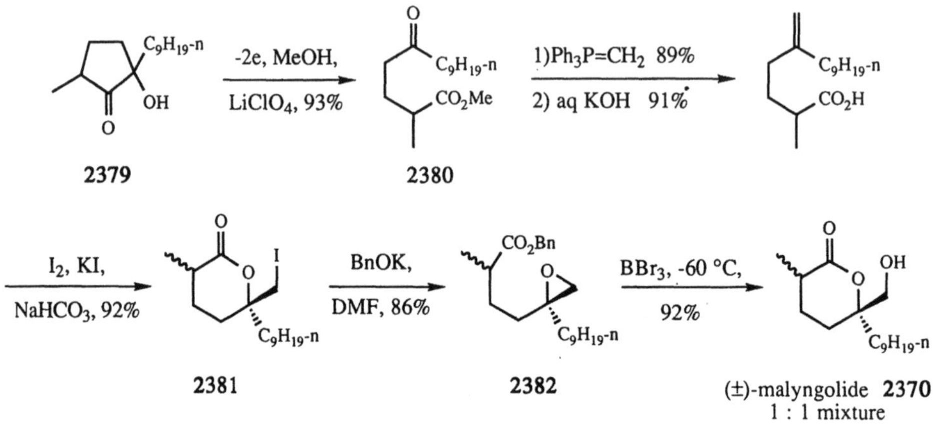

Scheme 425. Torii Synthesis of (±)-Malyngolide

tetraol **2396**. Hydrogenation and hydrolysis of **2396** affords a 1:1 mixture of diastereomers that was separated after lactonization to give lactone **2397**. Oxidative cleavage of **2397** and reduction with diphenyltin hydride produces (−)-malyngolide in eight steps and 16% overall yield.

The strategy of Eliel [664] for the enantio- and diastereospecific synthesis of malyngolide involves the coupling of optically pure Grignard reagent **2399** with optically pure ketone **2401** (Scheme 432). Grignard reagent **2399** is prepared in five steps and approximately 55% yield from *N*-crotyl-(−)-ephedrine (Scheme 431). Oxathiane **2400** is converted to ketone **2401** by addition of its lithio anion to decanal followed by oxidation. Addition of Grignard reagent **2399** to **2401** proceeds with 98% stereoselectivity, giving **2402** in 96% yield. Alcohol **2402** is converted to δ-hydroxy acid **2404** which cyclizes to (−)-malyngolide upon standing. A yield of 20% is achieved over 12 steps. In a similar manner, the other three isomers of (−)-malyngolide can be obtained.

Another enantiospecific synthesis of malyngolide (**2370**) that does not exhibit diastereoselectivity is that of Ho [665] (Scheme 433). The synthesis begins with mannofuranose **2406** available from D-mannose. Conversion of **2406** to epoxide **2407** occurs in two steps and 80% overall yield. Copper-catalyzed opening of epoxide **2407** with *n*-nonylmagnesium bromide gives diol **2408**. Removal of the secondary hydroxyl requires five steps to yield lactol **2409**, after hydrolysis. Wittig olefination, hydrogenation and saponification produces diol **2410** as a 1:1 mixture of diastereomers at C3. Lactonization gives **2370** and epimalyngolide as a separable mixture of diastereomers. Malyngolide is obtained in approximately 12% yield over 12 steps.

Hagiwara's [666] synthesis (Scheme 434) of malyngolide (**2370**) is one of the few to solve the problem of diastereoselectivity. Condensation of β-ketoester **2411** with THP protected α-hydroxy ketone **2412** provides lactone **2413**. Reduction of the ketone and elimination give α,β-unsaturated lactone **2414**.

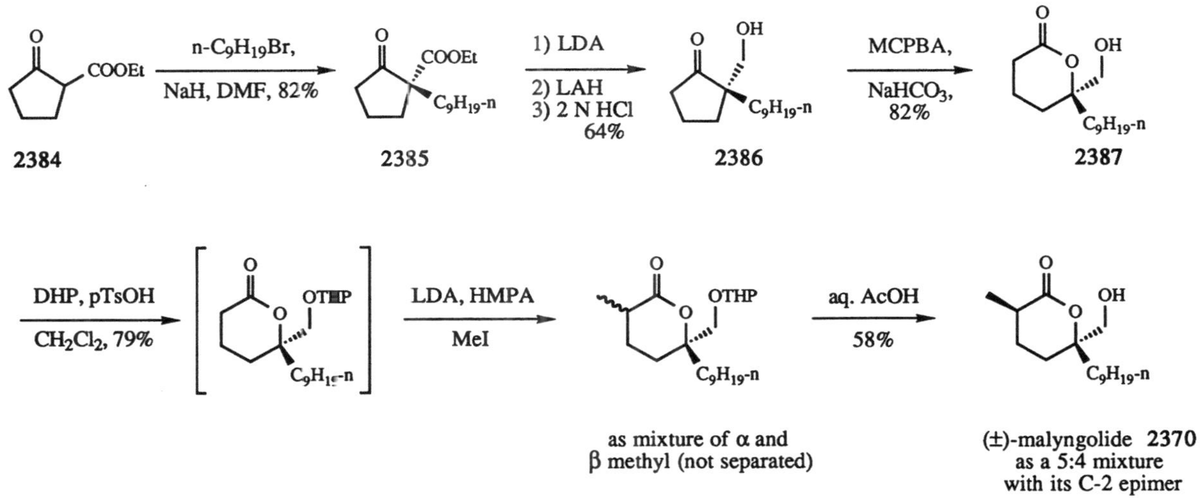

Scheme 426. Matsuo Synthesis of (±)-Malyngolide

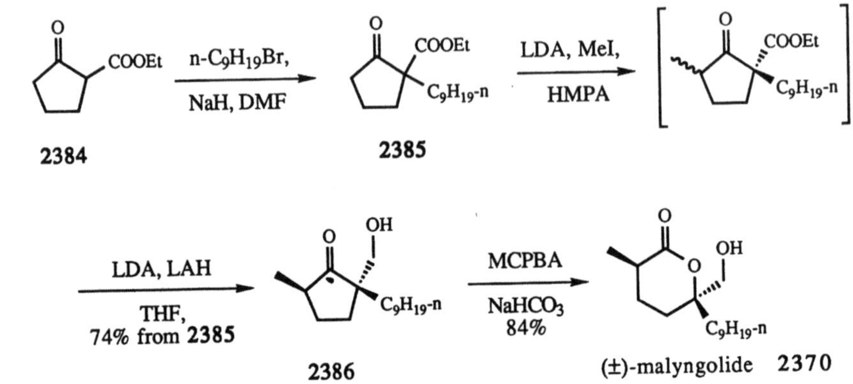

Scheme 427. Matsuo Improved Synthesis of (±)-Malyngolide

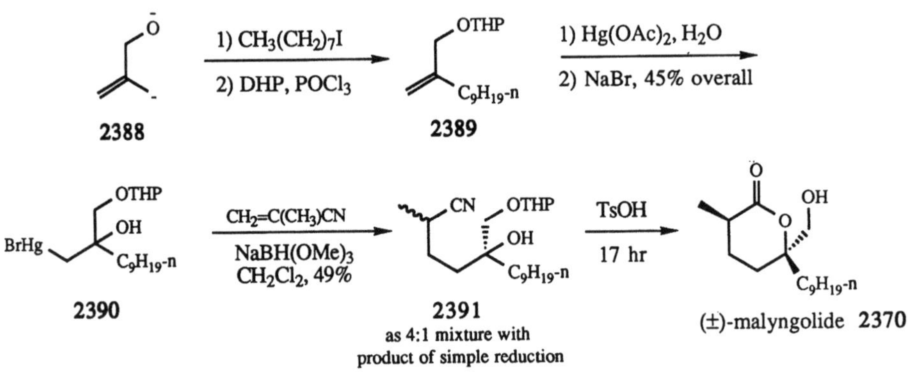

Scheme 428. Kozikowski Synthesis of (±)-Malyngolide

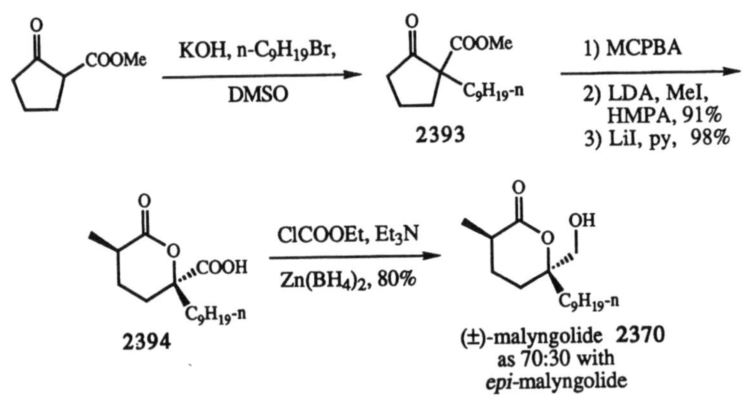

Scheme 429. Kim Synthesis of Racemic Malyngolide

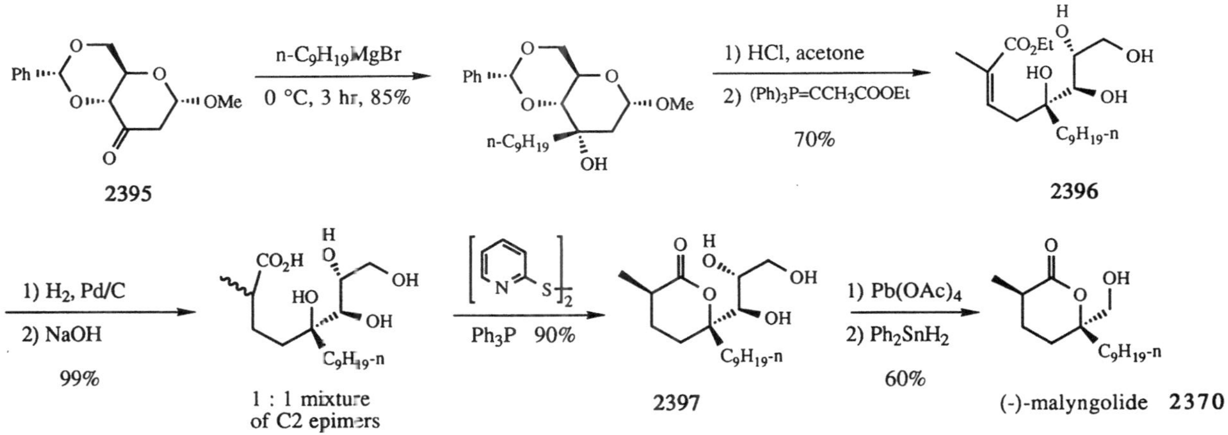

Scheme 430. Sinay Synthesis of (–)-Malyngolide

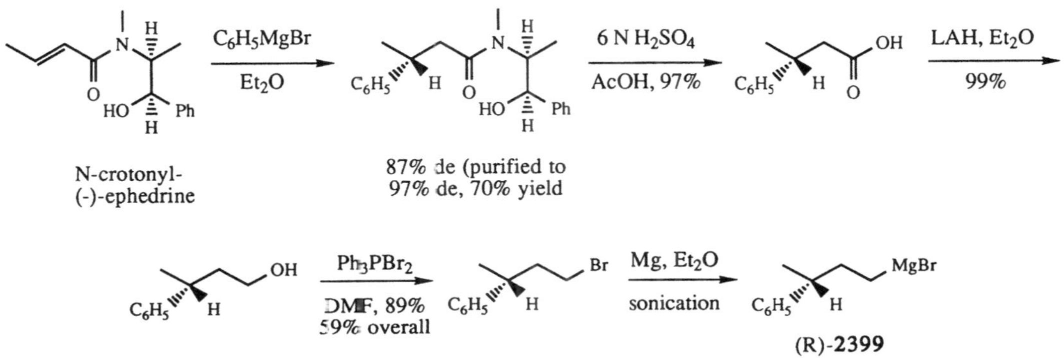

Scheme 431. Synthesis of Grignard Reagent **2399** (Eliel)

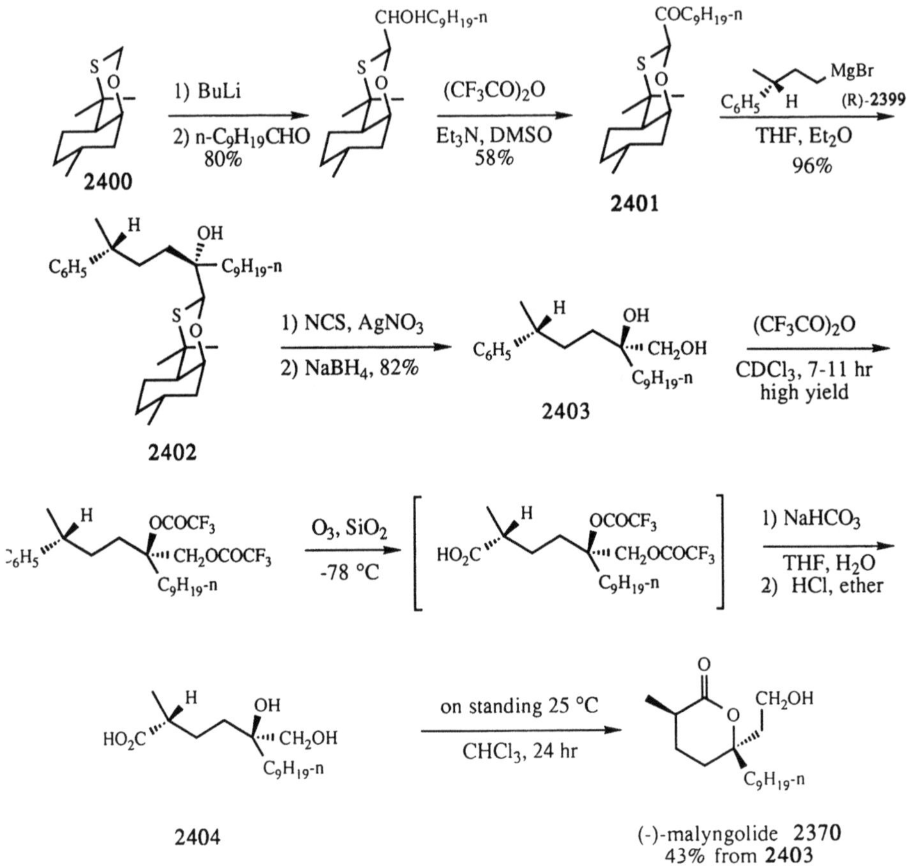

Scheme 432. Eliel Synthesis of (+)- and (−)-Malyngolide

Reduction and hydrolysis of **2414** (the lower pathway) affords malyngolide and 2-epi-malyngolide in approximately a 1:1 ratio. However, if the sequence is reversed and the THP group is removed prior to hydrogenation, participation of the free hydroxyl in the hydrogenation process gives malyngolide in a 15:1 ratio with 2-epi-malyngolide. Choice of solvent is crucial; the highest diastereoselectivity is obtained using hexane. Racemic malyngolide is obtained in five steps and 34% overall yield.

Another enantiospecific, but not diastereospecific synthesis of malyngolide is that of Nagano [667] shown in Scheme 435. Alkylation of the acetonide of dimethyl tartrate gives a 1:4 mixture of the α and β diastereomers **2417a** and **2417b**, along with 20–30% of the dialkylated product. Reduction of the major isomer gives triol **2418** which is converted to **2420** by transformation to the iodide **2419** and alkylation with diethyl methylmalonate. Deprotection, decarboxylation and cyclization of **2420** gives (+)-malyngolide (**2370**) along with 30% of 2-epi-malyngolide. Malyngolide is obtained in 4.3% yield over 11 steps.

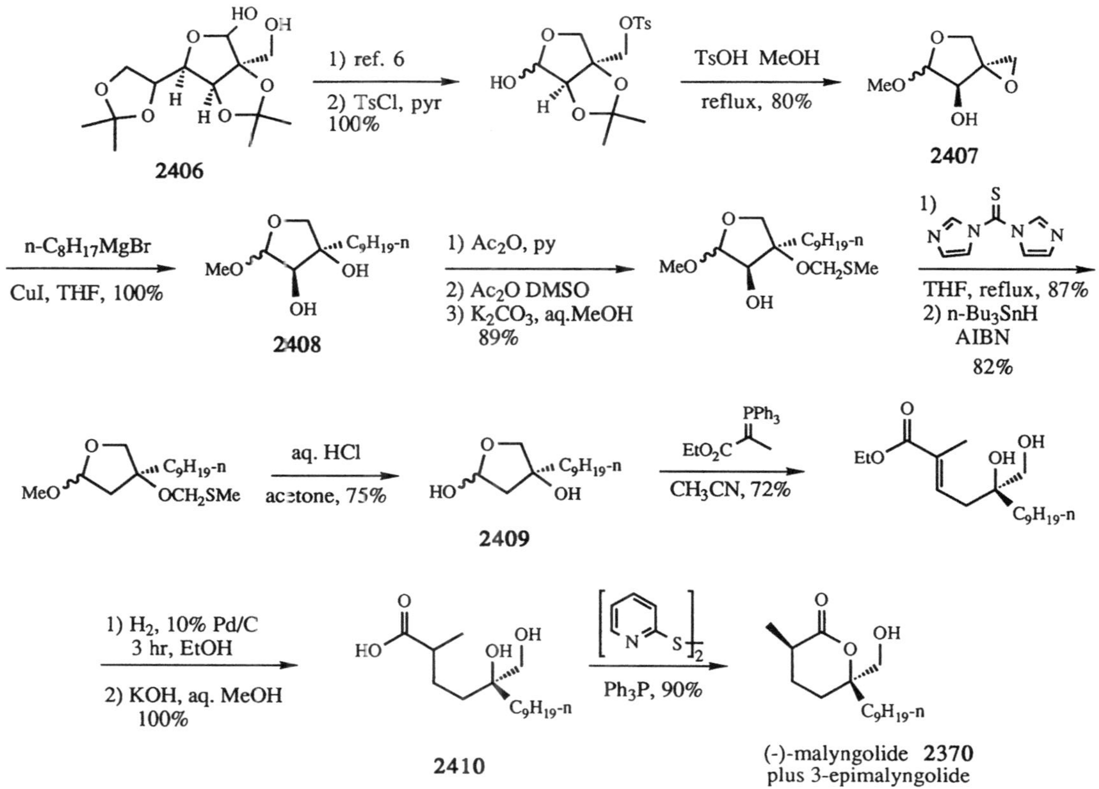

Scheme 433. Ho Synthesis of (−)-Malyngolide

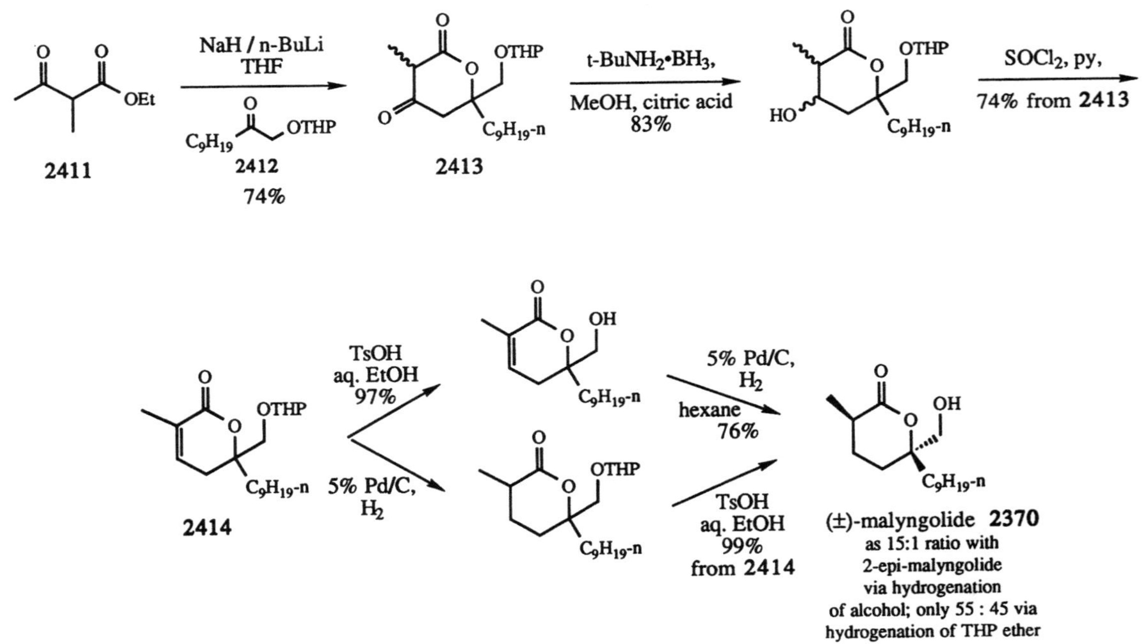

Scheme 434. Hagiwara Synthesis of (±)-Malyngolide

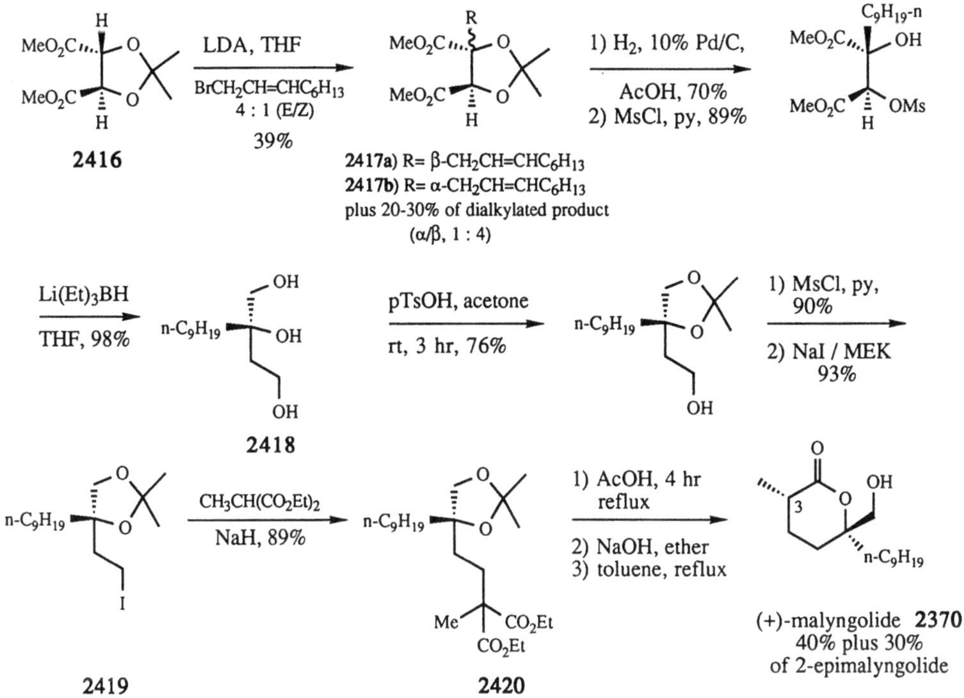

Scheme 435. Nagano Synthesis of (+)-Malyngolide

The synthesis of malyngolide (**2370**) by Wuts [668] is outlined in Scheme 436. The key step involves carbon-carbon bond fragmentation of acid **2422** via oxidative electrolysis. Acid **2422** is obtained through standard transformations, begining with Birch reduction of 3-methoxybenzoic acid (**2421**), in six steps. Electrolysis of **2422** gives unsaturated ester **2423** which is alkylated to give compound **2424**. Oxidation and cyclization provides malyngolide as a 1.8:1 mixture with 2-epi-malyngolide, in a total of 11 steps.

Giese's synthesis [669] of malyngolide provides all four stereoisomers in approximately 10% overall yield for each (Scheme 437). Sharpless epoxidation of allylic alcohol **2426** provides epoxide **2427** in 41% yield and >96% ee. Light-initiated radical coupling of iodide **2428** with methyl methacrylate in the presence of sodium borohydride and catalytic amounts of tri-*n*-butyltin chloride gives ester **2429** as a mixture of all four diastereomers. Hydrogenation and solvolysis of **2429** gives an approximately 1:1 mixture of (−)-malyngolide and 2-epi-malyngolide (**2370**) in a total of 6 steps. Use of the enantiomer of epoxide **2427** gives (+)-malyngolide in a similar manner.

One of the few syntheses of malyngolide (**2370**) that is both enantio- and diastereospecific is that of Fujisawa [670] shown in Scheme 438. Incubation of β-keto thiolester **2431** with baker's yeast in aqueous sucrose solution produces β-hydroxy ester **2432** in 88% yield (100% de and > 96% ee). Transesterification

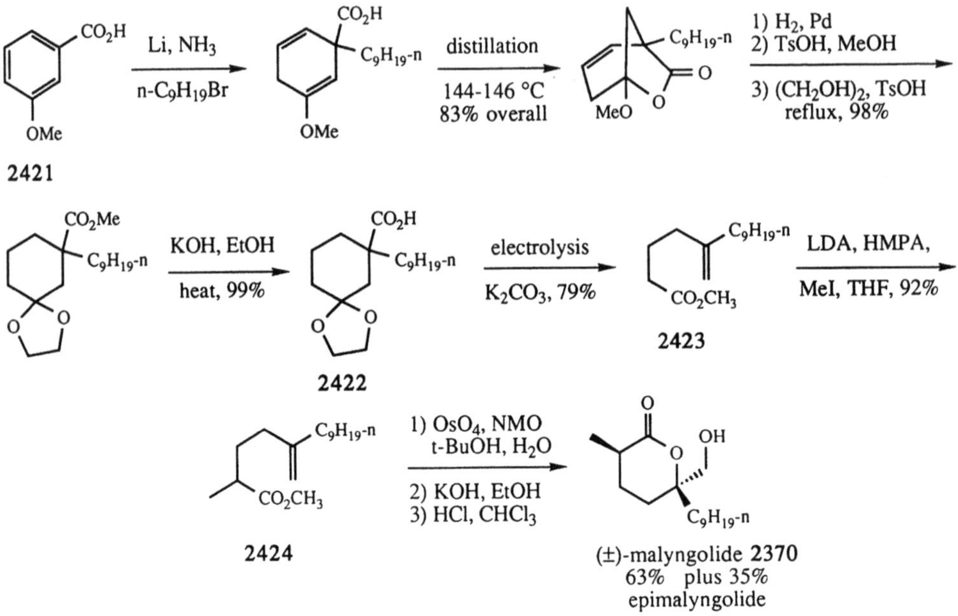

Scheme 436. Wuts Synthesis of (±)-Malyngolide

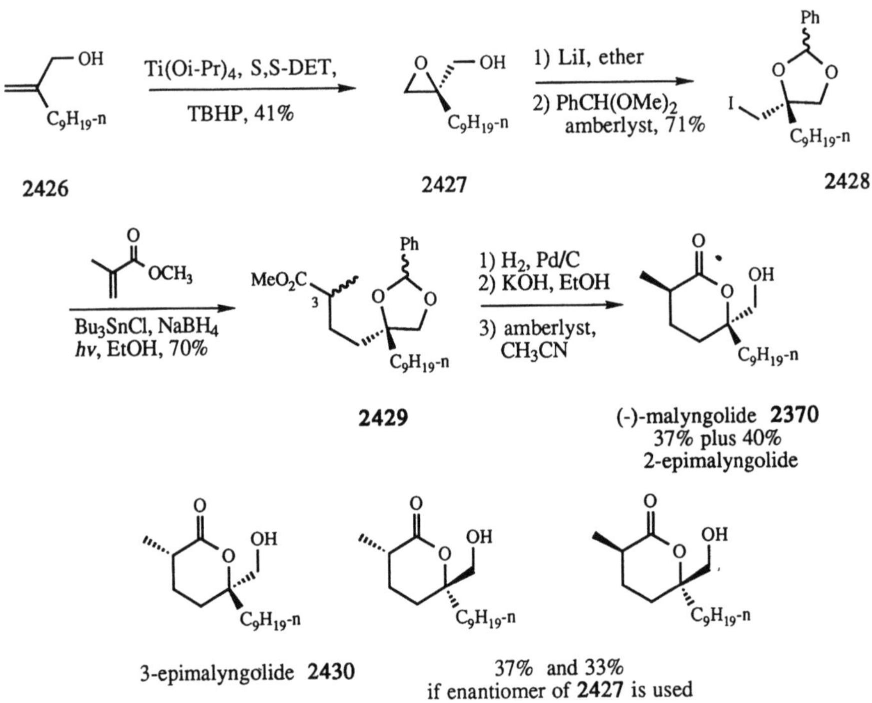

Scheme 437. Giese Synthesis of (−)-Malyngolide And Its Three Stereoisomers

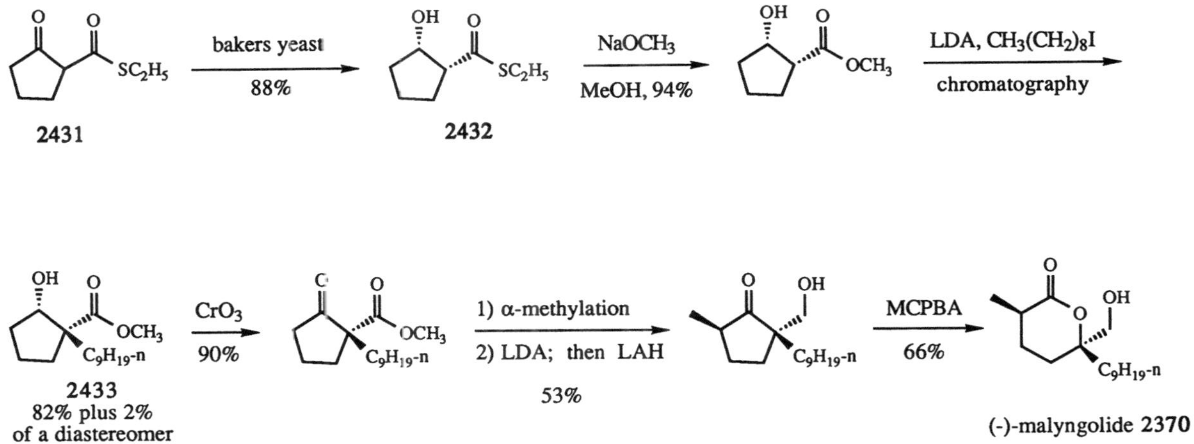

Scheme 438. Fujisawa Synthesis of (−)-Malyngolide

and alkylation provides **2433** in 82% yield, along with 2% of its diastereomer. The synthesis is completed by following the procedure of Matsuo (Scheme 427) to give (−)-malyngolide in seven steps and 21% overall yield.

6.3.17 Okadaic Acid

The potent antitumor agent okadaic acid (Scheme 439) was first isolated from sponges of the genus *Halichondria* and was shown to possess structure **2435** by X-ray crystallography [671]. The structurally similar metabolites acanthafolicin [672] and dinophysistoxins 1–3 [673] as well as **2435** have been isolated from other organisms including dinoflagellates, suggesting that these metabolites may be synthesized by symbiotic microorganisms. Metabolites **2435–2437** are unique in that they possess 3 spiroketal arrangements within their complex skeleta. In the solid state, each of the three spiroketals appear to be anomerically maximized (axial spiro C–O bonds), suggesting that the relative stereochemistry of the spiro carbons might be established in acid-promoted thermodynamic spirocyclizations.

Isobe and coworkers took a classical approach to the problem [674], culminating in the only synthesis of okadaic acid. The molecule was divided into three segments (**2447**, **2461** and **2472**) which were synthesized in optically pure form using chiral pool elements and then coupled using sulfone anion additions to aldehydes.

Segment **2447** was assembled in 15 steps from the sugar derivative **2438** (Scheme 440). Although **2438** contains 3 stereogenic centers, only the one at C4 was incorporated into **2447**. The exocyclic stereocenter at C2 was established via diastereoselective oxymercuration of the *E*-alkene **2442** giving **2443**. This was converted to lactone **2444** which was combined with the optically pure acetylide anion **2445** giving ketone **2446**. Conjugate addition of methyl cuprate to the alkyne results in a (*Z*)- α,β-unsaturated ketone which is spirocyclized to the key segment **2447**.

Segment **2461** is characterized by the presence of 5 contiguous stereocenters bearing C–O bonds (C22–C26). Carbon Ferrier rearrangement of the glucal **2449** (Scheme 441) followed by ester saponification and acetalization led to **2450** in which 3 of these C–O stereocenters were established. The C23–C24 olefin was eventually epoxidized and opened with sodium pheylmethanolate to establish the stereocenters at these two carbons resulting in **2454** after protection. Again, an optically active sulfone was added to an aldehyde **2455** leading to the cyclization precursor **2458**. Closure was achieved by hydrogenolysis of the benzyl ether in the presence of catalytic HOAc. Functional group manipulation led from **2459** to aldehyde segment **2461**.

Segment **2472** was also assembled from a carbohydrate (Scheme 442). D-Glucose led in three steps to tetraacetate **2462**. Ferrier rearrangement led to **2463** which, when treated with a methyl cuprate reagent led to **2465**. This isomer

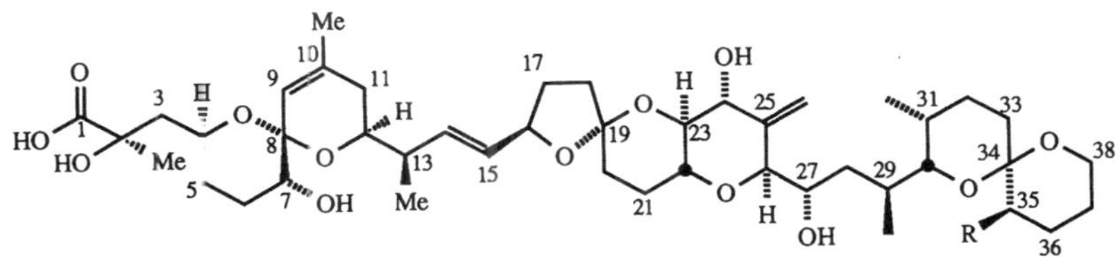

2435 okadaic acid R = H
2436 dinophysistoxin-3 (R = CH_3)
2437 acanthafolicin (R = H, 9,10-α-episulfide)

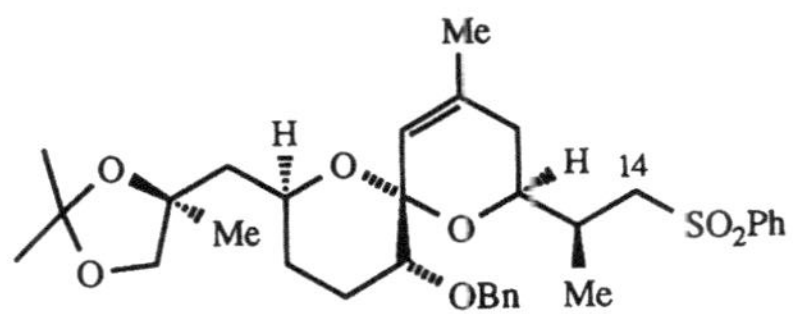

2447

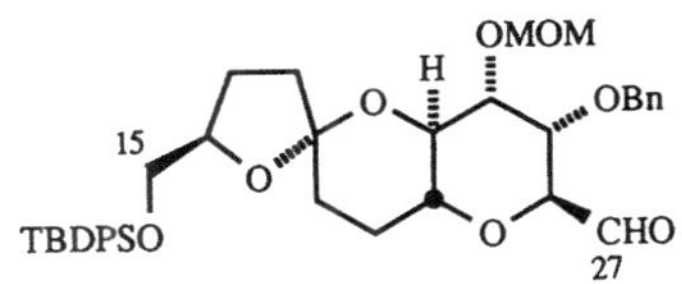

2461

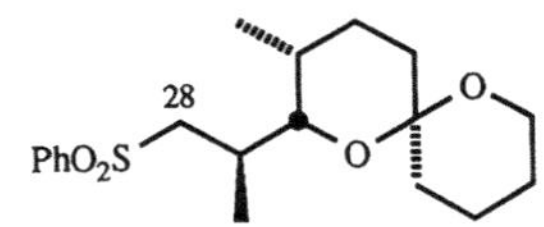

2472

Scheme 439. Okadaic Acid Synthesis Strategy (Isobe)

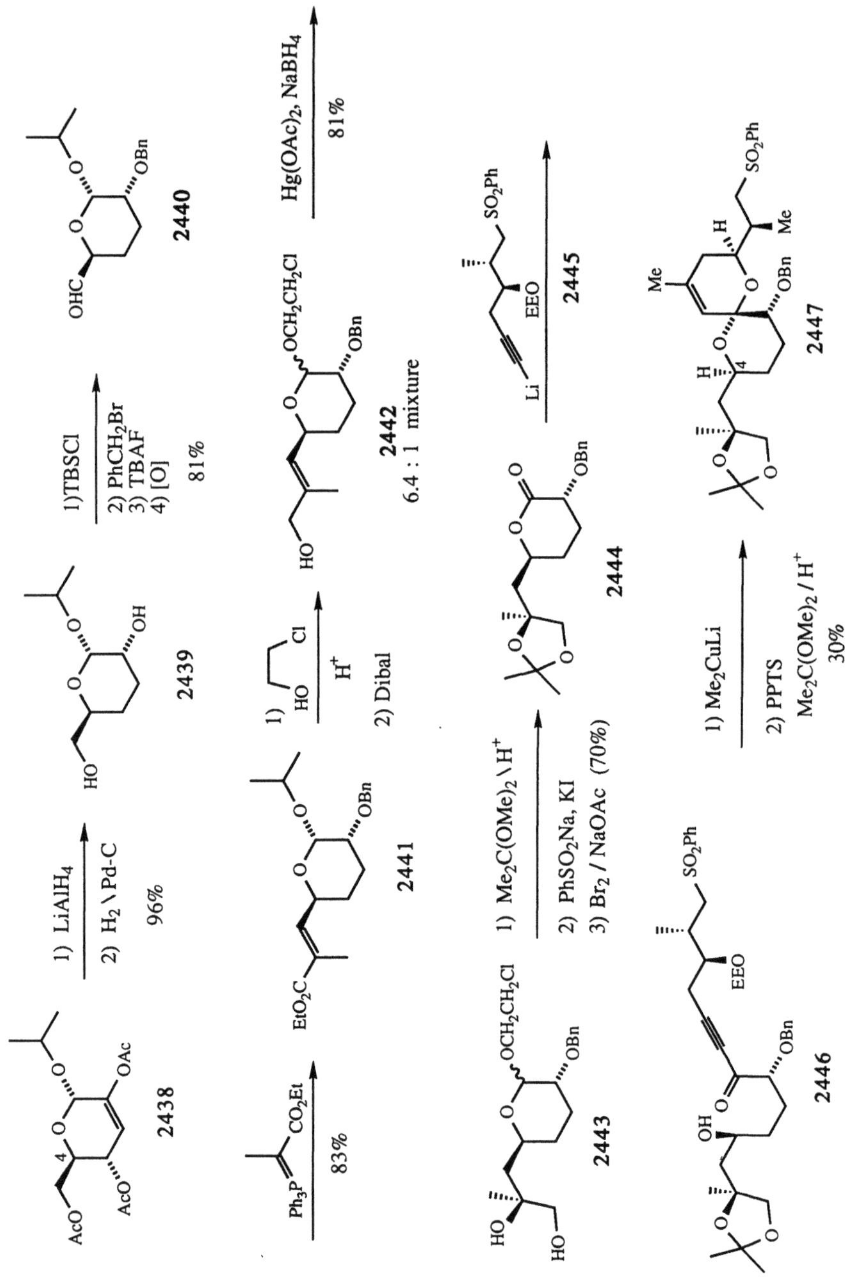

Scheme 440. Synthesis of Okadaic Acid Segment **2447** (Isobe)

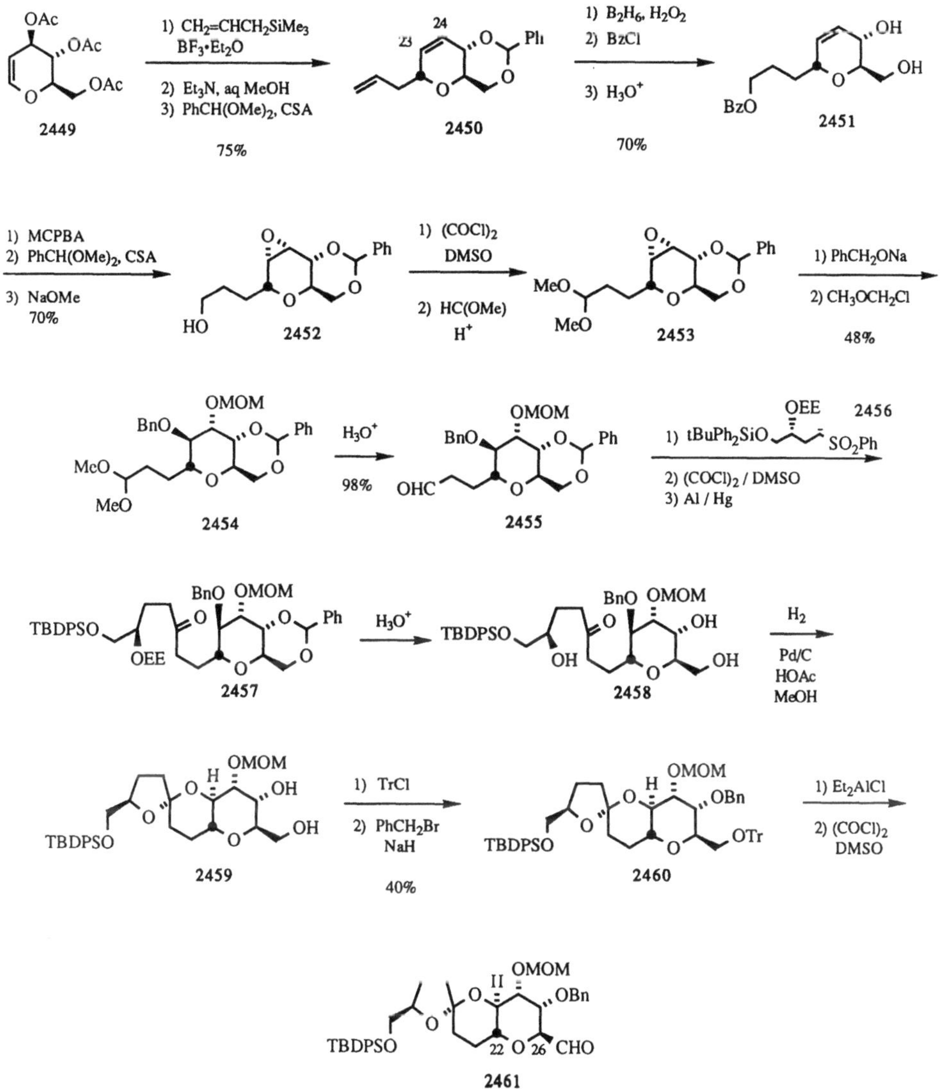

Scheme 441. Synthesis of Okadaic Acid Fragment **2461** (Isobe)

arises via elimination to enone **2464** which then underwent axial addition of the cuprate reagent giving **2465**. Five steps led to lactone **2467** which was combined with sulfone **2468** leading to **2469**. Further manipulation led to **2471**. Addition of MeLi to the α,β-unsaturated sulfone led eventually to **2472**. The stereocontrol in this reaction is thought to be due to chelation of the alkyllithium with a tetrahydropyran oxygen delivering the reagent to one face of the olefin (**2473** → **2474**). Segment **2472** was thus available in 16 steps from **2462**.

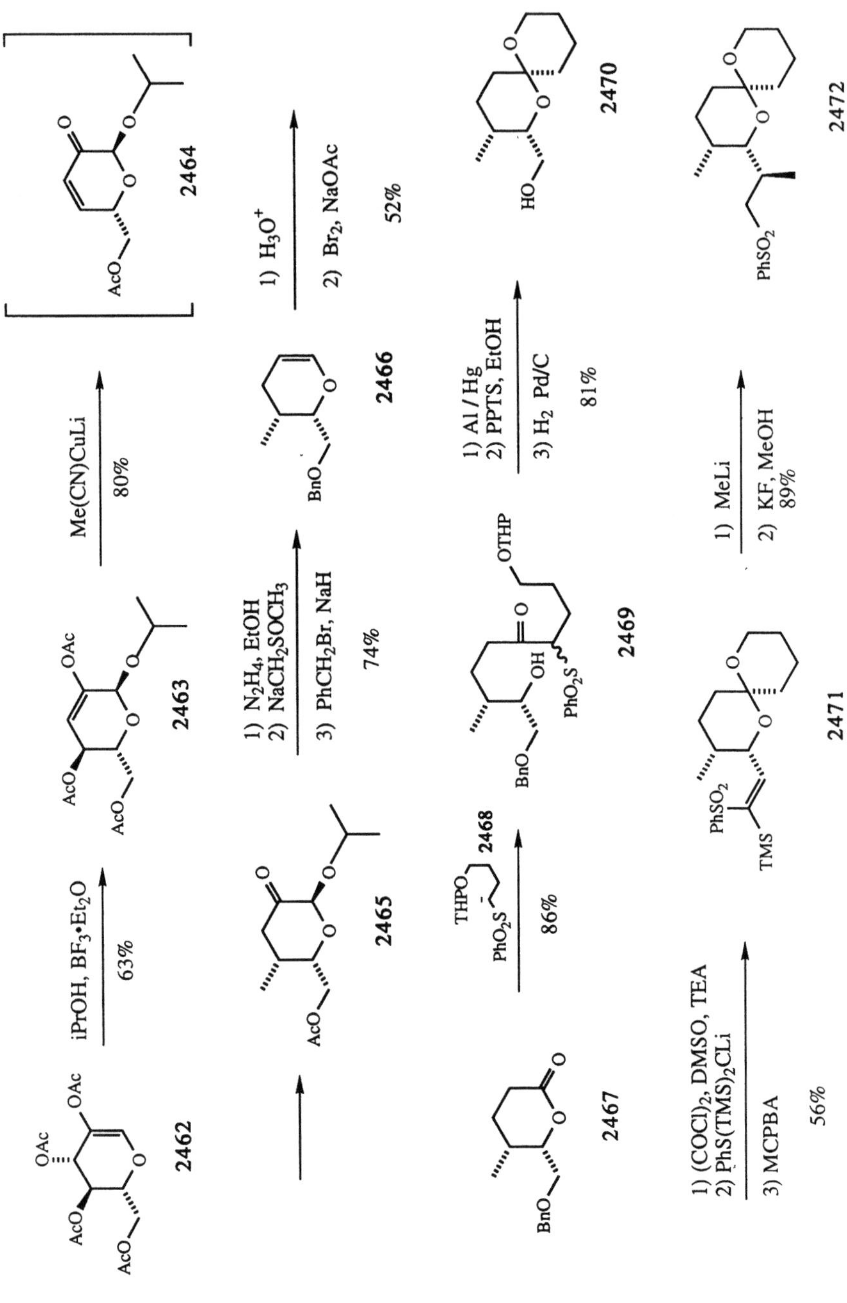

Scheme 442. Synthesis of Okadaic Acid Fragment **2472** (Isobe)

2473 2474

The segments were combined in a B + C → BC → ABC sequence (Scheme 443). The anion of sulfone segment **2472** was joined with the aldehyde of segment **2461** generating **2476**. The resulting diastereomeric mixture was oxidized to the ketone and the sulfone was reductively removed to give an intermediate ketone which was reduced to give the correct configuration of the alcohol at C27 (**2477**). The reduction proceeded stereospecifically with $NaBH_4$ or $LiAlH_4$:$Zn(BH_4)_2$ (85:15, desired:undesired) and DIBAL (70:30) reduced with less selectivity. These results were rationalized via the two transition state models shown in Scheme 444. It was stated that the $NaBH_4$ and $LiAlH_4$ reductions proceeded via **2483** while the "low or opposite-dominant selectivities" were caused by "chelative interaction .... at the transition state." Experimental support for these hypotheses other than the product structures and ratios was not given. Attention was next turned to adding the C41 methylene carbon which was easily accomplished in three steps from **2478** with the parent Wittig reagent $Ph_3P{=}CH_2$ performing the methylenation. From here, three steps were required to convert C15 to aldehyde **2480**, ready for coupling to segment **2447**. The reaction sequence used to form the C14–C15 olefin entailed sulfone anion addition to the aldehyde, acetylation of the resulting alcohol and reductive elimination with sodium amalgam, resulting in a *trans* olefin (**2481**) in 32% yield for the 3 steps. At this juncture, the entire carbon skeleton was complete leaving only the C1 carboxylic acid terminus to be established. This was accomplished by deprotection of the C1–C2 diol and oxidation in 2 stages to the carboxylic acid via the aldehyde **2482**. Final removal of 3 benzyl groups with Li/NH_3 provided 1.7 mg of crystalline okadaic acid (**2435**).

An analysis of the course of the synthesis is shown in Scheme 445. The three starting materials were simple sugar derivatives and led in nearly linear fashion to each of the three main segments **2447**, **2461** and **2472**. The assembly of the segments took an additional 19 steps, resulting in an overall 54 step process in the longest linear sequence. A total of 106 separate operations (minimum) were required to assemble okadaic acid.

6.3.18 Debromoaplysiatoxin

A series of complex hemi-spiroketals were isolated from the marine aplysiid *Stylocheilus longicauda* by Scheuer in the early 1970s [675]. The aplysiatoxin-oscillatoxin metabolites were eventually traced to the blue-green alga *Lyngbya majuscala* and other sources [676]. A partial listing of the metabolites is shown in Scheme 446. The relative and absolute configurations of this metabolite series

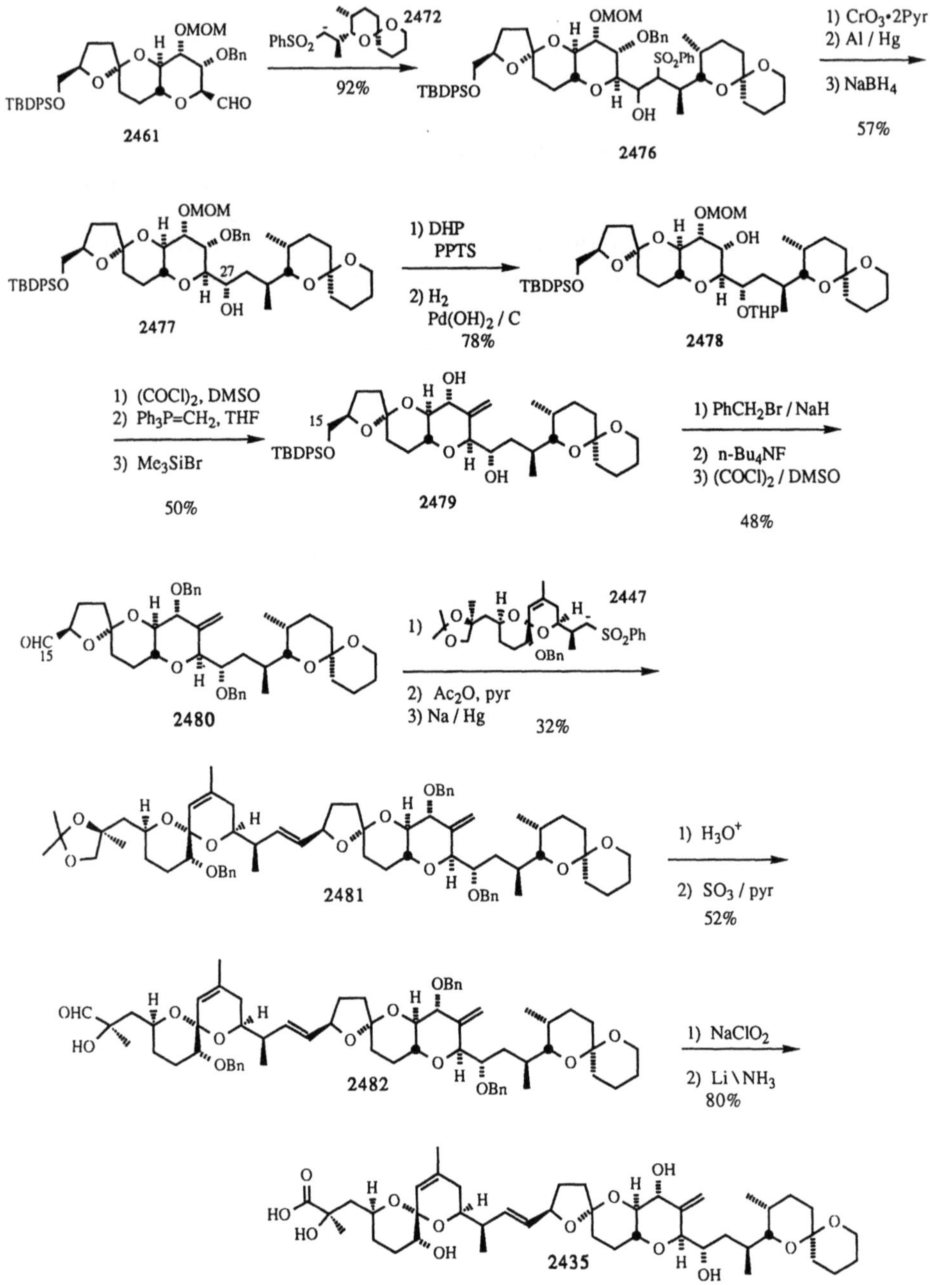

Scheme 443. Assembly of Okadaic Acid Segments (Isobe)

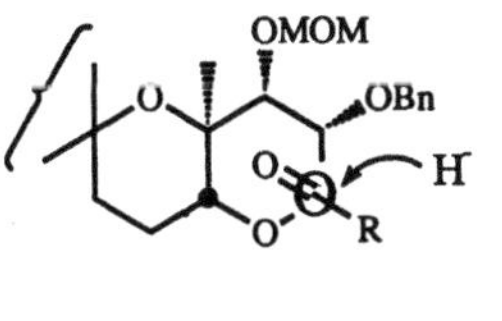

2483

Stereoelectronically-controlled anti-periplanar addition ($LiAlH_4$, $NaBH_4$)

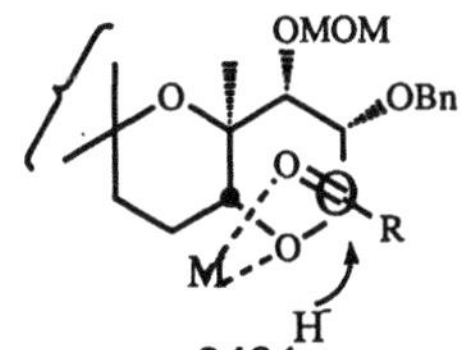

2484

Chelation -controlled addition (Dibal, $Zn(BH_4)_2$, B_2H_6)

Scheme 444. Transition State Hypotheses Concerning the Reduction of the C27 Ketone

were eventually confirmed by Moore and coworkers using extensive spectroscopic and chemical degradation techniques, in addition to X-ray crystallography [677]. Compounds in this family exhibit a variety of physiological activities, ranging from tumor promotion to a peculiar form of contact dermatitis [678]. The compounds possess potentially mobile hemi-spiroketal arrangements due to a hydroxyl at C3, although open forms of the spiroketal ring system have not been observed. In addition, the spiroketal configurations are not anomerically maximized (see **2490**).

The approach of Kishi is the only successful synthesis of a metabolite in this series to-date [679]. The fragments used to assemble debromoaplysiatoxin **2487** are shown in Scheme 447. The synthesis of fragments **2492**, **2494**, **2496** and **2498** was straightforward and proceeded from optically active starting materials. Fragment **2497** is a common reagent. The combining of the fragments is an instructive exercise in manipulation of functionality utilizing protecting group technology (Schemes 449 and 450). The anion of sulfone **2492** was alkylated with the epoxide **2496** and the sulfone was removed by reduction providing **2500**. Diol deprotection and conversion to the epoxide **2501** was straightforward. Epoxide opening with the dithiane **2494** led to **2502** in "almost quantitative" yield. Esterification with **2498** followed by protecting group manipulation led to **2503**. Note that the last two steps in the **2502** → **2503** sequence exchanged the TBDPS group for BOM. Apparently a BOM group already installed on **2498** in place of TBDPS was not satisfactory in the esterification. Elaboration of **2504** to the carboxylic acid **2507** proceeded via a two stage oxidation. Coupling with the magnesium salt **2497** and removal of the MPM groups gave the cyclization substrate **2509** in reasonable overall yield. It is noteworthy that **2509** does not exist in a cyclic form. Conversion of **2509** → **2510** with silver trifluoroacetate suggests that macrocyclization facilitates formation of the hemispiroketal and further suggests that the natural products exist as the most thermodynamically stable isomer at C3 and C7. Removal of the protecting groups from **2510** produces debromoaplysiatoxin **2487**. Aplysiatoxin has already been produced from **2487** by bromination. The synthesis of **2487** proceeds in 18 steps from the fragments pictured in Scheme 447 in 2–3% overall yield.

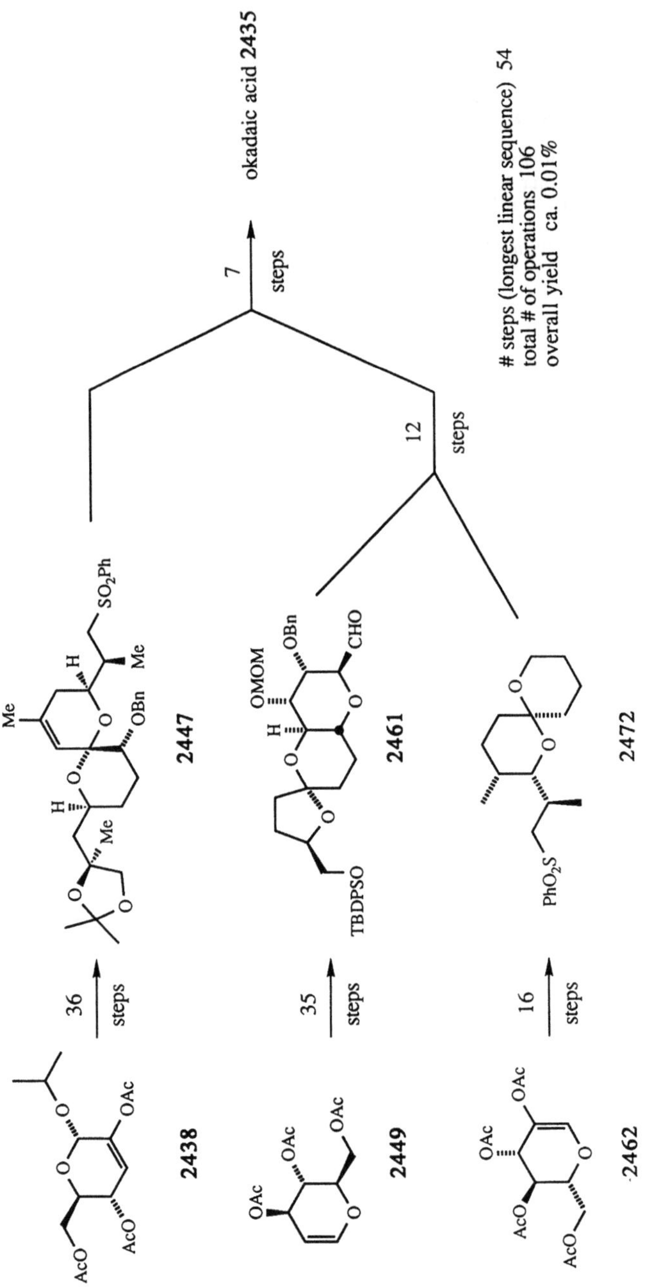

Scheme 445. Summary of Okadaic Acid Synthesis (Isobe)

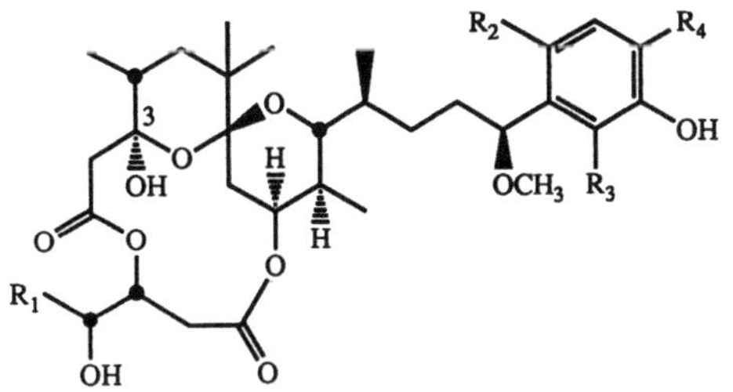

	R1	R2	R3	R4
2486 Aplysiatoxin	CH_3	Br	H	H
2487 Debromoaplysiatoxin	CH_3	H	H	H
2488 Oscillatoxin A	CH_3	Br	Br	Br
2489 19,21-dibromaplysiatoxin	H	H	Br	Br

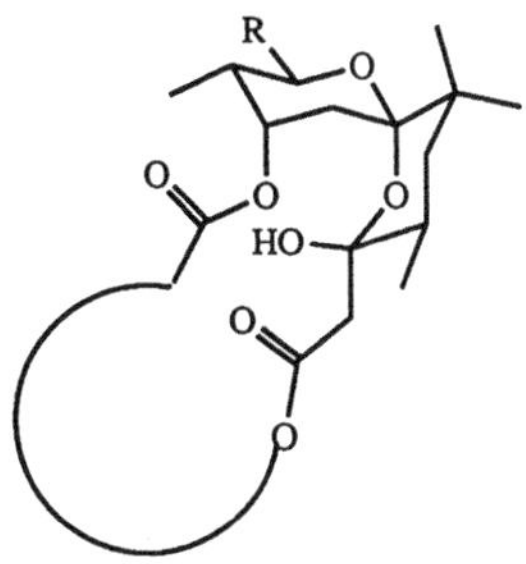

2490 solid state spiroketal conformation

Scheme 446. Some Metabolites of *Lyngbya majuscala*

Although not culminating in a completed metabolite synthesis, the work of Ireland in this area is worthy of note [680]. Utilizing extensive studies on the hetero Diels-Alder reaction performed by this group, the C3 nor-hydroxy compound **2532** was assembled (Scheme 452). The key step involved the [4 + 2] cycloaddition of optically active partners **2517** and **2521**. The vinyl ether **2517** was prepared from the lactone **2516** by reaction with the Tebbe reagent (Scheme 450). The lactone, in turn, was derived via standard manipulations from the alcohol **2511**, using the Sharpless epoxidation of **2512** to enter the desired enantiomeric series. Enone **2521** was produced from (*S*)-**2519** and was used as a mixture of epimers at C15 (Scheme 451). It was shown, however, that the absolute stereochemistry at C15 could be induced by reduction of the corresponding ketone using the Noyori chiral binaphthol reagent producing **2522**.

The cycloaddition between **2517** and **2521** proceeded to give **2523** which was manipulated to **2524** (Scheme 452). Spiroepimerization of **2524** was induced by HCl in $CHCl_3$ leading to **2525** possessing the maximum number of equatorial substituents. Apparently, steric considerations outweighed the stereoelectronic preference of the anomeric effect. After a series of standard transformations to give **2528**, the diastereomers were separated providing **2529**. Unfortunately, the key remote oxidation of **2529** → **2532** could not be realized under a variety of conditions (Scheme 453). Anticipating eventual success, compound **2529** was converted to **2533** featuring a DCC/DMAP mediated macrocyclization.

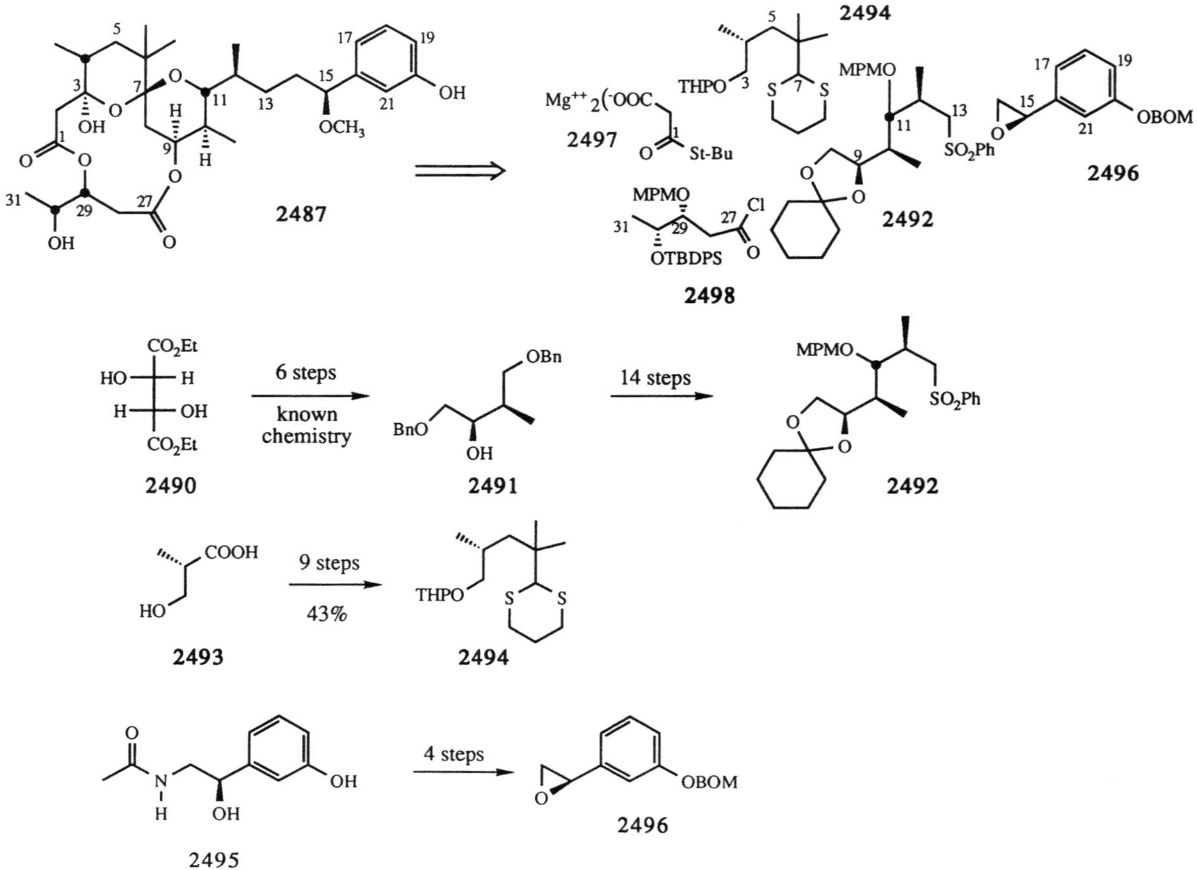

Scheme 447. Fragments used in the Kishi Debromoaplysiatoxin Synthesis

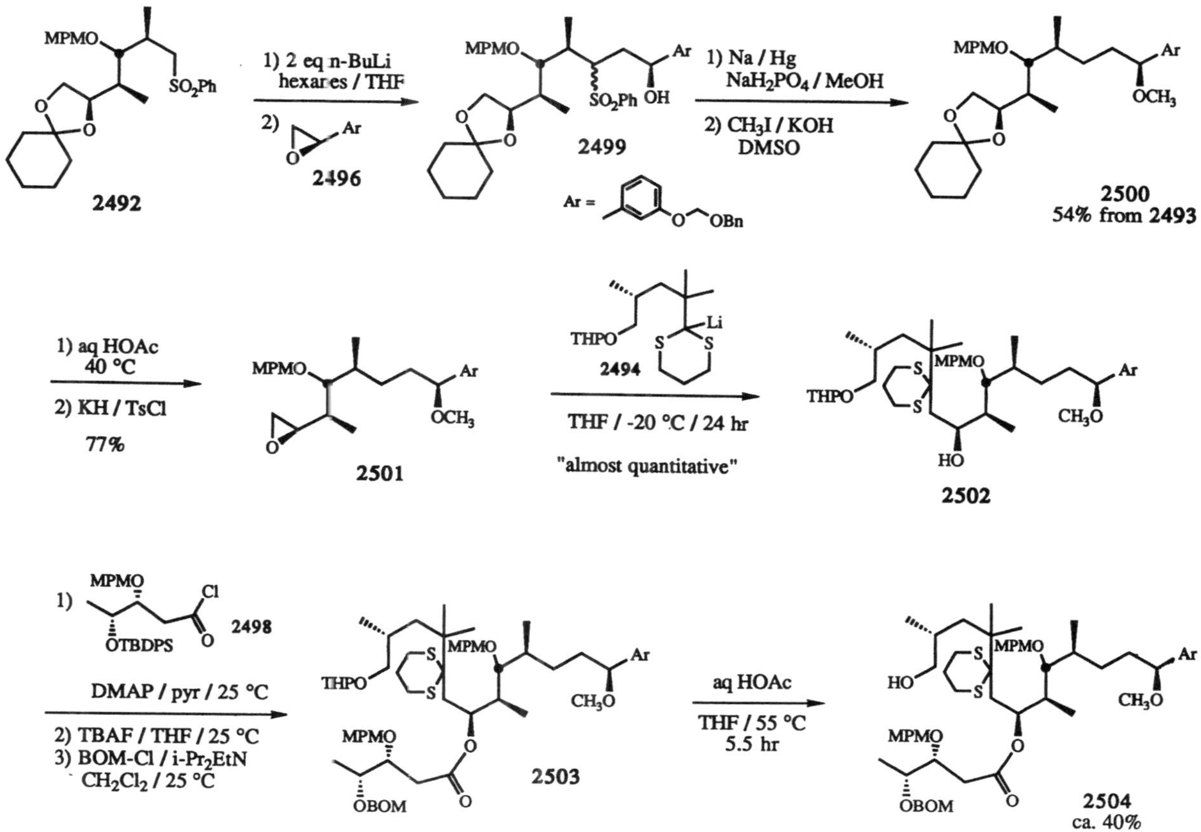

Scheme 448. Initial Assemby of Debromoaplysiatoxin Fragments

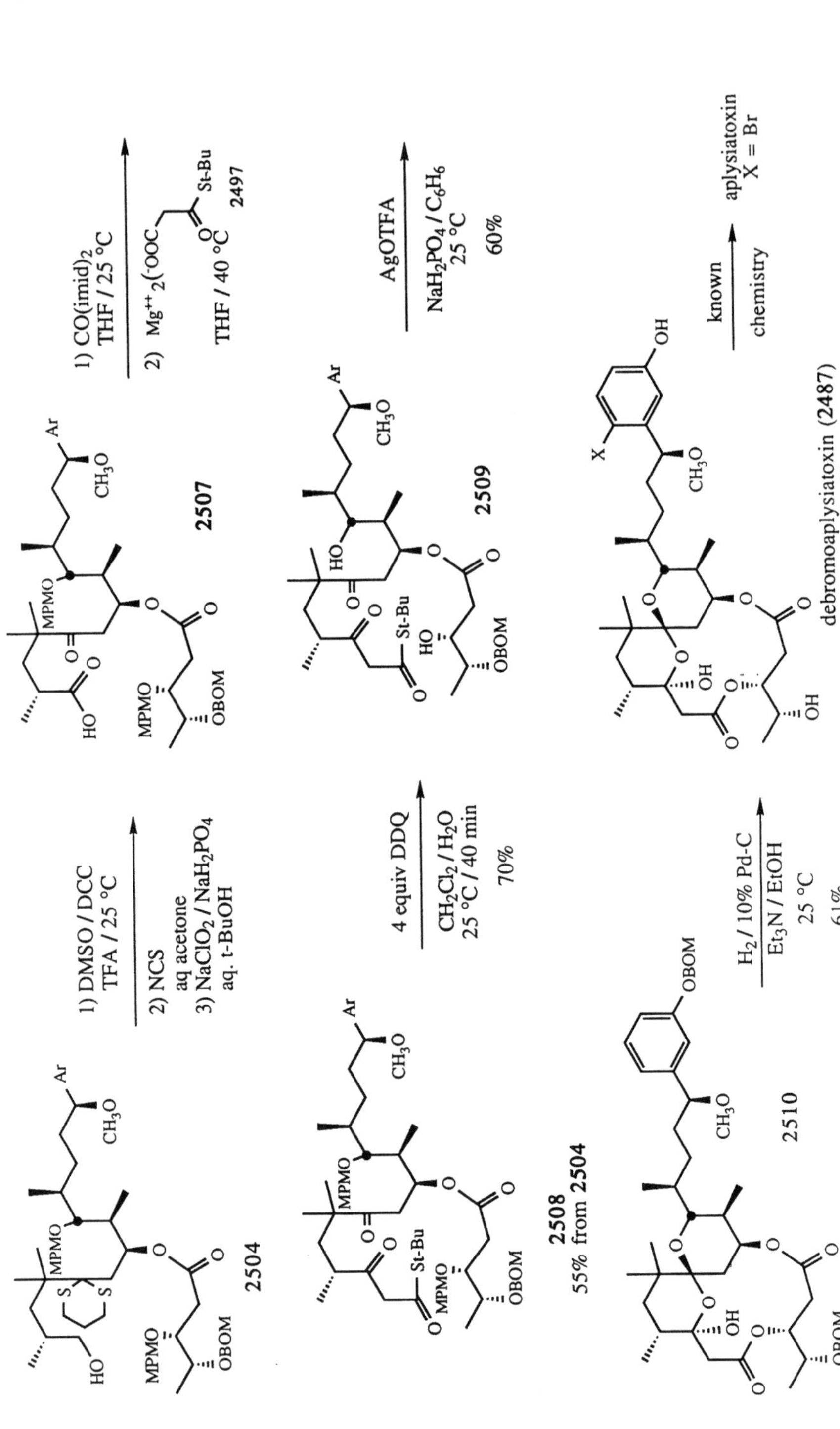

Scheme 449. Final Assembly of Debromoaplysiatoxin (Kishi)

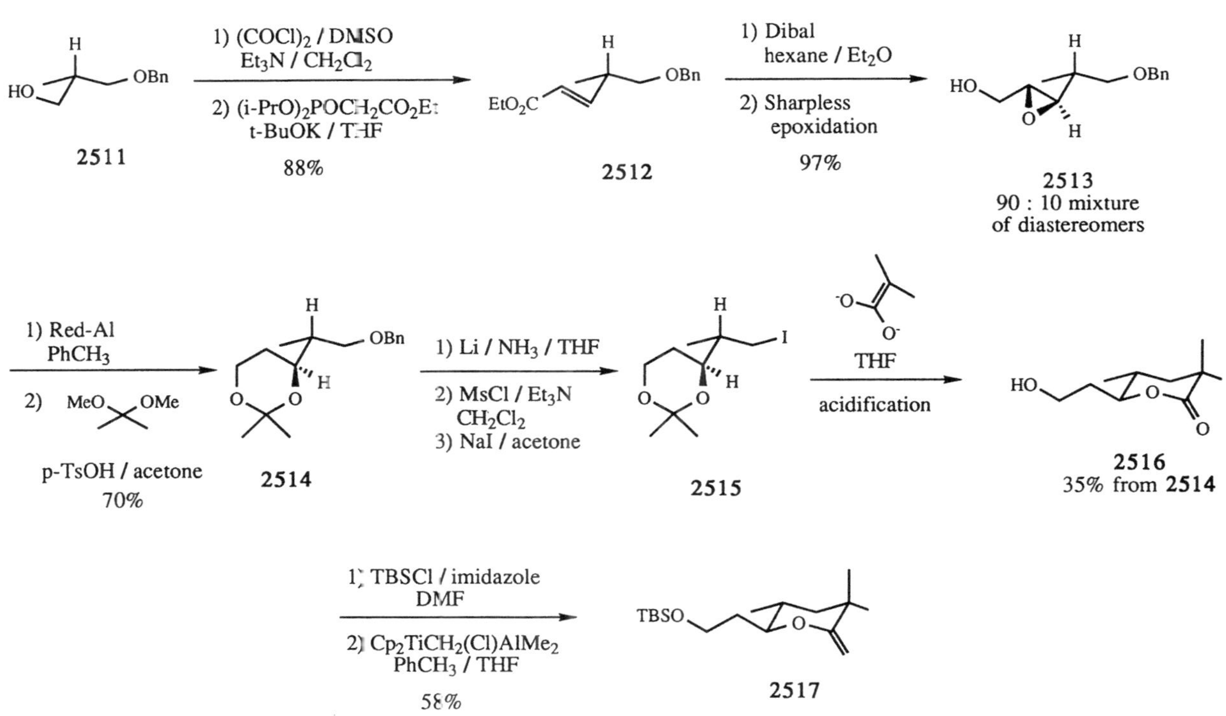

Scheme 450. Synthesis of Vinyl Ether **2517** (Ireland)

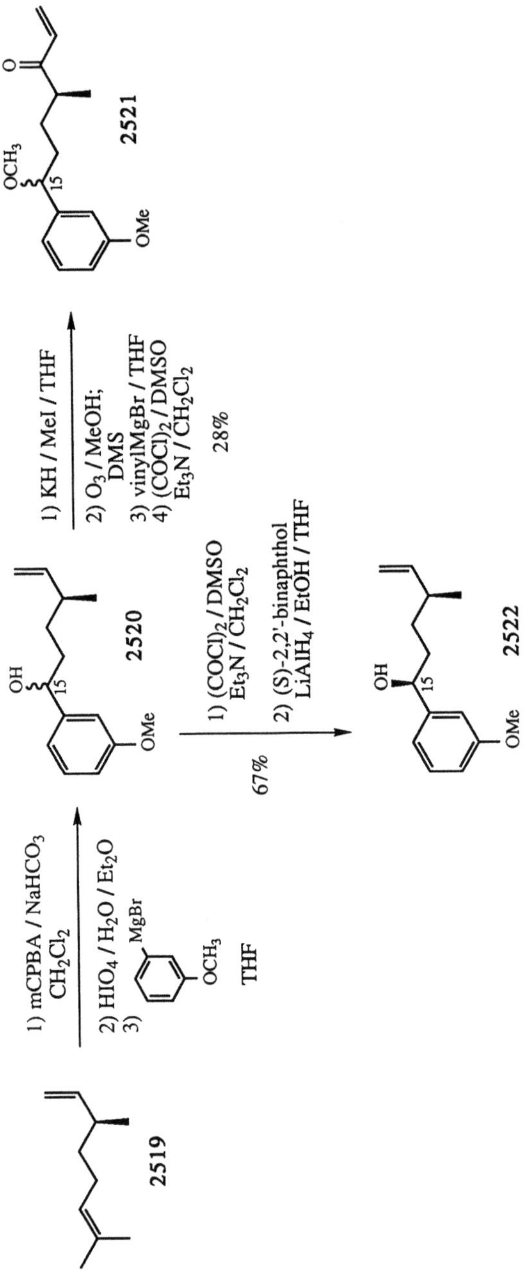

Scheme 451. Synthesis of Enone 2521

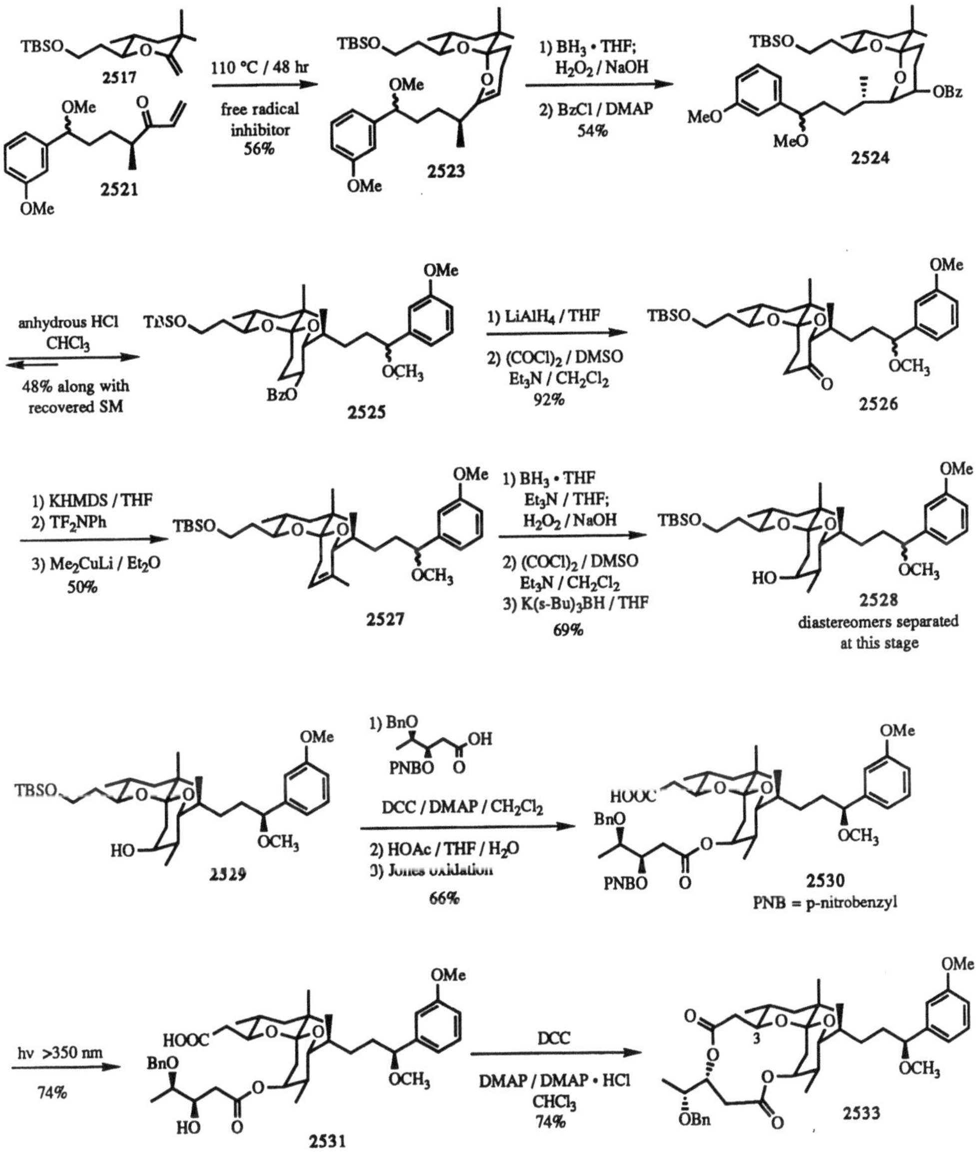

Scheme 452. Cycloaddition of **2517** with **2521** and Subsequent Transformations (Ireland)

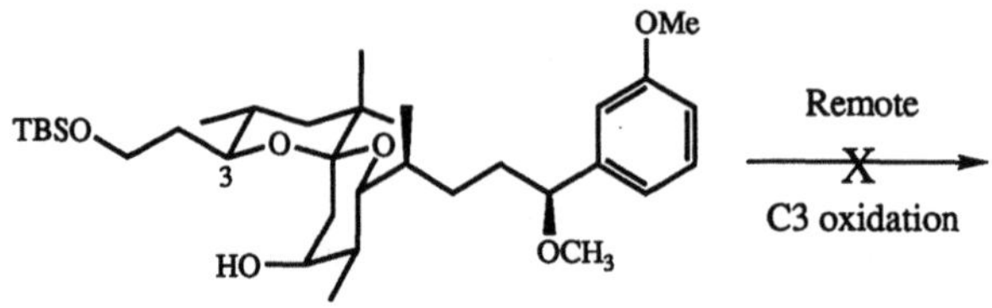

2529

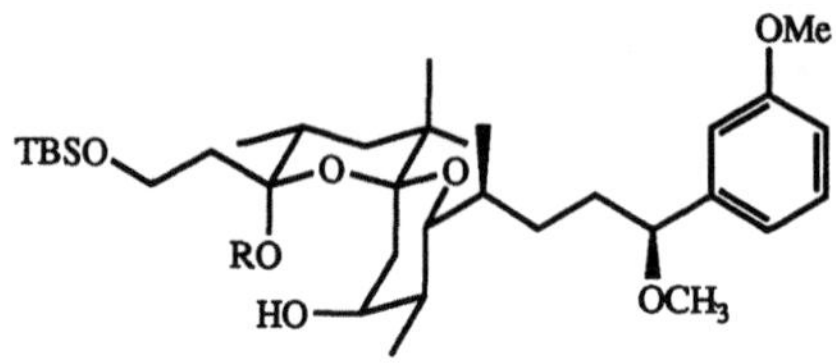

2532

Scheme 453. Attempted Remote Oxidation of **2529**→**2532** (Ireland)

7 Summary

Clearly the activity in marine natural products synthesis is deeper and broader than one might have expected and in every way mirrors not only synthesis of terrestrial metabolites, but organic synthesis as a whole. As long as the marine environment continues to be a source of biologically active substances and molecules of unprecedented architecture and complexity, it is anticipated that synthesis activity in this area will continue its remarkable growth.

8 References

312. Nakamoto WH, Shiba T (1984) Tetrahedron Lett 25:4411
313. Wakamiya T, Oda Y, Fujita H, Shiba T (1986) Tetrahedron Lett 27:2143
314. Holler TP, Spaltenstein A, Turner E, Klevit RE, Shapiro BM, Hopkins PB (1987) J Org Chem 52:4421
315. (a) Morimoto H (1955) J Pharm Soc Japan 75:766 and 937, (b) For leading references see: Watase H (1958) Bull Chem Soc Japan 31:932
316. Impellizzeri G, Mangiafico S, Oriente G, Piaatelli M, Sciuto S, Fattorusso E, Magno S, Santacroce C, Sica D (1975) Phytochemistry 14:1549
317. (a) Murukami S, Takemoto T, Shimizu Z (1953) J Pharm Soc Japan 73:1026, (b) Nilta I, Watase H, Tomiie Y (1958) Nature 181:761, (c) Maeda M, Kodama T, Tanaka T, Yoshizumi H, Takemoto T, Nomoto K, Fujita T (1987) Tetrahedron Lett 28:633
318. (a) McGeer EG, Olney JW, McGeer PL (eds) (1978) Kainic Acid as a Tool in Neurobiology. Raven Press, New York and references therein, (b) Goldberg O, Luini A, Teichberg VI (1983) J Med Chem 26:39
319. Oppolzer W, Andres H (1978) Tetrahedron Lett 3397
320. (a) Oppolzer W, Robbiani C, Battig K (1980) Helv Chim Acta 63:2015, (b) Oppolzer W, Robbiani C, Battig K (1984) Tetrahedron 40:1391
321. Oppolzer W, Andres H (1979) Helv Chim Acta 62:2282
322. Oppolzer W, Thirring K (1982) J Am Chem Soc 104:4978
323. Kraus GA, Nagy JO (1983) Tetrahedron Lett 24:3427
324. Kraus GA, Nagy JO (1985) Tetrahedron 41:3537
325. DeShong P, Kell DA (1989) Tetrahedron 27:3979
326. Cooper J, Knight DW, Gallagher PT (1987) J Chem Soc Chem Commun 1220
327. Baldwin JE, Li C-S (1987) J Chem Soc Chem Commun 166
328. (a) Takemoto T, Daigo K, Kondo Y, Kondo K (1966) J Pharm Soc Japan 86:1874, (b) Daigo K (1959) J Pharm Soc Japan 79:350
329. Ohfune Y, Tomita M (1982) J Am Chem Soc 104:3511
330. Andersen RJ (1978) Tetrahedron Lett 2541
331. (a) Schmidt U, Lieberknecht A, Griesser H, Bokens H (1982) Tetrahedron Lett 23:4911, (b) Schmidt U, Lieberknecht A, Griesser H, Bokens H (1985) Liebigs Ann Chem 785
332. (a) Hofheinz W, Oberhansli WE (1977) Helv Chim Acta 60:660, (b) Erickson KL,Wells RJ (1982) Aust J Chem 35:31
333. Baker JT (ed) (1981) Roche Research Institute of Marine Pharmacology: Final Research Report. Dee Why, Australia, p 30
334. De Laszlo S, Williard PG (1985) J Am Chem Soc 107:199
335. Ugi I (1962) Angew Chem Int Ed Engl 1:1
336. Sharma GM, Burkholder PR (1967) J of Antibiotics Ser A (Japan) 20:200
337. (a) Sharma GM, Vig B, Burkholder PR (1970) J Org Chem 35:2823, (b) Andersen RJ, Faulkner DJ (1973) In: Worthen, LR (ed) Food-Drugs Sea, Proc. [Conf.], 3rd 1972, Mar, Technol Soc, Washington DC, pp 111–115
338. Sharma BM, Burkholder PR (1967) Tetrahedron Lett 4147
339. Yamada Y, Kim J-S, Iguchi K, Suzuki M (1974) Chem Lett 1399
340. Evans DA, Wong RY (1977) J Org Chem 42:350
341. Fischer A, Henderson G (1983) Tetrahedron Lett 24:131
342. Norte M, Fernandez JJ (1987) Tetrahedron Lett 28:1693

343. (a) Fattorusso E, Minale L, Dodano G, Moody K, Thomson RH (1970) J Chem Soc Chem Commun 752, (b) Moody K, Thomson RH, Fattorusso E, Minale L, Sodano G (1972) J Chem Soc Perkin Trans 1 18, (c) Mcmillan JA, Paul IC, Goo YM, Rinehart Jr KL, Krueger WC, Pschigoda LA (1981) Tetrahedron Lett 22:39
344. Cimino G, De Rosa S, De Stefano S, Self R, Sodano G (1983) Tetrahedron Lett 24:3029
345. (a) Nishiyama S, Yamamura S (1983) Tetrahedron Lett 24:3351, (b) Nishiyama S, Yamamura S (1985) Bull Chem Soc Japan 58:3453
346. (a) Kazlauskas R, Lidgard RO, Murphy PT, Wells RJ (1980) Tetrahedron Lett 21:2277, (b) Kazlauskas R, Lidgard RO, Murphy PT, Wells RJ, Blount JF (1981) Aust J Chem 34:765
347. Nishiyama S, Yamamura S (1982) Tetrahedron Lett 23:1281
348. Nishiyama S, Suzuki T, Yamamura S (1982) Chem Lett 1851
349. Nishiyama S, Suzuki T, Yamamura S (1982) Tetrahedron Lett 23:3699
350. For leading references in this area see: Pettit GR (1980 + 1982) Synthetic Peptides. Elsevier, Amsterdam The Netherlands vols 5 + 6
351. (a) Yamamoto H (1987) J Chem Soc Perkin Trans 1 613, (b) Akaji K, Fujii N, Tokunaga F, Miyata T, Iwanaga S, Yajima H (1989) Chem Pharm Bull 37:2661
352. Rinehart KL Jr, Kishore V, Nagarajan S, Lake RJ, Gloer JB, Bozich FA, Li K-M, Maleczka RE Jr, Todsen WL, Munro MHG, Sullins DW, Sakai R (1987) J Am Chem Soc 109:6846. This manuscript contains references to the prior literature involving isolation and early synthesis studies
353. (a) Schmidt U, Kroner M, Griesser H (1988) Tetrahedron Lett 29:3057, (b) Schmidt U, Kroner M, Griesser H (1988) Tetrahedron Lett 29:4407
354. Hamada Y, Kondo Y, Shibata M, Shioiri T (1989) J Am Chem Soc 111:669
355. Jouin P, Poncet J, Dufour M-N, Pantaloni A, Castro B (1989) J Org Chem 54:617
356. Zabriskie TM, Klocke JA, Ireland CM, Marcus AH, Molinski TF, Faulkner DJ, Xu C, Clardy J (1986) J Am Chem Soc 108:3123
357. Crews P, Manes LV, Boehler M (1986) Tetrahedron Lett 27:2797
358. Grieco PA, Hon YS, Perez-Medrano A (1988) J Am Chem Soc 110:1630
359. Chan WR, Tinto WF, Manchand PS, Todaro LJ (1987) J Org Chem 52:3091
360. Grieco PA, Perez-Medrano A (1988) Tetrahedron Lett 29:4225
361. White JD, Amedio Jr JC (1989) J Org Chem 54:736
362. Hirai Y, Yokota K, Sakai H, Yamazaki T, Momose T (1989) Heterocycles 29:1865
363. Inanaga J, Hirata K, Saeki H, Katsuki T, Yamaguchi M (1979) Bull Chem Soc Japan 52:1989
364. (a) Ireland CM, Durso Jr AR, Newman RA, Hacker MP (1982) J Org Chem 47:1807, (b) Biskupiak JE, Ireland CM (1983) J Org Chem 48:2302
365. Wasylyk JM, Biskupiak JE, Costello CE, Ireland CM (1983) J Org Chem 48:4445
366. Ireland CM, Scheuer PJ (1980) J Am Chem Soc 102:5688
367. Hamada Y, Shibata M, Shioiri T (1985) Tetrahedron Lett 26:6501
368. Hamada Y, Shibata M, Shioiri T (1985) Tetrahedron Lett 26:5159
369. Schmidt U, Griesser H (1986) Tetrahedron Lett 27:163
370. Sugiura T, Hamada S, Shioiri T (1987) Tetrahedron Lett 28:2251
371. Schmidt U, Gleich P (1985) Angew Chem Int Ed Engl 24:569
372. Schmidt U, Weller D (1986) Tetrahedron Lett 27:3495
373. Kato S, Hamada S, Shioiri T (1986) Tetrahedron Lett 27:2653
374. Hamamoto Y, Endo M, Nakagawa M, Nakanishi T, Mizukawa K (1983) J Chem Soc Chem Commun 323
375. Kato S, Hamada S, Shioiri T (1985) Tetrahedron Lett 26:3223
376. Pettit GR, Kamano Y, Herald CL, Tuinman AA, Boettner FE, Kizu H, Scmidt JM, Baczynskzyj L, Tomer KB, Bontems RJ (1987) J Am Chem Soc 109:6883
377. Pettit GR, Singh SB, Hogan F, Lloyd-Williams P, Herald DL, Burkett DD, Clewlow PJ (1989) J Am Chem Soc 111:5463
378. Stonard RJ, Andersen RJ (1980) J Org Chem 45:3687
379. Schmidt U, Wild J (1984) Angew Chem Int Ed Engl 23:991
380. For leading references see; Muratake H, Natsume M (1987) Tetrahedron 28:2265. For the absolute configuration of lyngbyatoxin A see: Sakai S, Hitotsuyanagi Y, Aimi N, Fujiki H (1986) Tetrahedron Lett 27:5219
381. Cardellina II JH, Marner F-J, Moore RE (1979) Science 204:193
382. Murataka H, Natsume M (1987) Tetrahedron Lett 28:2265
383. Kigoshi H, Shizuri Y, Niwa H, Yamada K (1981) Tetrahedron Lett 22:4729

384. Kurosawa E, Fukuzawa A, Irie T (1972) Tetrahedron Lett 2121
385. Gonzalez AG, Martin JD, Martin VS, Norte M, Perez R, Ruano Z, Drexler SA, Clardy J (1982) Tetrahedron 38:1009
386. (a) Irie T, Suzuki M, Masamune T (1965) Tetrahedron Lett 1091, (b) Irie T, Suzuki M, Masamune T (1968) Tetrahedron 24:4193
387. Gonzalez AG, Martin JD, Martin VS, Norte M, Perez R, Ruano Z, Drexler SA, Clardy J (1982) Tetrahedron 38:1009
388. Kinnel R, Duggan AJ, Eisner T, Meinwald J, Miura I (1977) Tetrahedron Lett 3913
389. Waraszkiewicz SM, Sun HH, Erickson KL, Finer J, Clardy J (1978) J Org Chem 43:3194
390. Kurosawa E, Fukuzawa A, Irie T (1972) Tetrahedron Lett 2121
391. Anorbe B, Martin VS, Palazon JM, Trujillo JM (1986) Tetrahedron Lett 27:4991
392. Fukuzawa A, Sato H, Miyamoto M, Masamune T (1986) Tetrahedron Lett 27:2901
393. Palazon JM, Martin VS (1988) Tetrahedron Lett 29:681
394. Mukaiyama T, Shoda S, Watanabe S (1977) Chem Lett 383
395. (a) Murai A, Murase H, Matsue H, Masamune T (1977) Tetrahedron Lett 2507, (b) Masamune T, Matsue H (1975) Chem Lett 895
396. Overman LE, Thompson AS (1988) J Am Chem Soc 110:2248
397. Holmes AB, Jennings-White CLD, Kendrick DA (1983) J Chem Soc Chem Commun 415
398. Holmes AB, Jennings-White CLD, Kendrick DA (1984) J Chem Soc Chem Commun 1594
399. Feldman KS, Mechem CC, Nader L (1982) J Am Chem Soc 104:4011
400. Hunt DJ, Jackson SD, Moyes RB, Wells PB, Whyman R (1982) J Chem Soc Chem Commun 84
401. Feldman KS (1982) Tetrahedron Lett 23:3031
402. (a) Kikuchi H, Tsukitani Y, Iguchi K, Yamada Y (1982) Tetrahedron Lett 23:5171, (b) Kikuchi H, Tsukitani Y, Iguchi K, Yamada Y (1983) Tetrahedron Lett 24:1549, (c) Iguchi K, Yamada Y, Kikuchi H, Tsukitani T (1983) Tetrahedron Lett 24:4433, (d) Kobayashi M, Yasuzawa T, Yoshihara M, Akutsu H, Kyogoku Y, Kitagawa I (1982) Tetrahedron Lett 23:5331, (e) Kobayashi M, Yasuzawa T, Son BW, Kyogoku Y, Kitagawa I (1983) Chem Pharm Bull 31:1440
403. Fukushima M, Kato T, Ota K, Yamada Y, Kikuchi H, Kitagawa I (1983) Proc Jap Cancer Assoc 42:243
404. Kikawa Y, Naarumiya S, Fukushima M, Wakatsuka H, Hayashi O (1984) Proc Natl Acad Sci USA 81:1317
405. (a) Corey EJ (1983) Experientia 39:1084, (b) Corey EJ, Schmidt G, Shimoji K (1983) Tetrahedron Lett 24:3196
406. Corey EJ, Mehrotra MM (1984) J Am Chem Soc 106:3384
407. Shibasaki M, Ogawa Y (1985) Tetrahedron Lett 26:3841
408. Hashimoto S, Arai Y, Hamanaka N (1985) Tetrahedron Lett 26:2679
409. Nagaoka H, Miyakoshi T, Yamada Y (1984) Tetrahedron Lett 25:3621
410. Nagaoka H, Iguchi K, Miyakoshi T, Yamada N, Yamada Y (1986) Tetrahedron Lett 27:223
411. Corey EJ, Matsuda SPT (1987) Tetrahedron Lett 28:4247
412. Corey EJ, Xiang YB (1988) Tetrahedron Lett 29:995
413. Baker BJ, Okuda RK, Yu PTK, Scheuer PJ (1985) J Am Chem Soc 107:2976
414. Mori K, Takeuchi T (1988) Tetrahedron 44:333
415. Sasai H, Shibasaki M (1987) Tetrahedron Lett 28:333
416. Suzuki M, Morita Y, Yanagisawa A, Baker BJ, Scheuer PJ, Noyori R (1988) J Org Chem 53:286
417. Nagaoka H, Miyaoka H, Miyakoshi T, Yamada Y (1986) J Am Chem Soc 108:5019
418. Higgs MD, Mulheirn LJ (1981) Tetrahedron 37:4259
419. Corey EJ, De B (1984) J Am Chem Soc 106:2735
420. (a) Moore RE (1976) Lloydia 39:181, (b) Moore RE Acc Chem Res 1977 10 40
421. Yamada K, Tan H, Tatematsu H (1979) J Chem Soc Chem Commun 572
422. Moore RE, Pettus Jr JA, Doty MS (1968) Tetrahedron Lett 4787
423. Pettus Jr JA, Moore RE (1970) J Chem Soc Chem Commun 1093
424. (a) Muller DG, Clayton MN, Gassmann G, Boland W, Marner F-J, Jaenicke L (1984) Experientia 40:211, (b) Muller DG, Clayton MN, Gassmann G, Boland W, Marner F-J, Schotten T, Jaenicke L (1985) Naturwissenschaften 72:97
425. Moore RE, Pettus Jr JA, Mistysyn J (1974) J Org Chem 39:2201
426. Kajiwara T, Kodama K, Hatanaka A (1980) Bull Jap Soc Sci Fish 771
427. Muller DG, Jaenicke L, Donike M, Akintobi T (1971) Science 171:815

428. Ohloff G, Pickenhagen W (1969) Helv Chim Acta 52:882
429. Das KC, Weinstein B (1969) Tetrahedron Lett 3459
430. Billups WE, Chow WY, Cross JH (1974) J Chem Soc Chem Commun 252
431. Yamada K, Tan H, Hirota K (1980) Tetrahedron Lett 21:4873
432. Colobert F, Genet J-P (1985) Tetrahedron Lett 26:2779
433. Schotten T, Boland W, Jaenicke L (1985) Helv Chim Acta 68:1186
434. Schotten T, Boland W, Jaenicke L (1986) Tetrahedron Lett 27:2349
435. Ali A, Sarantakis AD, Weinstein B (1971) J Chem Soc Chem Commun 940
436. Schotten T, Boland W, Jaenicke L (1985) Helv Chim Acta 68:1186
437. Schneider MP, Goldbach M (1980) J Am Chem Soc 102:6114
438. Dorsch D, Kunz E, Helmchen G (1985) Tetrahedron Lett 26:3319
439. Billups WE, Chow WY, Cross JH (1974) J Chem Soc Chem Commun 252
440. Colobert F, Genet J-P (1985) Tetrahedron Lett 26:2779
441. Schotten T, Boland W, Jaenicke L (1986) Tetrahedron Lett 27:2349
442. Jaenicke L, Akintobi T, Muller DG (1971) Angew Chem Internat Ed Engl 10:492
443. Janeicke L, Akintobi T, Marner F-J (1973) Liebigs Ann Chem 1252
444. Ali A, Sarantakis A D, Weinstein B (1971) J Chem Soc Chem Commum 940
445. Schneider MP, Goldbach M (1980) J Am Chem Soc 102:6114
446. Moore RE, Yost G (1973) J Chem Soc Chem Commun 937
447. Asaoka M, Takenouchi K, Takei H (1988) Chem Lett 921
448. Jaenicke L, Muller DG, Moore RE (1974) J Am Chem Soc 96:3324
449. Boland W, Jaenicke L (1978) Chem Ber 111:3262
450. Boland W, Jaenicke L (1979) J Org Chem 44:4819
451. Boland W, Jaenicke L, Muller DG (1981) Liebigs Ann Chem 2266
452. Boland W, Mertes K, Jaenicke L, Muller DG, Folster E (1983) Helv Chim Acta 66:1905
453. Crouse GD, Paquette LA (1981) J Org Chem 46:4272
454. Burks Jr JE, Crandall JK (1984) J Org Chem 49:4663
455. (a) Muller DG, Peters A, Gassmann G, Boland W, Marner F-J, Jaenicke L (1982) Naturwissenschaften 69:290, (b) Muller DG, Boland W, Marner F-J, Gassmann G (1982) Naturwissenschaften 69:501
456. Boland W, Jakoby K, Jaenicke L (1982) Helv Chim Acta 65:2355
457. Boland W, Niedermeyer U, Jaenicke L (1985) Helv Chim Acta 68:2062
458. Marner F-J, Muller B, Jaenicke L (1984) Z Naturforsch 39c:689
459. Jaenicke L, Muller DG, Moore RE (1974) J Am Chem Soc 96:3324
460. Marner F-J, Jaenicke L (1975) Chem Ber 108:2202
461. Boland W, Niedermeyer U, Jaenicke L (1985) Helv Chim Acta 68:2062
462. Jakel E, Schneider MP (1987) J Chem Soc Chem Commun 733
463. Muller DG, Jaenicke L (1973) FEBS Lett 30:137
464. (a) Muller DG (1968) Planta 81:160, (b) Muller DG, Jaenicke L, Donike M, Akintobi T (1971) Science 171:815
465. Jaenicke L, Seferiadis K (1975) Chem Ber 108:225
466. Wiedenmann M, Hopf H (1977) Z Naturforsch 32b:119
467. Schneider MP, Goldbach M (1980) J Am Chem Soc 102:6114
468. Goldbach M, Jakel E, Schneider MP (1987) J Chem Soc Chem Commun 1434
469. Boland W, Schroer N, Sieler C, Feigel M (1987) Helv Chim Acta 70:1025
470. (a) Muller DG, Marner F-J, Boland W, Jaenicke L, Gassmann G (1981) Naturwissenschaften 68:478, (b) Marner F-J, Boland W, Jaenicke L (1982) Liebigs Ann Chem 579
471. Naf F, Decorzant R, Escher SD (1982) Tetrahedron Lett 23:5043
472. Endo M, Nakagawa M, Hamamoto Y, Nakanishi T (1983) J Chem Soc Chem Commun 322
473. Suntory Ltd, Jpn Kokai Tokyo Koho Jpn Pat 5,829,737 [83 20737], cf (1983) Chem Abstracts 99:70287s
474. Endo M, Nakagawa M, Hamamoto Y, Nakanishi T (1983) J Chem Soc Chem Commun 980
475. (a) Urech R (1984) J Chem Soc Chem Commun 989, (b) Urech R (1986) Aust J Chem 39:443
476. Still IWJ, Shi Y (1987) Tetrahedron Lett 28:2489
477. Carter GT, Rinehart Jr KL (1978) J Am Chem Soc 100:4302
478. Boukouvalas J, Golding BT, McCabe RW, Slaich PK (1983) Angew Chem Int Ed Engl 22:618
479. (a) Blunt JW, Munro MHG, Yorke SC (1982) Tetrahedron Lett 23:2793, (b) Yorke SC, Blunt JW, Munro MHG, Cook JC, Rinehart Jr KL (1986) Aust J Chem 39:447
480. Boylan DB, Scheuer PJ (1967) Science 155:52

481. Yoshikawa M, Sugimura T, Tai A (1989) Agric Biol Chem 53:37
482. Anderson R, Kates M, Volcani BE (1978) Biochim Biophys Acta 528:89
483. Ohashi K, Yamagiwa Y, Kamikawa T, Kates M (1988) Tetrahedron Lett 29:1185
484. Hayashi A, Matsuura F (1978) Chem Phys Lipids 22:9
485. Ohashi K, Kosai S, Arizuka M, Watanabe T, Fukunaga M, Monden K, Uchikoda T, Yamagiwa Y, Kamikawa T (1988) Tetrahedron Lett 29:1189
486. Ravi BN, Wells RJ (1982) Aust J Chem 35:105
487. Ortuno RM, Bigorra J, Font J (1988) Tetrahedron 44:5139
488. Warren RG, Wells RJ, Blount JF (1980) Aust J Chem 33:891
489. Williams DR, Harigaya Y, Moore JL, D'sa A (1984) J Am Chem Soc 106:2641
490. Hatakeyama S, Sakurai K, Saijo K, Takano S (1985) Tetrahedron Lett 26:1333
491. Raederstorff D, Shu AYL, Thompson JE, Djerassi C (1987) J Org Chem 52:2346
492. Mena PL, Pilet O, Djerassi C (1984) J Org Chem 49:3260
493. Cimino G, De Stefano S (1977) Tetrahedron Lett 1325
494. Thaller V, De Marquez MD (1985) J Chem Res (S) 104 and (1985) J Chem Res (M) 1268
495. (a) Carle JS, Christophersen C (1979) J Am Chem Soc 101:4012, (b) Carle JS, Christophersen C (1980) J Org Chem 45:1586, (c) Carle JS, Christophersen C (1981) J Org Chem 46:3440, (d) Wulff P, Carle JS, Christophersen C (1981) J Chem Soc Perkin Trans 1 2985
496. Hino T, Tanaka T, Matsuki K, Nakagawa M (1983) Chem Pharm Bull 31:1806
497. (a) Kosuge T, Tsuji K, Hirai K, Yamaguchi K, Okamoto T, Iitaka Y (1981) Tetrahedron Lett 22:3417, (b) Kosuge T, Tsuji K, Hirai K (1982) Chem Pharm Bull 30:3255
498. (a) Inoue S, Okada K, Tanino H (1984) Tetrahedron Lett 25:4407, (b) Okada K, Tanino H, Hashizume K, Mizuno M, Kakoi H, Inoue S (1984) Tetrahedron Lett 25:4403
499. Inoue S, Okada K, Tanino H, Kakoi H (1986) Tetrahedron Lett 27:5225
500. Okada K, Mizuno Y, Tanino H, Kakoi H, Inoue S (1989) Chem Lett 703
501. Inoue S, Okada K, Tanino H, Kakoi H (1988) Tetrahedron Lett 29:1547
502. Capon RJ, MacLeod JK, Scammels PJ (1986) Tetrahedron 42:6545
503. MacLeod JK, Monahan LC (1988) Tetrahedron Lett 29:391
504. Muratake H, Natsume M (1989) Tetrahedron Lett 30:5771
505. Yasukouchi T, Kanematsu K (1989) Tetrahedron Lett 30:6559
506. Guella G, Mancini I, Zibrowius H, Pietra F (1989) Helv Chim Acta 72:1444
507. Heitz S, Durgeat M, Guyot M, Brassy C, Bachet B (1980) Tetrahedron Lett 21:1457
508. Hogan IT, Sainsbury M (1984) Tetrahedron 40:681
509. Brennan MR, Erickson KL (1978) Tetrahedron Lett 1637
510. Ohta T, Somei M (1989) Heterocycles 29:1663
511. Dellar G, Djura P, Sargent MV (1981) J Chem Soc Perkin Trans 1 1679
512. Cardellina II JH, Kirkup MP, Moore RE, Mynderse JS, Seff K, Simmons CJ (1979) Tetrahedron Lett 4915
513. (a) Kano S, Sugino E, Hibino S (1980) J Chem Soc Chem Commun 1241, (b) Kano S, Sugino E, Shibuya S, Hibino S (1981) J Org Chem 46:3856
514. Takano S, Suzuki Y, Ogasawara K (1981) Heterocycles 16:1479
515. Kawasaki T, Nonaka Y, Sakamoto M (1989) J Chem Soc Chem Commun 43
516. (a) Moody CJ, Shah P (1989) J Chem Soc Perkin Trans 1 376, (b) Moody C, Shah P (1989) J Chem Soc Perkin Trans 1 2463
517. (a) Rinehart Jr KL, Kobayashi J, Harbour GC, Hughes RG, Mizak SA, Scahiel TA (1984) J Am Chem Soc 106:1524, (b) Kobayashi J, Harbour GC, Gilmore J, Rinehart Jr KL (1984) J Am Chem Soc 106:1526, (c) Rinehart Jr KL, Kobayashi J, Harbour GC, Gilmore J, Mascal M, Holt RG, Shield LS, Lafargue F (1987) J Am Chem Soc 109:3378
518. Murakami Y, Takahashi H, Nakazawa Y, Koshimizu M, Watanabe T, Yokoyama Y (1989) Tetrahedron Lett 30:2099
519. Hino T, Lai Z, Seki H, Hara R, Kuramochi T, Nakagawa M (1989) Chem Pharm Bull 37:2596
520. Rinehart KL, Kobayashi J, Harbour GC, Gilmore J, Mascal M, Holt TG, Shield LS, Lafargue F (1987) J Am Chem Soc 109:3378
521. Van Wagenen BC, Cardellina II JH (1989) Tetrahedron Lett 30:3605
522. Wasserman HH, Kelly TA (1989) Tetrahedron Lett 30:7117
523. Still IWJ, McNulty J (1989) Heterocycles 29:2507
524. Nakagawa M, Liu J-J, Hino T (1989) J Am Chem Soc 111:2721
525. Sakai R, Kohmoto S, Higa T, Jefford CW, Bernardinelli G (1987) Tetrahedron Lett 28:5493
526. Torisawa Y, Hashimoto A, Nakagawa M, Hino T (1989) Tetrahedron Lett 30:6459

527. Shi L, Xia W, Yang J, Wen X, Huang YZ (1987) Tetrahedron Lett 28:2155
528. Coval SJ, Scheuer PJ (1985) J Org Chem 50:3024
529. Burke SD, Piscopio AD, Buchanan JL (1988) Tetrahedron Lett 29:2757
530. Oppolzer W, Dupuis D, Poli G, Raynham TM, Bernardinelli G (1988) Tetrahedron Lett 29:5885
531. Sugahara T, Iwata T, Yamaoka M, Takano S (1989) Tetrahedron Lett 30:1821
532. Evans DA, Chapman KT, Bisaha J (1988) J Am Chem Soc 110:1238
533. Kobayashi J, Cheng J, Nakamura H, Ohizumi Y, Hirata Y, Sasaki T, Ohta T, Nozoe S (1988) Tetrahedron Lett 29:1177
534. Bracher F (1989) Heterocycles 29:2093
535. (a) Nakamura H, Kobayashi J, Ohizumi Y (1982) Tetrahedron Lett 23:5555, (b) Ohizumi Y, Kajiwara A, Nakamura H, Kobayashi J (1984) J Pharm Pharmacol 36:785, (c) (1984) Chem Abstracts 100:83029y, (d) Ninth International Congress on Heterocyclic Chemistry Tokyo (1983) Abstract G-66
536. (a) Pelletier JC, Cava MP (1985) Tetrahedron Lett 26:1259, (b) Pelletier JC, Cava MP (1987) J Org Chem 52:616
537. Kelly TR, Maguire MP (1985) Tetrahedron 41:3033
538. Sakamoto T, Miura N, Kondo Y, Yamanaka H (1986) Chem Pharm Bull 34:2760
539. Bassoli A, Maddinelli G, Rindone B, Tollari S, Chioccara F (1987) J Chem Soc Chem Commun 150
540. Andrew RG, Raphael RA (1987) Tetrahedron 43:4803
541. Schmitz FJ, Agarwal SK, Gunasekera SP, Schmidt PG, Shoolery JN (1983) J Am Chem Soc 105:4835
542. For a review see: Arai T, Kubo A (1983). In: Brossi A (ed) The Alkaloids. Academic, New York, vol 21 (ch 3)
543. Kubo A, Nakahara S (1988) Heterocycles 27:2095
544. Echavarren AM, Stille JK (1988) J Am Chem Soc 110:4051
545. Prager RH, Tsopelas C (1989) Heterocycles 29:847
546. Chevolot L (1981). In: Scheuer PJ (ed) Marine Natural Products: Chemical and Biological Perspectives. Academic, New York, vols 4 (ch 2)
547. Yasumoto T, Yasumura D, Yotsu M, Michishita T, Endo A, Kotaki Y (1986) Agric Biol Chem 50:793
548. Kishi Y, Nakatsubo F, Aratani M, Goto T, Inoue S, Kakoi H, Sugiura S (1970) Tetrahedron Lett 5127
549. Kishi Y, Nakatsubo F, Aratani M, Goto T, Inoue S, Kakoi H (1970) Tetrahedron Lett 5129
550. Kishi Y, Aratani M, Fukuyama T, Nakatsubo F, Goto T, Inoue S, Tanino H, Sugiura S, Kakoi H (1972) J Am Chem Soc 94:9217
551. Kishi Y, Fukuyama T, Aratani M, Nakatsubo F, Goto T, Inoue S, Tanino H, Sugiura S, Kakoi H (1972) J Am Chem Soc 94:9219
552. For leading references see: Ghazarossian VE, Schantz EJ, Schnoes HK, Strong FM (1974) Biochem Biophys Res Commun 59:1219
553. Shimizu Y, Alam M, Oshima Y, Fallon WE (1975) Biochem Biophys Res Commun 66:731
554. Schantz EJ, Mold JD, Stanger DW, Shavel J, Riel FJ, Bowden JP, Lynch JM, Wyler RS, Riegel B, Sommer HJ (1957) J Am Chem Soc 79:5230 and 5235
555. (a) Schantz EJ, Ghazarossian VE, Schnoes HK, Strong FM, Springer JP, Pezzanite JO, Clardy J (1975) J Am Chem Soc 97:1238, (b) Bordner J, Thiessen WE, Bates HA, Rapoport H (1975) J Am Chem Soc 97:6008
556. Tanino H, Nakata T, Kaneko T, Kishi Y (1977) J Am Chem Soc 99:2818
557. Jacobi PA, Martinelli MJ, Polanc S (1984) J Am Chem Soc 106:5594
558. Harbour GC, Tymiak AA, Rinehart Jr KL, Shaw PD, Hughes Jr RG, Mizsak SA, Coats JH, Zurenko GE, Li LH, Kuentzel SL (1981) J Am Chem Soc 103:5604
559. Snider BB, Faith WC (1983) Tetrahedron Lett 24:861
560. Snider BB, Faith WC (1984) J Am Chem Soc 106:1443
561. Hassner A, Keshava Murthy KS (1986) Tetrahedron Lett 27:1407
562. Uyehara T, Furuta T, Kabasawa Y, Yamada J-I, Kato T (1986) J Chem Soc Chem Commun 539
563. (a) Walts AE, Roush WR (1985) Tetrahedron 41:3463, (b) Roush WR, Walts AE (1984) J Am Chem Soc 106:721
564. Forenza S, Minale L, Riccio R, Fattorusso R (1971) J Chem Soc Chem Commun 1129

565. Garcia EE, Benjamin LE, Fryer I (1973) J Chem Soc Chem Commun 78
566. de Nanteuil G, Ahond A, Poupat C, Thoisin O, Potier P (1986) Bull Soc Chim France 813
567. Foley LH, Büchi G (1982) J Am Chem Soc 104:1776
568. Ahond A, Bedoya Zurita M, Colin M, Fizames C, Laboute P, Lavelle F, Laurent D, Poupat C, Pusset J, Pusset M, Thoison O, Potier P (1988) C R Seances Acad Sci Paris series 2 307:145
569. Kato Y, Fusetani S, Matsunaga S, Hashimoto K (1985) Tetrahedron Lett 26:3483
570. Meade EA, Krawczyk SH, Townsend LB (1988) Tetrahedron Lett 29:4073
571. Ayer SW, Andersen RJ, He C, Clardy J (1984) J Org Chem 49:3869
572. Chan AT, Andersen RJ, LeBlanc MJ, Harrison PJ (1980) Marine Biology 59:7
573. Hirota K, Kubo K, Kitade Y, Maki Y (1985) Tetrahedron Lett 26:2355
574. Cimino G, De Giulio A, De Rosa S, De Stefano S, Puliti R, Mattia CA, Mazzarella L (1985) J Nat Prod 48:523
575. Fujii T, Saito T, Mori S (1988) Heterocycles 27:1145
576. For isolation of 1-methylisoguanosine and initial report of its synthesis see: (a) Quinn RJ, Gregson RP, Cook AF, Bartlett RT (1980) Tetrahedron Lett 21:567. For reports on the pharmacological activity see:, (b) Baird-Lambert J, Marwood JF, Davies LP, Taylor KM (1980) Life Sci 26:1069, (c) Davies LP, Taylor KM, Gregson RP, Quinn RJ (1980) Life Sci 26:1079, (d) Davies LP, Cook AF, Poonian M, Taylor KM (1980) Life Sci 26:1089
577. Cook AF, Bartlett RT (1980) J Org Chem 45:4020
578. Chern JW, Townsend LB (1985) Tetrahedron Lett 26:6419
579. (a) Arai T, Yazawa K, Mikami Y, Kubo A, Takahashi K (1976) J Antibiotics (Tokyo) 29:398, (b) Mikami Y, Yokoyama K, Omi A, Arai T (1976) J Antibiotics (Tokyo) 29:406
580. (a) Frincke JM, Faulkner DJ (1982) J Am Chem Soc 104:265, (b) McIntyre DE, Faulkner DJ, Van Engen D, Clardy J (1979) Tetrahedron Lett 4163
581. (a) Fukumi H, Kurihara H, Hata T, Tamura C, Mishima H, Kubo A, Arai T (1977) Tetrahedron Lett 3825, (b) Fukumi H, Kurihara H, Mishima H (1978) Chem Pharm Bull 26:2175
582. Parker KA, Casteel DA (1988) J Org Chem 53:2847
583. McKillop A, Brown SP (1987) Synth Commun 17:657
584. Frincke JM, Faulkner DJ (1982) J Am Chem Soc 104:265
585. (a) Parker KA, Cohen ID, Padwa A, Dent W (1984) Tetrahedron Lett 25:4917, (b) Padwa A, Chen Y-Y, Dent W, Nimmesgern H (1985) J Org Chem 50:4006
586. Kubo A, Nakahara S, Inaba K, Kitahara Y (1985) Chem Pharm Bull 33:2582
587. Danishefsky S, Berman E (1980) Tetrahedron Lett 21:4819
588. (a) Kubo A, Nakahara S (1981) Chem Pharm Bull 29:595, (b) Kubo A, Nakahara S, Inaba K, Kitahara Y (1986) Chem Pharm Bull 34:4056
589. (a) Braun M, Büchi G (1976) J Am Chem Soc 98:3049, (b) Braun M, Büchi G, Bushey DF (1978) J Am Chem Soc 100:4208
590. (a) Cariello L, Crescenzi G, Prota G, Giordano F, Mazzarella L (1973) J Chem Soc Chem Commun 99, (b) Cariello L, Crescenzi G, Capasso F, Giordano F, Mazzarella L (1974) Tetrahedron 30:3281, 3611 and 4191, (c) Cariello L, Crescenzi G, Prota G, Zanetti L (1974) Experientia 30:849
591. Cimino G, De Stefano S, Minale L (1975) Experientia 31:1387
592. Stierle DB, Faulkner DJ (1980) J Org Chem 45:4980
593. Muchowski JM, Naef R (1984) Helv Chim Acta 67:1168
594. Bowden BF, Clezy PS, Coll JC, Ravi BN, Tapiolas D (1984) Aust J Chem 37:227
595. Laatsch H, Pudleiner H (1989) Liebigs Ann Chem 863
596. Ballantine JA, Psaila AF, Pelter A, Murray-Rust P, Ferrito V, Schembri P, Jaccarini V (1980) J Chem Soc Perkin Trans 1 1080
597. (a) Baltzer F (1925) Publ Staz Zool Napoli 6:223, (b) De Silva FBD (1934) Varkbl Biol 15:221
598. Nigrelli RF, Stempien NF, Ruggieri GD, Berenbaum MC (1981) J Chem Soc Perkin Trans 1 3135
599. Battersby AR, Dutton CJ, Fookes CJR (1988) J Chem Soc Perkin Trans 1 1569
600. Yvin JC, Chevolot-Magueur A-M, Chevolot L, Lallemand J-Y, Potier P, Guilhem J (1982) J Am Chem Soc 104:4497
601. Elsner H, Liedmann A, Oppers K (1938) Naunyn-Schmiedebergs Arch Exp Pathol Pharmakol 190:510
602. Seebach D, Dust M, Naef R, Banziger M (1984) Angew Chem Int Ed Engl 23:530
603. Poss AJ, Belter RK (1988) J Org Chem 53:1535

604. Glombitza K-W, Sukopp I, Wiedenfeld H (1985) Planta Med 437
605. Lu PW, Lillard DW, Seib PA, Kramer KJ, Liang YT (1984) J Agric Food Chem 32:21
606. Peters JE, Catalfomo P, Constantine GH, Kirk PW (1975) J Pharm Sci 64:176
607. Schiehser GA, White JD, Matsumoto G, Pezzanite JO, Clardy J (1986) Tetrahedron Lett 27:5587
608. (a) Pallenberg AJ, White JD (1986) Tetrahedron Lett 27:5591, (b) White JD, Badger RA, Kezar III HS, Pallenberg AJ, Schiehser GA (1989) Tetrahedron 45:6631
609. Rollin P (1987) Tetrahedron Lett 28:3813
610. Higa T, Scheuer PJ (1977) In: Faulkner DJ, Fenical WH (eds) Marine Natural Products Chemistry, Plenum, New York, p 35
611. Iguchi M, Nishiyama A, Etoh H, Okamoto K, Yamamura S, Kato Y (1986) Chem Pharm Bull 34:4910
612. (a) Sharma GM, Vig B, Burkholder PR, "Food-Drugs From the Sea" in (1969) Proc Mar Technol Soc 307, (b) Carte B, Faulkner DJ (1981) Tetrahedron 37:2335, (c) Wells RJ (1979) Pure Appl Chem 51:1829
613. Capon R, Ghisalberti EL, Jefferies PR, Skelton BW, White AH (1981) J Chem Soc Perkin Trans 1 2464
614. Francesconi KA, Ghisalberti EL (1985) Aust J Chem 38:1271
615. Doyle MP, Siegfried B, Dellaria Jr JF (1977) J Org Chem 42:2426
616. Sharma GM, Vig B (1972) Tetrahedron Lett 1715
617. Rose AF, Pettus JA, Sims JJ (1977) Tetrahedron Lett 1847
618. Beissner RS, Guilford WJ, Coates RM, Hager LP (1981) Biochemistry 20:3724
619. (a) Kazlauskas R, Murphy PT, Quinn RJ, Wells RJ (1977) Tetrahedron 37, (b) Pettus JA, Wing RM, Sims JJ (1977) Tetrahedron Lett 41
620. Beechan CM, Sims JJ (1979) Tetrahedron Lett 1649
621. Nakayama M, Fukuoka Y, Nozaki H, Matsuo A, Hayashi S (1980) Chem Lett 1243
622. Boschelli D, Smith III AB, Stringer OD, Jenkins Jr RH, Davis FA (1981) Tetrahedron Lett 22:4385
623. Simmons CJ, Marner F-J, Cardellina II JH, Moore RE, Seff K (1979) Tetrahedron Lett 2003, (b) Cardellina II JH, Moore RE (1979) Tetrahedron Lett 2007
624. James GD, Pattenden G, Mills S (1985) Tetrahedron Lett 26:3617
625. (a) Neeman I, Fishelson L, Kashman Y (1975) Mar Biol 30:293, (b) Kashman Y, Groweiss A, Shmueli U (1980) Tetrahedron Lett 21:3629, (c) Spector I, Shochet NR, Kashman Y, Groweiss A (1983) Science (Washington DC) 493, (d) Groweiss A, Shmueli U, Kashman Y (1983) J Org Chem 48:3512, (e) Kashman Y, Groweiss A, Lidor R, Blasberger D, Carmely S (1985) Tetrahedron 41:1905
626. Zibuck R, Liverton NJ, Smith III AB (1986) J Am Chem Soc 108:2451
627. Kubodera N, Nagano H, Takagi M, Matsunaga I (1982) Heterocycles 18:259
628. Kameyama T, Takahashi A, Kurasawa S, Ishizuka M, Okami Y, Takeuchi T, Umezawa H (1987) J Antibiot 40:1664 and 1671
629. Bergeron RJ, McManis JS (1989) Tetrahedron 45:4939
630. Gerwick WH, Lopez A, Van Duyne GD, Clardy J, Ortiz W, Baez A (1986) Tetrahedron Lett 27:1979
631. Alonso R, Brossi A (1988) Tetrahedron Lett 29:735
632. Ayyangar NR, Khan RA, Deshpande VH (1988) Tetrahedron Lett 29:2347
633. Cardellina II JH, Hendrickson RL, Manfredi KP, Strobel SA, Clardy J (1987) Tetrahedron Lett 28:727
634. Brehm M, Dauben WG, Kohler P, Lichtenthaler FW (1987) Angew Chem Int Ed Engl 26:1271
635. Uemura D, Toya Y, Watanabe I, Hirata Y (1979) Chem Lett 1481
636. Jarglis P, Lichtenthaler FW (1982) Angew Chem Int Ed Engl 21:141
637. Williard PG, De Laszlo SE (1984) J Org Chem 49:3489
638. Hofheinz W, Oberhansli WE (1977) Helv Chim Acta 60:660
639. Kohler H, Gerlach H (1984) Helv Chim Acta 67:1783
640. Nozaki H, Minohara K, Miyazaki I, Kondo H, Shirane F, Makayama M (1980) Agric Biol Chem 52:3229
641. Ohira S, Shirane F, Nozaki H, Yahiro S, Nakayama M (1989) Bull Chem Soc Japan 62:2427
642. (a) Kurata K, Amiya T (1980) Bull Chem Soc Japan 53:2020, (b) Kurata K, Amiya T (1977) Chem Lett 1435, (c) Kurata K, Amiya T (1980) Chem Lett 297, (d) Kurata K, Amiya T (1980) Phytochemistry 19:141

643. Mori T, Bando H, Kanaiwa Y, Amiya T, Kurata K (1983) Chem Pharm Bull 31:1754
644. Lindquist N, Fenical WH, Sesin DF, Ireland CM, Van Duyne GD, Forsyth CJ, Clardy J (1988) J Am Chem Soc 110:1308
645. Forsyth C, Clardy J (1988) J Am Chem Soc 110:5911
646. (a) McAdam DP, Stick RV (1986) Tetrahedron Lett 27:251, (b) McAdam DP, Perera AMA, Stick RV (1987) Aust J Chem 40:1901
647. Edmonds JS, Francesconi KA (1981) Nature 289:602
648. Edmonds JS, Francesconi KA, Healy PC, White AH (1982) J Chem Soc Perkin Trans 1 2989
649. Okaichi T, Hashimoto Y (1962) Agric Biol Chem 26:224
650. Hagiwara H, Numata M, Konishi K, Oka Y (1965) Chem Pharm Bull 13:253
651. Asato AE, Moore RE (1973) Tetrahedron Lett 4941
652. (a) Sleeper HL, Fenical W (1977) J Am Chem Soc 99:2367, (b) Fenical W, Sleeper H, Paul VG, Stallard MO, Sun HH (1979) Pure Appl Chem 51:1865
653. Clelland J, Knox GR (1983) J Chem Soc Chem Commun 1219
654. Cardellina II JH, Moore RE, Arnold EV, Clardy J (1979) J Org Chem 44:4039
655. Sakito Y, Tanaka S, Asami M, Mukaiyama T (1980) Chem Lett 1223
656. Babler JH, Invergo BJ, Sarussi SJ (1980) J Org Chem 45:4241
657. Cardillo G, Orena M, Porzi G, Sandri S (1981) J Org Chem 46:2439
658. Torii S, Inokuchi T, Yorikata K (1981) J Org Chem 46:5030
659. Matsuo K, Kinuta T, Tanaka K (1981) Chem Pharm Bull 29:3047
660. Matsuo K, Tanaka K (1981) Chem Pharm Bull 29:3070
661. Kozikowski AP, Nieduzak TR, Scripko J (1982) J Am Chem Soc 103:2301
662. Kim S, Hong CY, Moon YC (1982) J Org Chem 47:4350
663. Pougny J-R, Rollin P, Sinay P (1982) Tetrahedron Lett 23:4929
664. Kogure T, Eliel E (1984) J Org Chem 49:578
665. Ho P-T, Wong S (1985) Can J Chem 63:2221
666. Hagiwara H, Uda H (1985) J Chem Soc Perkin Trans 1 1157
667. Tokunga Y, Nagano H, Shiota M (1986) J Chem Soc Perkin Trans 1 581
668. Wuts PGM, Cheng M-C (1986) J Org Chem 51:2844
669. Giese B, Rupaner R (1977) Liebigs Ann Chem 231
670. Sato T, Maeno H, Noro T, Fujisawa T (1988) Chem Lett 1739
671. Tachibana K, Scheuer PJ, Tsukitani Y, Kikuchi H, Van Engen D, Clardy J, Gopichand Y, Schmitz FJ (1981) J Am Chem Soc 103:2469
672. Schmitz FJ, Prasad RS, Gopichand Y, Hossain MB, Van der Helm D (1981) J Am Chem Soc 103:2467
673. (a) Yasumoto T, Murata M, Oshima Y, Sano M, Matsumoto G, Clardy J (1985) Tetrahedron 41:1019, (b) Murata M, Sano M, Iwashita T, Naoki H, Yasumoto T (1986) Agric Biol Chem 50:2693
674. (a) Isobe M, Ichikawa Y, Bai D-L, Goto Y (1985) Tetrahedron Lett 26:5203, (b) Isobe M, Ichikawa Y, Goto Y (1985) Tetrahedron Lett 26:5199, (c) Isobe M, Ichikawa Y, Masaki H, Goto T (1984) Tetrahedron Lett 25:3607, (d) Isobe M, Ichikawa Y, Goto T (1985) Tetrahedron Lett 27:963, (e) Ichikawa Y, Isobe M, Goto T (1984) Tetrahedron Lett 25:5049, (f) Ichikawa Y, Isobe M, Bai D-L, Goto T (1987) Tetrahedron 43:4737, (g) Ichikawa Y, Isobe M, Goto T (1987) Tetrahedron 43:4749, (h) Ichikawa Y, Isobe M, Masaki H, Kawai T, Goto T, Katayama C (1987) Tetrahedron 43:4759, (i) Isobe M, Ichikawa Y, Bai D-L, Masaki H, Goto T (1987) Tetrahedron 43:4767
675. Scheuer PJ, Kato Y (1975) Pure Appl Chem 41:1, (b) Scheuer PJ, Kato Y (1976) Pure Appl Chem 48:29
676. (a) Mynderse JS, Moore RE, Kashiwagi M, Norton TR (1977) Science (Washington DC) 538, (b) Moore RE (1981). In: Scheuer PJ (ed) Marine Natural Products: Chemical and Biological Perspectives. Academic, New York, vol 4 (ch 1)
677. (a) Moore RE, Blackman AJ, Cheuk CE, Mynderse JS, Matsumoto G, Clardy J, Woodard RW, Craig JC (1984) J Org Chem 49:2484, (b) Moore RE, Mynderse JS (1978) J Org Chem 43:2301
678. Serdula M, Bartolini G, Moore RE, Gooch J, Wiebenga N (1982) Hawaii Med J 41:200
679. Park P-U, Broka CA, Johnson BF, Kishi Y (1987) J Am Chem Soc 109:6205
680. Ireland RE, Thaisrivongs S, Dussault PH (1988) J Am Chem Soc 110:5768

Subject Index

MIX
Papier aus verantwortungsvollen Quellen
Paper from responsible sources
FSC® C105338

If you have any concerns about our products,
you can contact us on
ProductSafety@springernature.com

In case Publisher is established outside the EU,
the EU authorized representative is:
Springer Nature Customer Service Center GmbH
Europaplatz 3, 69115 Heidelberg, Germany

Printed by Libri Plureos GmbH
in Hamburg, Germany